Remote Sensing of Coastal Aquatic Environments

T0181369

Remote Sensing and Digital Image Processing

VOLUME 7

The titles published in this series are listed at the end of this volume

REMOTE SENSING OF COASTAL AQUATIC ENVIRONMENTS

Technologies, Techniques and Applications

edited by

RICHARD L. MILLER
NASA, Earth Science Applications Directorate, Stennis Space Center, MS, U.S.A.

CARLOS E. DEL CASTILLO
NASA, Earth Science Applications Directorate, Stennis Space Center, MS, U.S.A.

and

BRENT A. MCKEE
Tulane University, New Orleans, LA, U.S.A.

A C.I.P. Catalogue record for this book is available from the Library of Congress.

ISBN 978-90-481-6792-0 (PB)
ISBN 978-1-4020-3100-7(e-book)

Published by Springer,
P.O. Box 17, 3300 AA Dordrecht, The Netherlands.

Sold and distributed in North, Central and South America
by Springer,
101 Philip Drive, Norwell, MA 02061, U.S.A.

In all other countries, sold and distributed
by Springer,
P.O. Box 322, 3300 AH Dordrecht, The Netherlands.

Printed on acid-free paper

Cover illustration:
A comparison of data acquired at 50m (left) and then resampled to a
spatial resolution of 1km that is typical of many satellite-based instruments (right) -
also see p. 54, figure 3.

Printed in the Netherlands.

Contents

Chapter 1
INTRODUCTION TO RADIATIVE TRANSFER   1

J. RONALD V. ZANEVELD, MICHAEL J. TWARDOWSKI,
ANDREW BARNARD AND MARLON R. LEWIS

Chapter 2
AN INTRODUCTION TO SATELLITE SENSORS, OBSERVATIONS AND TECHNIQUES   21

CHRISTOPHER W. BROWN, LAURENCE N. CONNOR,
JOHN L. LILLIBRIDGE, NICHOLAS R. NALLI AND RICHARD V. LEGECKIS

Chapter 3
OPTICAL AIRBORNE REMOTE SENSING 51

JEFFREY S. MYERS AND RICHARD L. MILLER

Chapter 4
IN-WATER INSTRUMENTATION AND PLATFORMS FOR OCEAN COLOR REMOTE SENSING APPLICATIONS   69

MICHAEL S. TWARDOWSKI, MARLON R. LEWIS,
ANDREW H. BARNARD AND J. RONALD V. ZANEVELD

Chapter 5
THE COLOR OF THE COASTAL OCEAN AND APPLICATIONS IN THE SOLUTION OF RESEARCH AND MANAGEMENT PROBLEMS   101

FRANK E. MULLER-KARGER, CHUANMIN HU,
SERGE ANDRÉFOUËT, RAMÓN VARELA, AND ROBERT THUNELL

Chapter 6
BIO-OPTICAL PROPERTIES OF COASTAL WATERS   129

EURICO J. D'SA AND RICHARD L. MILLER

Chapter 7
REMOTE SENSING OF ORGANIC MATTER IN COASTAL WATERS   157

CARLOS E. DEL CASTILLO

Chapter 8
HYPERSPECTRAL REMOTE SENSING   181

ZHONGPING LEE AND KENDALL L. CARDER

Chapter 9
COMPUTATIONAL INTELLIGENCE AND ITS APPLICATION
IN REMOTE SENSING   205

HABTOM RESSOM, RICHARD L. MILLER,
PADMA NATARAJAN, AND WAYNE H. SLADE

Chapter 10
MODELING AND DATA ASSIMILATION

JOHN R. MOISAN, ARTHUR J. MILLER,
EMANUELE DI LORENZO AND JOHN WILKIN

Chapter 11
MONITORING BOTTOM SEDIMENT RESUSPENSION AND SUSPENDED SEDIMENTS IN SHALLOW COASTAL WATERS

RICHARD L. MILLER, BRENT A. MCKEE, AND EURICO J. D'SA

Chapter 12
REMOTE SENSING OF HARMFUL ALGAL BLOOMS

RICHARD P. STUMPF AND MICHELLE C. TOMLINSON

Chapter 13
MULTI-SCALE REMOTE SENSING OF CORAL REEFS 297

SERGE ANDRÉFOUËT, ERIC J. HOCHBERG, CHRISTOPHE CHEVILLON,
FRANK E. MULLER-KARGER, JOHN C. BROCK AND CHUANMIN HU

Chapter 14
REAL-TIME USE OF OCEAN COLOR REMOTE SENSING
FOR COASTAL MONITORING 317

ROBERT A. ARNONE AND ARTHUR R. PARSONS

LIST OF CONTRIBUTORS

SERGE ANDREFOUËT
Institut de Recherche pour le Développement (IRD), Nouvelle Calédonie

ROBERT A. ARNONE
Naval Research Laboratory, Code 7330, Ocean Sciences Branch, Stennis Space Center, MS, 39529 USA

ANDREW BARNARD
WET Labs, Inc., 620 Applegate St., Philomath, OR, 97370 USA

JOHN C. BROCK
USGS Center for Coastal and Watershed Studies, 600 4th Street South, St. Petersburg, FL, 33701 USA

CHRISTOPHER W. BROWN
Cooperative Research Programs, Office of Research and Applications (ORA), National Environmental Satellite, Data and Information Service (NESDIS), National Oceanic and Atmospheric Administration (NOAA), College Park, MD, 20742 USA

KENDALL L. CARDER
College of Marine Science, University of South Florida, St. Petersburg, FL, 33701 USA

CHRISTOPHE CHEVILLON
Institut de Recherche pour le Développement, BP A5, 98848 Nouméa, New Caledonia

LAURENCE N. CONNOR
Oceanic Research and Applications Division, ORA, NESDIS, NOAA, Camp Springs, MD, 20746 USA

CARLOS E. DEL CASTILLO
National Aeronautics and Space Administration, Earth Science Applications Directorate, Stennis Space Center, MS, 39529 USA

EMANUELE DI LORENZO
Scripps Institution of Oceanography, University of California, San Diego, La Jolla, CA, 92093 USA

EURICO J. D'SA
Lockheed Martin Space Operations, Remote Sensing Directorate, Stennis Space Center, MS, 39529 USA

ERIC J. HOCHBERG
University of Hawaii, School of Ocean and Earth Science and Technology, Hawaii Institute of Marine Biology, P.O. Box 1346, Kaneohe, HI, 96744 USA

CHUANMIN HU
Institute for Marine Remote Sensing, College of Marine Science, University of South Florida 140 7th Ave. S., St. Petersburg FL, 33701 USA

ZHONGPING LEE
Naval Research Laboratory, Code 7333, Stennis Space Center, MS, 39529 USA

RICHARD V. LEGECKIS
Oceanic Research and Applications Division, ORA, NESDIS, NOAA, Silver Spring, MD, 20910 USA

MARLON R. LEWIS
Dalhousie University, Halifax, Nova Scotia B3H 4J1, CANADA

JOHN L. LILLIBRIDGE
Oceanic Research and Applications Division, ORA, NESDIS, NOAA, Silver Spring, MD, 20910 USA

BRENT A. MCKEE
Department of Earth and Environmental Sciences, Tulane University, 208 Dinwiddie Hall, New Orleans, LA, 70118 USA

ARTHUR J. MILLER
Scripps Institution of Oceanography, University of California, San Diego, La Jolla, CA, 92093 USA

RICHARD L. MILLER
National Aeronautics and Space Administration, Earth Science Applications Directorate, Stennis Space Center, MS, 39529 USA

JOHN R. MOISAN
NASA, Goddard Space Flight Center, Wallops Flight Facility, Wallops Island, VA, 23337 USA

FRANK E. MULLER-KARGER
Institute for Marine Remote Sensing, College of Marine Science, University of South Florida 140 7th Ave. S., St. Petersburg FL, 33701 USA

JEFFREY S. MYERS
SAIC, Airborne Sensor Facility, Ames Research Center, Moffett Field, CA, 94035 USA

NICHOLAS R. NALLI
Cooperative Institute for Research in the Atmosphere, Colorado State University, Camp Springs, MD, 20746 USA

PADMA NATARAJAN,
Department of Electrical and Computer Engineering, University of Maine, Orono, ME, 04473 USA

ARTHUR R. PARSONS
Naval Research Laboratory, Code 7330, Ocean Sciences Branch, Stennis Space Center, MS, 39529 USA

HABTOM RESSOM
Lombardi Comprehensive Cancer Center, Biostatistics Shared Resource, Department of Oncology, Georgetown University Medical Center, Washington, DC, 20057 USA

WAYNE H. SLADE
Department of Electrical and Computer Engineering, University of Maine, Orono, ME, 04473 USA

RICHARD P. STUMPF
National Oceanic and Atmospheric Administration, National Ocean Service, Silver Spring, MD 20910 USA

ROBERT THUNELL
University of South Carolina, Columbia, SC, 29208 USA

MICHELLE C. TOMLINSON
National Oceanic and Atmospheric Administration, National Ocean Service, Silver Spring, MD 20910 USA

MICHAEL J. TWARDOWSKI
WET Labs, Inc., 165 Dean Knauss Dr., Narragansett, RI, 02882 USA

RAMÓN VARELA
Fundación La Salle de Ciencias Naturales, Venezuela

JOHN WILKIN
Rutgers University, Douglas Campus, Institute of Marine and Coastal Sciences, 71 Dudley Road, New Brunswick, NJ, 08901 USA

J. RONALD V. ZANEVELD
WET Labs, Inc., 620 Applegate St., Philomath, OR, 97370 USA

PREFACE

Coastal waters are important ecological systems and vital assets for many nations. Estuaries, bays, and coastal margins are among the most productive natural systems on Earth. Coastal productivity supports multiple trophic levels as well as recreational and commercial fisheries. Coastal waters are active areas for the processing of nutrients and carbon and therefore play a roll in the flux, cycling, and fate of atmospheric CO_2. Coastal waters link land and ocean systems primarily through the discharge of rivers. Rivers serve as a major conduit for the delivery of significant amounts of dissolved and particulate materials from terrestrial environments to the coastal ocean. As a consequence, human activities within a watershed can be expressed within coastal waters by changes in water quality and overall system response. Because the majority of the human population lives within 60 km of a coast, coastal waters are critical areas for recreation, commerce, and national defense.

Coastal waters are also complex, dynamic environments. A vast array of coupled biological, chemical, geological, and physical processes occur over multiple time and space scales. The optical environment of coastal waters is particularly complex. The propagation and utilization of light within the water column varies over time and space scales corresponding to changes in concentrations in optically active materials (e.g., phytoplankton, colored dissolved organic matter, and suspended particulates).

There is considerable interest in studying coastal waters to gain a better understanding of Earth system processes for climate change research or environmental factors for management decisions. Unfortunately, the dynamic nature of coastal waters renders most traditional field measurements and sampling protocols ineffective in capturing the range and variability of many coastal processes. In contrast, remote sensing from aircraft and space-based platforms offers unique large-scale synoptic data to address the complex nature of coastal waters. The utility of remote sensing to a wide range of disciplines is well documented. However, to apply remote sensing to a particular application, especially within a dynamic coastal environment, new users are often challenged to find appropriate reference material to gain an adequate understanding of remote sensing in a quick and effective manner.

The motivation to write this book was borne from our early experiences, and frustrations, to use remote sensing to address research problems in our major disciplines, which are related to coastal marine processes. We were faced with the challenge of learning a new technology laden with new and confusing terminology, data, and methods of processing and analysis. Although we were primarily interested in the application of remote sensing to our research, we were forced to learn even basic principles by scouring countless technical manuals, reports, and often-cryptic scientific papers. Simply, a book on the use of remote sensing for coastal aquatic environments did not exist. Hence, the major goal of writing this book was to produce a valuable resource for students, researchers, and decision makers involved in various studies of coastal waters that are interested in applying remote sensing to their work. However, a book such as this cannot be totally comprehensive and cover all aspects of the remote sensing of coastal waters. The primary focus of this book is on optical remote sensing using passive instruments, with a major emphasis on the visible and near-Infrared regions of the electromagnetic spectrum. A discussion of active instruments is largely omitted.

This book is presented as a series of chapters to address the technologies, techniques, and applications of remote sensing related specifically to the study of coastal aquatic environments. Although each chapter was written to 'stand on its own', the content of each chapter was carefully selected to provide an integrated approach to basic terminology and principles. This book is therefore intended as a comprehensive reference on the subject as well. We also encouraged an overlap in discussions of key topics between chapters. However, the authors were asked to cast their discussion of these topics within the context of the main theme of their chapter. In this way, the reader is provided different ways of considering the same concept or application and we hope that this approach will help clarify some difficult topics and accelerate a reader's understanding of key concepts. The book is loosely organized into three sections corresponding to the main theme of a chapter: technologies (Chapters 2 – 4); techniques (Chapters 5 – 10); and, applications (Chapters 11 – 14). Chapter 1 provides an excellent overview of the basic terms and principles of radiative transfer theory to serve as a foundation for the book.

We are indebted to all the authors that contributed to this book. We were extremely fortunate to enlist recognized leaders in coastal remote sensing that enthusiastically embraced the goal of this challenging project, despite their full schedules and monumental workloads. Everyone worked very hard and was motivated to create a book that we all wished was available when we began to explore the exciting world of remote sensing.

As with many efforts to produce a book, friends and relatives often endure the thrills and disappointments that the authors encounter along the way. This book is no exception. Our families provided endless support and encouragement and always reminded us of the contribution that we could make. Richard Miller is forever grateful to his wife Jo Ann for her countless contributions, his daughter Virginia for raising the bar with her academic achievements and piercing discussions, and most importantly, his son Richard Jr. for forfeiting time with his dad so that he could work endless hours on this project. Carlos Del Castillo thanks his wife Mayra for her patience and help during many long hours of writing and reviewing chapters. Brent McKee thanks his family (Becky and Buck) for their encouragement, support and understanding during the long hours and late nights of this project.

The work by Miller and Del Castillo on this book was accomplished outside their official duties with the National Aeronautics and Space Administration (NASA). Hence, the views, opinions, and findings contained in this chapter are those of the authors and should not be construed as an official NASA or U.S. Government position, policy, or decision.

Richard L. Miller
Carlos E. Del Castillo
Brent A. McKee

Chapter 1

INTRODUCTION TO RADIATIVE TRANSFER

[1]J. RONALD V. ZANEVELD, [2]MICHAEL J. TWARDOWSKI,
[1]ANDREW BARNARD AND [3]MARLON R. LEWIS

[1]*WET Labs, Inc., 620 Applegate St., Philomath, OR, 97370 USA*
[2]*WET Labs, Inc., 165 Dean Knauss Dr., Narragansett, RI, 02882 USA*
[3]*Dalhousie University, Halifax, Nova Scotia B3H 4J1, CANADA*

1. Introduction

For the purpose of this paper, we define radiative transfer as the study of the change in direction and intensity of radiation in the atmosphere and ocean, due to absorption, scattering, fluorescence, inelastic scattering, and air-sea interface effects. This is a very large topic and many books have been written on this subject. It is therefore impossible to cover the subject comprehensively in a short article such as this. For detailed presentations on various aspects of this subject the reader is referred to books such as Preisendorfer (1965, 1976), Jerlov (1974), Jerlov and Steemann-Nielsen (1974), Liou (1980), Shifrin (1988), Dera (1992), Mobley (1994), and Thomas et al. (1999). A very readable introduction to the subject is the first half of Kirk (1994). Detailed definitions of optical and radiometric parameters and protocols for their measurement can be found in Mueller et al. (2003)

Energy is generated in the sun by thermonuclear reactions. Gamma rays resulting from these reactions diffuse toward the surface of the sun. On their way to the surface, the gamma rays are scattered, absorbed and reemitted by nuclei and electrons and are changed to x-rays, then to ultraviolet rays and finally emerge mainly as visible light from the surface. After approximately eight minutes of travel, a small fraction of the solar photons reach the outer atmosphere of the earth. The amount of solar radiation reaching the top of the atmosphere depends on the motion of the earth around the sun. The planet orbits about the sun in a slightly elliptical orbit. It also rotates about its axis, which is tilted relative to the plane of the planet about the sun. These motions cause the irradiance incident on the top of the atmosphere to fluctuate on a daily and seasonal basis. Longer term effects such as precession, obliquity and eccentricity are also present; these effects give rise to the various well-known Milankovitch modes of variability in incident solar irradiance.

The atmosphere consists primarily of a mixture of gases, some of which are relatively permanent and some of which vary in concentration. Nitrogen and oxygen dominate the permanent gases, while water vapor and ozone dominate the variable constituents. These gases are distributed non-linearly as a function of height above the planet. Each of the gases has a particular absorption spectrum. Molecular scattering by the gases diffuses the collimated solar photons. In addition to the gases, aerosols (particles) are present. These particles have various origins, such as deserts, ice, fossil fuel burning, forest fires, etc. The aerosols also absorb and scatter light.

The light that arrives at the sea surface is already considerably modified from that which reached the top of the atmosphere. The intensity and spectrum are changed by the

1

R.L. Miller et al. (eds.), Remote Sensing of Coastal Aquatic Environments, 1–20.

absorption and scattering of the gases and aerosols, and the directionality has been changed by the scattering. In addition to direct sun light there is now also a large amount of diffuse light, the intensity and spectrum of which depends on the solar zenith angle and the nature and distribution of aerosols and gases. The sea surface can be characterized by a directional slope spectrum of the waves. This spectrum determines the refraction and reflection patterns at the surface, and so the directional pattern of the light beneath the surface. In addition to the slope spectrum, the sea surface can generate white caps and bubbles, which also affect the redistribution of light.

The sea water itself has certain scattering and absorption characteristics. In addition, the water contains particles of various origins and dissolved materials, each with its own optical characteristics. A fraction of the light that reaches the sea surface and penetrates into the water is reflected back into the atmosphere. This upward light is again modified by the atmosphere and its constituents before some of it reaches spacecraft orbiting the earth. In shallow waters some light may reach the bottom and be reflected upwards.

The formidable task of radiative transfer is to quantify all the processes described above. The forward problem for remote sensing is to predict the spectral intensity of the light reaching the satellite sensor based on a quantitative description of all the absorption, scattering and reflection characteristics of the optical components in the atmosphere and ocean. The inverse problem is the determination of the concentration of the atmospheric and oceanic constituents (including the bottom) when the quantity and spectral characteristics of the light received at the space- or airborne sensor are known.

The above tasks require the exact definition of radiometric quantities, definitions of the inherent optical properties (IOP, the scattering and absorption of the various atmospheric and oceanic constituents) and apparent optical properties (AOP) that describe the attenuation characteristics of the light field. All of these parameters must then be combined with the incident radiance distribution in physical relations that allow mathematical calculations.

Here we have taken a narrative approach, introducing parameters as they are needed. The usual approach is to define all the radiometric quantities, and the inherent and apparent optical properties, and then proceed from there. We have taken the inverse approach for readability. A more complete discussion of the various properties is given at the end of the chapter.

2. The Equation of Radiative Transfer

In this section it is assumed that all parameters are measured at a single wavelength, λ, which is omitted for brevity. The intensity of a narrow beam of light emanating from a source can be described by its radiance: energy per unit area per unit solid angle (the cone into which the light radiates, or is measured, units of steradian, sr). Units of radiance are therefore W/m^2sr, denoted by the symbol L. Imagine this narrow beam being attenuated by absorption and scattering in a medium. It makes sense for the loss of radiance ΔL due to attenuation over short distances Δr to be proportional to the distance traveled and radiance itself. We then get: $\Delta L = -cL\Delta r$, where c is a proportionality constant, called the attenuation coefficient, units of m^{-1}. Taking the limit, we obtain $dL/dr = -cL$, so that $L(r) = L(0) \exp(-cr)$. This equation is used in attenuation meters which contain highly collimated sources to obtain the attenuation coefficient, c. This coefficient is both a function of wavelength and location; c should thus be properly given by $c(\lambda, \vec{x})$, where $\vec{x}$ is the position vector (x,y,z).

If we look at the attenuation of radiance in the case of sunlight in a scattering medium, we not only have attenuation of the radiance as it travels from one location $\vec{x}_1$

to another, $\vec{x}_2$, but light from other directions can be scattered into the direction specified by the two locations $\vec{x}_1$ and $\vec{x}_2$. We thus need to quantify this process. In general we can specify a direction by (θ,ϕ), where θ is the solar zenith angle and ϕ the azimuthal angle in a given reference frame. If we look at the rate of change of L, along a path r in the direction (θ,ϕ), due to scattered light entering the beam, we would expect this increase to be proportional to the radiance coming from another direction (θ',ϕ') multiplied by a small solid angle dω, to get the energy per unit area perpendicular to the incoming beam. Thus, dL/dr α Ldω. We now define a proportionality function, the volume scattering function, with units of $m^{-1}sr^{-1}$ to relate dL/dr to Ldω. Clearly this function depends on location, the incoming direction, and the outgoing direction; it is designated by $\beta(\vec{x},\theta,\phi,\theta',\phi')$: dL$(\vec{x},\theta,\phi)$/dr $= \beta(\vec{x},\theta,\phi,\theta',\phi')$ L$(\vec{x},\theta',\phi')$dω. This is the rate of increase of radiance in direction (θ,ϕ) at location $\vec{x}$, with wavelength λ (not explicitly specified), due to scattered light from direction (θ',ϕ'), with the same location and wavelength. Clearly there can be light coming from all directions (θ',ϕ'), we thus need to integrate over all these directions (there are 4π steradians in a sphere) to get the total amount of radiance that is scattered from all directions into (θ,ϕ) :

$$dL(\vec{x},\lambda,\theta,\phi)/dr = \int_0^{4\pi} \beta(\vec{x},\lambda,\theta,\phi,\theta',\phi') \, L(\vec{x},\lambda,\theta',\phi')d\omega'. \tag{1}$$

We thus have found the rate of increase in the radiance along a path due to scattering. Earlier, we had found the rate of decrease due to attenuation. We combine these two, losses due to attenuation and gains due to scattering to get:

$$dL(\vec{x},\lambda,\theta,\phi)/dr = -c(\vec{x},\lambda) \, L(\vec{x},\lambda,\theta,\phi) + \int_0^{4\pi} \beta(\vec{x},\lambda,\theta,\phi,\theta',\phi') \, L(\vec{x},\lambda,\theta',\phi')d\omega'. \tag{2}$$

This is the equation of radiative transfer (ERT) without internal sources such as fluorescence or Raman scattering. Many books have been written regarding solutions to the ERT. The classical Legendre function solution is by Chandrasekar (1960). Other solutions are given in Preisendorfer (1965). The most common approach in oceanography is to assume that horizontal gradients in radiance and IOP are much smaller than vertical ones, so that horizontal structure is ignored. Taking cosθ dr= dz, leads to:

$$\cos\theta \, dL(z,\lambda,\theta,\phi)/dz = -c(z) \, L(z,\lambda,\theta,\phi) + \int_0^{4\pi} \beta(z,\lambda,\theta,\phi,\theta',\phi') \, L(z,\lambda,\theta',\phi')d\omega'. \tag{3}$$

This is the ERT for the so-called plane parallel assumption without internal sources and is widely applied. Discussions on various solutions to this equation can be found in Mobley (1994) and Thomas and Stamnes (1999). Mobley's numerical solution is commercially available (Hydrolight, Sequoia, Inc.) and is widely used. In the surface zone of the ocean, in the presence of waves, the plane parallel assumption is incorrect, however (e.g., Zaneveld et al., 2001), as it is in the atmosphere in the presence of clouds. In fact, waves form a formidable obstacle to the correct measurement of components for, and the verification of, solutions and inversions of the ERT. If we think of the light field at a given location as being made up of all radiances at that location, we can readily surmise that due to refraction at the surface the light field in the presence of waves is not the same as that for a flat surface. Horizontal variations in the light field occur on the scales of the smallest capillary waves which have wavelengths on the order of a cm to length scales of hundreds of meters for open ocean swell. We can horizontally average Eq. (2), leading to

$$\int_0^D \int_0^D dL(\vec{x},\theta,\phi)dxdy /dr = -\int_0^D \int_0^D c(\vec{x}) L(\vec{x},\theta,\phi)dxdy +$$

$$\int_0^{4\pi} \int_0^D \int_0^D \beta(\vec{x},\theta,\phi,\theta',\phi') L(\vec{x},\theta',\phi')dxdyd\omega', \tag{4}$$

where D is an appropriate length scale. We can write an equivalent ERT (and use numerical solutions) for a horizontally averaged light field $\overline{L(\vec{x},\lambda,\theta,\phi)}$ that is the equivalent of Eq. (3), only if $c(\vec{x})$ and $\beta(\vec{x},\theta,\phi,\theta',\phi')$ are constant horizontally over the length scale D:

$$\cos\theta \int_0^D \int_0^D dL(\vec{x},\lambda,\theta,\phi)dxdy /dz = - c(z) \int_0^D \int_0^D L(\vec{x},\lambda,\theta,\phi)dxdy +$$

$$\int_0^{4\pi} \beta(z,\lambda,\theta,\phi,\theta',\phi') \int_0^D \int_0^D L(\vec{x},\lambda,\theta',\phi')dxdy \, d\omega'. \tag{5}$$

Note that the attenuation and scattering parameters can now be functions of depth only. While it is likely that the attenuation and scattering properties are constant over a few cm, this is highly unlikely over hundreds of meters. For example, Farmer and Li (1995) found bubble clouds aligned with the wind caused by Langmuir circulation on the scale of tens of meters and Colbo and Li (1999) observed that Langmuir circulation affected horizontal and vertical distribution of other particles on these scales. Barth (1999) using a SeaSoar undulating towed device, found sub-km scale variations in optical properties. Hence, it will be necessary to make a number of small scale transects of the inherent optical properties and radiance distribution in the ocean to properly assess the error due to the plane parallel assumption in different oceanic regimes.

The bottom can be included in equations such as the above. In that case one treats the bottom as a special case of scattering function $\beta(\vec{x},\theta,\phi,\theta',\phi')$ in which all downward values are infinite. What one is left with is the bi-directional reflectance distribution function, the BRDF (Voss et al., 2000; Zhang et al., 2003). The BRDF describes the transfer of radiance from a downwelling direction to an upwelling direction, where the amount of radiance transferred is a function of both the incoming and scattered directions. A common assumption for the directional scattering of the bottom is the Lambertian. In that case the scattered radiance is independent of direction.

In order to include fluorescence or Raman effects source terms must be added to the ERT. This poses no particular problem for numerical methods, such as Hydrolight. Thus fluorescence by phytoplankton pigments such as chlorophyll and phycoerythrin can be accommodated, in addition to Raman scattering which influences the radiance intensity (Marshall and Smith, 1990).

We have thus seen that plane parallel programs such as Hydrolight can be used in a horizontally average sense for the radiance, provided the IOP are constant horizontally. Refractive effects due to waves can then be included in an average sense. If one wants to investigate small scale effects of the full three dimensional ERT (Eq. (2) one can use so-called Monte Carlo methods. Reviews are provided in Mobley (1994) and Thomas and Stamnes (1999). In these methods, in principle, one follows the paths of multiple photons from the time they are inserted into the ocean to the moment they are absorbed, or leave the ocean. One calculates the probability of occurrence of a given scattering or attenuation event derived from the IOP. Using random numbers based on the probability distribution function of events, one obtains paths for the photons. After many calculations the photon density and direction provides the desired underwater or water leaving light field. This method has the advantage that any structure of IOP and interfaces can be used. In addition, internal sources such as fluorescence and Raman scattering can readily be handled. The method is computationally intensive, but with faster computers it is likely that this will become the dominant method in the future for obtaining solutions to the ERT.

The ERT provides a link between the radiance distribution and the IOP. In principle, if the radiance distribution at the boundary and the IOP are known, we can solve for the radiances in the interior. The radiance distribution is difficult to measure although it has been done (Smith et al., 1970; Voss, 1989; Voss et al., 2003). Similarly the volume scattering function is difficult to measure underwater in its entirety, although again, a few examples exist (Petzold, 1972; Lee and Lewis, 2003). Integration over the radiance distribution provides links between more readily measured parameters as we shall see below.

3. Gershun's Equation

Integrated parameters are usually easier to measure as they contain fewer variables. If in Eq. (3) we assume the IOP to be homogeneous and we integrate over all directions, we obtain:

$$d[\cos\theta \ L(z,\theta,\phi)d\omega]/dz = -c(z) \int_{0}^{4\pi} L(z,\theta,\phi)d\omega +$$

$$\int_{0}^{4\pi} \int_{0}^{4\pi} \beta(z,\theta,\phi,\theta',\phi') \ L(z,\theta',\phi') \ d\omega d\omega'. \tag{6}$$

The above leads us to define the following radiometric quantities:

$$E(z) = [\int_{0}^{4\pi} \cos\theta \ L(z,\theta,\phi)d\omega], \text{ and} \tag{7a}$$

$$E_0(z) = [\int_{0}^{4\pi} L(z,\theta,\phi)d\omega]. \tag{7b}$$

The first quantity, $E(z)$, is called the plane irradiance, because it is a measure of the flux of energy through a plane perpendicular to the z –direction. Its units are W/m^2. Note that $E(z)$ consists of the difference between the downwelling plane irradiance, $E_d(z)$ and the upwelling plane irradiance, $E_u(z)$, which represent the weighted integration over the upper $(\theta < \pi/2)$ and lower hemisphere $(\theta > \pi/2)$ respectively. The second quantity, $E_0(z)$, called the scalar irradiance, measures the total energy flux through a point. Its units are W/m^2. Inserting these quantities into Eq. (6) leads to:

$$dE(z)/dz = -c(z) \, E_0(z) + b(z) \, E_0(z). \qquad (8)$$

This is Gershun's equation. The attenuation coefficient c is the sum of the scattering and absorption coefficients (units m^{-1}) so that Eq. (8) can be rewritten as:

$$dE(z)/dz = -a(z) \, E_0(z). \qquad (9)$$

This provides an interesting link between the radiometric quantities (based on radiance) and the IOP. Let us define a diffuse attenuation coefficient, units of m^{-1}, similar to the beam attenuation coefficient:

$$K(z) = - dE(z)/ E(z)dz, \text{ so that}$$

$$E(z) = E(0)\exp[- \int_0^z K(z)dz]. \qquad (10a)$$

Similar attenuation coefficients can be defined for the radiances:

$$k(z,\theta,\phi)) = - d \, L(z,\theta,\phi)/ L(z,\theta,\phi) \, dz, \text{ and}$$

$$L(z,\theta,\phi) = L(0,\theta,\phi)\exp[- \int_0^z k(z,\theta,\phi)dz]. \qquad (10b)$$

The similarity of the definition of $K(z)$ and $k(z,\theta,\phi)$ to the inherent optical property, beam attenuation coefficient, leads us to call parameters such as $K(z)$, apparent optical properties (AOP, see section 6.2). These AOP are differential properties of the light field that are used to indicate the rate of change of radiometric properties with depth. Substitution of Eq. (9a) into (8) leads to:

$$\frac{a(z)}{K(z)} = \frac{E(z)}{E_0(z)} = \bar{\mu}(z). \qquad (11)$$

The quantity $\bar{\mu}(z)$ is called the average cosine (dimensionless) of the light field. This name follows from the definitions of $E(z)$ and $E_0(z)$. The average cosine is a useful concept as it connects the IOP $a(z)$ with the AOP $K(z)$. In the case of optical remote sensing it is reasonable to assume that the light field above the sea surface is dominated by the sun. The light field in the ocean tends to be oriented nearly vertically since the maximum intensity is due to the refracted image of the sun, and refraction reduces the solar zenith angle in water compared to that in air. The average cosine thus tends to vary in a narrow range of values from 0.7 to 0.9. An interesting aspect of the average cosine is that it converges to an asymptotic value $\bar{\mu}_\infty$ (Preisendorfer, 1976; Højerslev and Zaneveld, 1977) that is a function of the IOP only. That is, as one goes down into the ocean the light field reaches a constant shape, and the absolute values of the radiances decrease exponentially with a coefficient of K_∞, the asymptotic diffuse attenuation coefficient, which is an IOP, since its value is independent of the radiance distribution at the surface. Zaneveld (1989) and Berwald et al. (1995) have provided functional dependencies of $\bar{\mu}_\infty$ on the IOP in terms of ω_0 (the single scattering albedo = b/c, dimensionless).

4. Inversions and Remote Sensing

In section 2 we have sketched the methods with which one can obtain the radiance distribution if the IOP are known. The IOP can be measured directly (see Twardowski et al., chapter 4, this volume). In the case of remote sensing, however, we have an airborne or spaceborne radiance detector, typically at several wavelengths. The inverse problem of radiative transfer is then to derive IOP from the remotely sensed radiance. This is a problem for which there is no exact solution, so various approximations must always be made. Here we will use several approaches to the inversion problem. First we will use an intuitive approach, to get an idea of the parameters involved. Secondly, we will use an inversion from the ERT to see how far we can take an analytical approach before we must resort to approximations. The value of inversions from remote sensing lies in the determination of IOP which can then be further translated, if desired, into other less exact parameters such as chlorophyll, particle concentration, etc.

First we will derive from first principles the remote sensing reflectance just beneath the sea surface. We wish to derive equations for the upwelling light just beneath the surface, if the downwelling plane irradiance and the IOP are known. Zaneveld and Pegau (1998) and Zaneveld et al. (1998) have derived steady state two flow equations that can be adjusted to the present problem. From first principles one can think of a downwelling stream of photons that is attenuated on the way down. At all depths some of the photons are scattered into an upward direction. These upwelling photons are further attenuated on the way up. Finally only a small fraction of the upwelling photons are moving in a direction that can be detected by the sensor. The contribution to the upwelling nadir radiance, $L_u(z)$, (vertically upwelling, $\theta = \pi$, ϕ can be any value), at a given depth is the downwelling plane irradiance at that depth, $E_d(z)$, multiplied by a weighted integral of the volume scattering function in the backward direction, $\beta_b(z)$. This scattered light is a small fraction F_b of the backscattered light given by the backscattering coefficient b_b, as the remote sensing detector typically has a small solid angle of detection. Thus $\beta_b(z) = F_b b_b$. The vertical attenuation coefficient for downwelling irradiance is given by $K_d(z)$ and the vertical attenuation of the upwelling nadir radiance is given by $k_u(z)$. Integrating over all depths gives the nadir radiance just below the sea surface:

$$L_u(0^-) = E_d(0^-) \int_0^\infty (\beta_b(z) \, e^{-\zeta(z)} dz, \tag{12}$$

where

$$\zeta(z) = \int_0^\infty (K_d(z') + k_u(z')) \, dz'. \tag{13}$$

If the scattering and diffuse attenuation properties are assumed to be constant with depth,

$$L_u (0^-)/ E_d (0^-) = R_{rs}(0^-) = \frac{\beta_b(z)}{K_d(z') + k_u(z')},$$ (14)

where $R_{rs}(0^-)$ is the remote sensing reflectance just below the surface, with units of sr^{-1}. We had already seen that $\beta_b(z) = F_b\, b_b$. In addition, Eq. (11) showed that the diffuse attenuation coefficient was related to the absorption coefficient. We can thus reasonably expect the downwelling and upwelling diffuse attenuation coefficients to also be related to the absorption coefficient. Combining these arguments and applying them to Eq. (14) allows us to set:

$$L_u (0^-)/ E_d (0^-) = R_{rs}(0^-) = F_R \frac{b_b(z)}{a(z)}.$$ (15)

This relationship was first derived (using different approaches) by Gordon et al. (1975) and Morel and Prieur (1977). We thus found that the remote sensing reflectance is proportional to the backscattering coefficient and inversely proportional to the absorption coefficient. The proportionality factor F_R (sr^{-1}) depends on how the backscattered light relates to the backscattering coefficient, and therefore to the details of the volume scattering function in the backward direction and the radiance distribution. The relationship between the diffuse attenuation coefficients and the absorption coefficient also depends on the details of the radiance distribution and the IOP. Much radiative transfer is thus contained in the factor F_R. This factor has been studied in detail (for example Gordon et al., 1975; Gordon et al., 1988; Morel and Gentili, 1996).

Based on extensive Monte Carlo calculations, Gordon et al. (1975) derived for a collimated beam of irradiance:

$$R = E_u(0^-)/E_d(0^-) = 0.33\frac{b_b}{a + b_b}.$$ (16)

In order to relate the upwelling irradiance to the upwelling radiance, a commonly introduced factor is $Q = E_u(0^-) / Lu(0^-)$. Since typically backscattering is much less than absorption, this sets the parameter F_R in Eq. (15) equal to 0.33/Q. This parameterization is the starting point for many inversion algorithms, but it ignores the dependence of F_R on the shape of the volume scattering function and the radiance distribution. In order to obtain the dependence of F_R on these parameters, we must use a more analytical approach, which follows.

A more theoretical relationship (Zaneveld, 1982, 1995) can be derived from the ERT (in the form of Eq. 3) for the nadir radiance, L_u, for which $\cos\theta = -1$, and for which we can define an attenuation coefficient k_u (as in Eq. (10b)):

$$k_u(z)\, L_u(z) = -c(z)\, L_u(z) + \int_0^{4\pi} \beta(z,\pi,0,\theta',\phi')\, L(z,\theta',\phi')d\omega'.$$ (17)

We split the integration over $d\omega$ into integrations over θ and ϕ and split them into upwelling ($\theta > \pi/2$) and downwelling ($\theta < \pi/2$) components.

$$[k_u(z) + c(z)]\, L_u(z) = \int_0^{2\pi} \int_0^{\pi/2} \beta(z,\pi,0,\theta',\phi')L(\theta',\phi',z)\sin\theta\, 'd\theta\, 'd\phi' +$$

$$\int\limits_{0}^{2\pi} \int\limits_{\pi/2}^{\pi} \beta(z,\pi,0,\theta',\phi')L(\theta',\phi',z)\sin\theta \ 'd\theta \ 'd\phi' \ . \tag{18}$$

We now define shape factors f_b and f_L that relate the complex integral expressions above to simpler and more readily measured parameters.

$$f_b(z) = [\int\limits_{0}^{2\pi} \int\limits_{0}^{\pi/2} \beta(z,\pi,0,\theta',\phi')L(\theta',\phi',z)\sin\theta \ 'd\theta \ 'd\phi' \ ] / [\frac{b_b(z)}{2\pi} \ E_{od}(z)] \tag{19a}$$

$$f_L(z) = [\int\limits_{0}^{2\pi} \int\limits_{\pi/2}^{\pi} \beta(z,\pi,0,\theta',\phi')L(\theta',\phi',z)\sin\theta \ 'd\theta \ 'd\phi' \ ] / [b_f(z) \ L_u(z)] \ . \tag{19b}$$

The parameters are chosen this way because in Eq. (19a), all the scattering is in the backward direction. If the scattering were constant in the backward direction, we could take it outside of the integral, which would then reduce to $E_{od}(z)$, the downwelling scalar irradiance, defined as the unweighted integral over all radiances in the upper hemisphere. In Eq. (19b) the radiances are all upward, and if they were constant, we could take it outside of the integral, which would then become $b_f(z)$, the forward part of the scattering coefficient. The parameter $f_b(z)$ thus to a large extent describes how uniform the backscattering function is, and $f_L(z)$ describes primarily how uniform the upwelling radiance distribution is. Values of unity indicate uniformity. Substitution of Eq. (19) into (18) then gives:

$$[k_u(z) + c(z)] \ L_u(z) = f_b(z) \ E_{od}(z)\frac{b_b(z)}{2\pi} + f_L(z) \ L_u(z)b_f(z) \ , \text{ or} \tag{20}$$

$$L_u(z) / E_{od}(z) = \frac{f_b(z)\dfrac{b_b(z)}{2\pi}}{k_u(z) + c(z) - f_L(z)b_f(z)} \ . \tag{21}$$

This is an exact expression as it is only a rewrite of the ERT. We note that it contains the scalar irradiance rather than the plane irradiance. This equation can be the platform from which to obtain approximations for the remote sensing reflectance based on observations. Zaneveld (1995) has described a series of approximations that lead from more exact to more approximate expressions. The more approximate expressions contain measurable parameters, however, and so are needed for experimental work.

Zaneveld (1995) and Weidemann et al. (1995) have calculated values for $f_b(0^-)$ and $f_L(0^-)$ based on observations for homogeneous media. From Eq. (19b) we see that f_L is an integration over the forward scattering function multiplied by the upward radiance distribution. The upward radiance distribution is approximately uniform so that it was found that generally f_L was within a few percent of 1. We can assume as well that the backscattering coefficient and $(1- f_L \ b_f)$ are much less than the absorption coefficient, which would be the case in all but the most strongly scattering media. In addition, Zaneveld (1995) has argued, based on the observed nearly exact exponential shape of $L_u(z)$ near the surface, that $k_u(z)$ could be modeled as $a/\bar{\mu}_\infty$. Applying these approximations, Eq. (21) reduces to:

$$L_u(0^-) / E_{od}(0^-) \approx \frac{f_b}{2\pi (1 + 1/\bar{\mu}_\infty)} \frac{b_b}{a} . \tag{22}$$

Analogous to Eq. (11), we can define a downwelling average cosine, $\bar{\mu}_d (z) = E_d(z) / E_{od}(z)$. We can now obtain an expression analogous to Eq. (15):

$$L_u (0^-)/ E_d (0^-) = R_{rs}(0^-) \approx \frac{f_b}{2\pi \bar{\mu}_d (0^-) (1 + 1/\bar{\mu}_\infty)} \frac{b_b}{a} . \tag{23}$$

Clearly, through multiple approximations, it is possible to get closer to measurable parameters when proceeding from the purely theoretical expression given by Eq. (21). We also need an expression for f_b in terms of measurable parameters. Equation (19a) shows that f_b contains the backward part of the volume scattering coefficient and the forward part of the radiance distribution. We can assume that the downward radiance distribution near the surface forms a diffuse beam (due to forward scattering) along the refracted image of the sun. The refracted image of the sun, and the refracted detection angle of the remote sensing detector, form an angle θ_m. We can then hypothesize that most of the backward scattering will be between the diffuse forward beam and the diffuse backward beam, with an average volume scattering function of $\beta(\pi - \theta_m)$. If there is a significant amount of light that is backscattered twice before reaching the remote sensor, this hypothesis would be incorrect. However, even in the case of multiple scattering, the most likely scattering event will be in the very near forward direction. Multiple scattered light that reaches the remote sensor will thus most likely have undergone multiple forward scattering events and a single backscattering event. Even considering multiple scattering, the dominant backscattering will still be due to $\beta(\pi - \theta_m)$ except when scattering strongly dominates diffuse attenuation. Considering this, we can then write Eq. (19a) as:

$$f_b(0^-) \approx 2\pi \beta(\pi - \theta_m) / b_b. \tag{24}$$

Eq. (24) was tested extensively by Weidemann et al. (1995) and found to have typical errors of 5% and maximum errors of 12%. Note that this formulation makes no assumptions about the shape of the volume scattering function. Other formulations such as that by Morel and Gentili (1991, 1993) assume a constant shape for the particulate VSF. In any case, the reader should be aware that the relative orientation of the sun and the sensor has a potentially large impact on the remote sensing reflectance. Substitution of Eq. (24) into Eq. (23) then leads to:

$$L_u (0^-)/ E_d (0^-) = R_{rs}(0^-) \approx \frac{\beta(\pi - \theta_m)}{a \bar{\mu}_d (0^-) (1 + 1/\bar{\mu}_\infty)} . \tag{25}$$

Finally, we note that the average cosine of the near surface light field can be given by the cosine of the refracted solar zenith angle, $\cos\theta_s$, and that Zaneveld (1989) showed that $a/\bar{\mu}_\infty \approx c (1 - 0.52\omega_0 - 0.44\omega_0^2)$. We now have a complete model of the dependence of the remotely sensed reflectance on the IOP. We want to be able to invert relationships such as Eq. (25) for the IOP. In these models we have not specifically indicated wavelength, but for the purposes of inverting remotely sensed signals, one can assume that one has measured reflectance at N wavelengths. In principle one can then invert for N parameters. This is done by dividing the IOP into components (see section 6.1) and assigning spectral models for these components IOP, each with one or more parameters. One then minimizes the resultant N equations to obtain N or fewer parameters.

Examples of such methods are Roesler and Perry (1995), Garver and Siegel (1997), Hoge and Lyon (1999), Lee et al. (2002), and Roesler and Boss (2003).

5. Lidar

Light Detection And Ranging (Lidar) is a method whereby very short pulses of light from a laser are sent into the medium. By measuring the return as a function of time, profiles of optical properties can be determined. For oceanographic studies (Hoge, 1988), one typically uses a doubled YaG laser with a wavelength of 532nm.

5.1 ILLUMINATION AND DETECTION FOOTPRINTS

Here we will look at some of the oceanographic implications of illumination and detection footprints, or spot size. In a plane parallel situation, there is a very large illumination spot size (ISS) and a limited detection spot size (DSS). This is the classical passive remote sensing arrangement. For the active arrangement we can assume a small ISS and a large DSS. By means of a reciprocity argument, we can immediately say that this will give the same result as the passive arrangement. This reciprocity can be verified by a thought experiment. In the passive case, if we envision the detector beam, we know that if a photon leaves the confines of the beam, another photon will, statistically, take its place on the other side of the beam. No net photons are lost from this system due to course changing caused by scattering. We can then see the system as closed.

In the case of the active beam let us imagine the illumination beam to be exactly the same size as the detector beam in the passive case. The active detection beam is made large. Let us have the same number of photons per unit area per unit time enter the water. The same number of photons will wind up heading back to the satellite sensor as in the passive case. A number of photons will have left the source beam, but this doesn't matter as the detector beam is large and will catch them. Thus the two systems are equivalent as we started with the same number of photons (per unit area per unit time) and we get the same number of photons returned. The depth integrated return from an active system with a 1 km ISS and a larger DSS should then be the same as that of a passive sensor with a 1km DSS. This is a potentially very useful concept as we can now compare passive remote sensing signal strength such as that obtained with SeaWiFS or MODIS with depth-integrated Lidar signal strength.

The above construct simplifies life as strange things happen in the active source beam. On the way down photons leave the beam, so that K_d is not uniform within the beam. In the middle, if the beam is wider than 1/beam attenuation coefficient, we have a nearly plane parallel situation, and K_d is the same as in the case of sunlight, K_{dsun}. Towards the edges, light is leaking out of the beam and $K_d > K_{dsun}$. Just outside of the beam, however, we have only upwelled light near the surface. As we go down, we have more and more downward photons outside of the beam due to scattering thus, in that area, the light can be increasing with depth, leading to negative K_d. The beams will be interesting to model. On the way up we will also have non-uniform K_u's at the same depth. Using reciprocity we can avoid all that and study the passive case instead. We can then also use known models for diffuse K_d and K_u for radiance.

5.2 BACKSCATTERING SIGNAL

In this case, we are interested in the time dependent signal of the upwelling radiance, $L_u(0^-,t)$, just below the sea surface. In particular we wish to know how much energy is received in the time period (t, t+Δt). In that case, only photons having traveled distances

$(z, z+\Delta z) = c_w(t, t+\Delta t)$, where c_w is the speed of light in water, will be counted. If t is the round trip time to depth z, we can then approximate the signal from the layer due to backscattering as follows:

$$S_b(z,\Delta z) = \int_t^{t+\Delta t} L_u (0^-,t) \, dt = E_d (0^-) \int_z^{z+\Delta z} \beta_b(z') \, e^{-\zeta(z')} dz' , \quad (26)$$

where $\Delta z = (\Delta t/2)c_w$. If we assume a homogeneous ocean (i.e., β_b and ζ are constant with depth), and a light illuminating at 532 nm, we get:

$$S_b(z,\Delta z) = \int_t^{t+\Delta t} L_u (0^-,t) \, dt = E_d (\lambda, 0^-) \, F_b \frac{b_b(\lambda)}{\zeta(\lambda)} \, [e^{-\zeta(532)z} - e^{-\zeta(532)(z+\Delta z)}]. \quad (27)$$

5.3 STIMULATED FLUORESCENCE, EXCITATION WAVELENGTH 532 NM, EMISSION WAVELENGTH 685 NM

In the case of chlorophyll fluorescence, we note that the fluoresced light is equal to a quantum efficiency of conversion, $Q(532,685)$, multiplied by the light absorbed by pigments over the depth interval, less a small portion of the fluoresced light re-absorbed within the algal cells. Let us set the pigment absorption coefficient equal to $a_p(532)$. This fluoresced light must now be brought to the surface. Only half of the fluoresced light will go into upward directions and only a fraction of the fluoresced light is radiated into directions that can be detected by the satellite. We will call this fraction F_f (it should nearly be equal to $\Omega_s /2\pi$, where Ω_s is the detection solid angle of the satellite). The light that reaches the surface is attenuated by the diffuse attenuation coefficient for 685 nm light. The total signal at the surface due to fluoresced light in the interval $(z, z+\Delta z)$ is then:

$$S_f(685, z,\Delta z) = E_d (532, 0^-) \, F_f Q(532,685) \frac{a_p(532)/2}{K_d (532) + K_u(685)} \, [e^{-(K_d(532) + K_u(685))z} - e^{-(K_d(532) + K_u(685))(z+\Delta z)}] \quad (28)$$

5.4 RAMAN SCATTERING, EXCITATION WAVELENGTH 532 NM, EMISSION WAVELENGTH 651 NM

The emission is at around 651 nm (Marshall and Smith, 1990), so that we have to use the diffuse attenuation at 651nm for the upwelling light. We then get, similar to Eq. (5):

$$S_R(651, z,\Delta z) = E_d (532, 0^-) \, F_R \frac{b_R(532)/2}{K_d (532) + K_u(651)} \, [e^{-(K_d(532) + K_u(685))z} - e^{-(K_d(532) + K_u(651))(z+\Delta z)}], \quad (29)$$

where b_R is the Raman scattering coefficient.

5.5 SIGNAL STRENGTH ESTIMATIONS

We can now make some first order calculations as to the signal strengths of the scattered, Raman, and fluoresced light intensities. First, let us assume that $F_b = F_f = F_R = 1/2\pi$. If the upwelled scattered light, Raman, and fluoresced light are assumed to be totally diffuse, this would be reasonable. Note that by using the ratio, the signal will be in units of W/m^{2}sr. The incoming light is in irradiance units of W/m^2. The ratio will thus have units of sr^{-1}, and can be thought of as photons in over photons out per unit solid angle.

We now need to model $\beta_b(z)$, $K_d(z)$ and $k_u(z)$ (and hence $\zeta(z)$) at the various wavelengths. For simplicity we ignore the (λ) notation in the next few paragraphs. It was already shown that $k_u(z)$ can be modeled as (Zaneveld, 1995):

$$k_u(z) = a(z) / \overline{\mu}_\infty(z), \tag{30}$$

which was given as a function of $b(z)/c(z) = \omega_0(z)$ in section 4. This model is based on observations which show that in the backward direction the radiance has a constant attenuation coefficient with depth, even near the surface, whereas the forward radiance does not have a constant attenuation coefficient with depth.

Modeling $K_d(z)$ requires the inclusion of the depth dependence of the shape of the radiance distribution. This can be accomplished by using Gershun's equation (see section 3):

$$K(z) = a(z) / \overline{\mu}(z) \quad , \tag{31}$$

where $a(z)$ is the absorption coefficient and $\overline{\mu}(z)$ is the average cosine of the light field. $K_d(z)$ differs from $K(z)$ by only a few percent, so that we may set:

$$K_d(z) \approx a(z) / \overline{\mu}(z) \quad . \tag{32}$$

Berwald et al. (1995) have derived a parametric model for the dependence of $\overline{\mu}(z)$ on ω_0 for a vertical sun in a black sky. This fits our problem exactly. Barnard et al. (1998) have given global spectral slope averages for the particulate absorption and scattering coefficients and the yellow matter absorption all referenced to 488nm. Using these averages, we relate the absorption and scattering coefficients at 532nm to those at 651 and 685nm. In a few cases (e.g., Twardowski et al., 2001), we have measured backscattering coefficients. In general, only the total scattering coefficient is measured. For modeling purposes, in the latter case, one can use Petzold's (1972) average of: backscattering = 0.015*total scattering coefficient. In all cases we must add the pure water contribution since backscattering sensors are calibrated to exclude the pure water contribution.

The quantum efficiency for fluorescence is 2-5% (Sam Laney, pers. comm.) Marshall and Smith (1990) have determined that the Raman scattering coefficient is 2.6×10^{-4} m^{-1} at 488nm. Note that this is about one order of magnitude less than the scattering coefficient of pure water. Using a λ^{-4} dependence we calculate the Raman scattering coefficient at 532nm to be 1.8×10^{-4} m^{-1}.

We now have all the parameters in hand to model the range-gated backscattering, Raman and stimulated fluorescence signals.

6. Inherent, Radiometric, and Apparent Optical Properties

6.1 INHERENT OPTICAL PROPERTIES

Two fundamental IOPs are a (m^{-1}) and b (m^{-1}), the rates of radiant intensity loss over a fixed pathlength due to the processes of absorption and scattering, respectively. The beam attenuation coefficient, c (m^{-1}), is defined by their sum:

$$c = a + b. \tag{33}$$

There are many ways to decompose total or integrated IOPs into constituent IOPs. Scattering, for example, can be partitioned with respect to its angular distribution, the volume scattering function ($\beta(\theta)$, VSF), units of $m^{-1}sr^{-1}$. The VSF is defined by

$$\beta(\theta) = \frac{dI(\theta)}{EdV}, \tag{34}$$

where $dI(\theta)$ is the radiant intensity (w/sr) emanating into a small solid angle when a small volume dV is illuminated by an irradiance E. The beam attenuation coefficient and the volume scattering function are the IOPs that appeared in the equation of radiative transfer, Eq (3), and so form the connection between the particulate and dissolved materials and the remotely sensed radiances.

If we integrate the light emitted over all directions, we obtain the total scattering coefficient, b, units of m^{-1}. The total scattering coefficient can be divided into forward, b_f, and backward, b_b, components:

$$b_f = 2\pi \int_{0}^{\pi/2} \beta(\theta)\,\sin\theta\,d\theta \quad \text{and} \quad b_b = 2\pi \int_{\pi/2}^{\pi} \beta(\theta)\,\sin\theta\,d\theta \tag{35}$$

$$b = b_b + b_f. \tag{36}$$

The theoretical aspects of light scattering are treated extensively in van de Hulst (1981). For the various semi-analytical and analytical remote sensing algorithms (see section 4), we now have defined the two key IOPs relevant to the remote sensing reflectance, a and b_b. These IOPs are then often separated into operationally defined components such as the dissolved and particulate fractions and water:

$$a_t = a_g + a_p + a_w, \text{ and} \tag{37}$$

$$b_{bt} = b_{bp} + b_{bw}, \tag{38}$$

which applies to Eq. (29) as:

$$c_x = a_x + b_x \qquad \text{(where subscript } x = t, g, p, \text{ or } w). \tag{39}$$

The subscripts t, g, p, and w represent total, dissolved (historically called gelbstoff or gilvin), particulate, and water, respectively. Operationally, the dissolved fraction typically comprises all substances that pass through a 0.2 μm filter. Other commonly used parameters are a_{pg} and c_{pg}, defined as the quantities ($a_p + a_g$) and ($c_p + c_g$), respectively.

Equation (38) assumes that scattering from dissolved molecules in seawater will be negligible compared to the other terms. Another common assumption with errors typically less than 1% (Twardowski and Donaghay, 2001) is that $c_g \approx a_g$ because the total scattering from dissolved materials in natural waters, b_g, is sufficiently low relative to a_g. This may be disputed, however, in waters with a high content of fine clays, where colloidal material passing through a 0.2 μm filter may be detectable (Aas, 2000).

For algorithms that focus on the absorption and backscattering by phytoplankton, an additional partitioning of the particulate component of Eqs. (37) and (38) is often made:

$$a_p = a_\phi + a_d, \text{ and} \tag{40}$$

$$b_{bp} = b_{b\phi} + b_{bd}, \tag{41}$$

where the ϕ and d subscripts represent the algal and non-algal components, respectively. The non-algal component is comprised of non-living particulate organic material, living particles such as bacteria, inorganic minerals, and bubbles. The relative contribution of these different particle groups to particulate backscattering is poorly known, but recent progress has been made (Stramski et al., submitted). All the IOPs in Eqs. (33)-(41) have wavelength dependencies, examples of which can be found throughout the books by Shifrin (1988), Kirk (1994) and Mobley (1994).

Fluorescence is also an IOP that can be detected with passive and active remote sensing techniques. Common fluorophores in the dissolved fraction include humic substances (humic and fulvic acids), proteins, and hydrocarbons. Fluorescent phytoplankton pigments include chlorophyll, phycoerythrin, and phycocyanin.

The IOP fractional components discussed in this section can be related to several biogeochemical parameters (Twardowski et al., Chapter 4, Table 1). Algorithms exist to derive nearly all of these IOPs from passive remote sensing or active lidar platforms (for example, Garver and Siegel, 1997; Hoge and Lyon, 1999; Roesler and Boss, 2003) and, as a consequence, remote sensing algorithms have been developed for many of these biogeochemical properties. Excellent IOP reviews, including components and some biogeochemical associations, are given in Jerlov (1976), Shifrin (1988), Dera (1992), Kirk (1994), and Mobley (1994).

6.2 RADIOMETRY AND APPARENT OPTICAL PROPERTIES

The fundamental radiometric property is the radiance distribution ($L(\theta,\phi,\lambda)$, units of W m^{-2} sr^{-1} or quanta m^{-2} s^{-1} sr^{-1}), described as the radiant power in a specified zenith (θ) and azimuth (ϕ) direction per unit solid angle, per unit area normal to the incident beam at a given wavelength. The radiance distribution in the sea (and above it) can never be constant with depth, as it results from the modification of the incident radiance field by the sea-surface, the inherent optical properties of the ocean interior, and the reflectivity of the sea bottom.

All other radiometric quantities derive from this. In particular, the various irradiances are derived by weighted integration of the radiance field over defined solid angles. The downwelling ($E_d(\lambda)$, units of W m^{-2}) and upwelling ($E_u(\lambda)$, units of W m^{-2}) irradiances are given as the cosine-weighted integration of the radiance distribution over the upper (downwelling) and lower (upwelling) hemispheres, respectively (see section 3). These hemispheres are separated by a horizontal surface oriented normal to the vertical. The net downward irradiance ($E(\lambda)$, units of W m^{-2}) represents the vertical component of the irradiance vector and is given by the difference between the upward and downward irradiances, or the cosine weighted integral over all solid angles. A further quantity of biogeochemical and physical interest is the scalar irradiance ($E_o(\lambda)$, units of W m^{-2}) which results from the unweighted integration of radiance over all hemispheres.

A particularly useful relationship results from the integration of the radiative transfer equation to yield the Gershun equation as derived in section 3. Another derived quantity of interest results from spectral integration over the wavebands active in photosynthesis, generally taken as the interval from 350 or 400 nm to 700 or 750 nm. All of the above irradiances can be spectrally integrated, to provide a measurement of the so-called "Photosynthetically Available Radiation" (PAR, units of W m^{-2} or quanta s^{-1} m^{-2}). This is typically given as the integrated scalar irradiance.

6.3 THE AIR-SEA INTERFACE

For remote sensing, measurements of radiance and irradiance taken in water must be related to remotely sensed above-water radiances. To do this requires consideration of two factors, first the propagation of measurements taken at depth to the surface, and second the propagation of radiance across the sea-air boundary. The second is more straightforward than the first:

$$L_w(0^+,\theta,\phi) = L_u(0^-,\theta',\phi') \frac{1-\rho(\theta,\theta')}{n^2}, \tag{42}$$

where the water-leaving radiance just above the water, $L_w(0^+,\theta,\phi)$, in a given direction (θ,ϕ), derives from an upwelling below water (0^-) radiance stream, $L_u(0^-,\theta',\phi')$, of direction (θ',ϕ'). The two streams are related through Snell's law, $\theta' = \sin^{-1}[\sin\theta / n]$. The index of refraction, n, is actually an inherent optical property (the real part of the complex refractive index) and is dependent on salinity and (weakly) on temperature and pressure (Austin and Halikas, 1976). The reflection of the air water interface is given by ρ; note that $\rho = 1$ for incident angles greater than the critical angle ~ 48° for n = 1.34. Note as well that this relationship presumes no transpectral scattering (e.g., water Raman effects).

Most of the historical work has assumed nadir viewing geometry $(\theta = \pi)$. In this special case, $L_w(0^+,\pi,0,) \approx 0.55 \, L_u(0^-,\pi,0)$. However, most remote sensing instruments view the ocean surface at angles removed from nadir. Furthermore, the Fresnel scattering of downward radiance from the ocean surface upward is a strong function of illumination and viewing geometry. These so-called bi-directional characteristics of the radiance field incident on the sensor on orbit are therefore considerably more complicated. A rather complete theoretical analysis of this can be found in Morel and Gentilli (1996) and Mueller (2003); full evaluation of the bi-directionality of the radiance field below and above the sea-surface will require the routine measurement of the full radiance field (e.g., Morel et al., 1995; Voss et al., 2003).

A more problematic situation occurs when radiance measurements taken at depth are required to be propagated to the sea-surface to estimate $L_u(0^-,\pi,0)$. In practice, nadir viewing instruments are usually employed, but in principle, the full radiance distribution could be used as well. It is rarely possible or even feasible to measure L_u accurately near the sea-surface (i.e., 0^-), given the presence of surface waves of various scales, and typically, reliable measurements have only been made for depths $z > \sim$1-2 meters in the open ocean. More recent instruments provide accurate statistics of upwelling radiance at depths ~ 10 cm, at least for moderate sea-states (see Twardowski et al., Chapter 4).

Given an accurate measurement of radiance at a range of depths, the problem faced is the extrapolation to just below the sea surface. The usual approach is to assume homogeneity over the upper ocean in some sense, and compute $L_u(0^-,\pi,0)$ for $\theta=\pi$ in:

$$L(0^-,\theta,\phi) = L(z_0,\theta,\phi)\exp[\int_0^z k(z,\theta,\phi)dz], \tag{43}$$

where z_0 is a reference depth below the surface, and $k(z,\theta,\phi)$ is the diffuse spectral attenuation coefficient for radiance at depth z_0 (assumed constant over the interval 0^- to z_0) as defined in Eq. (9b). The diffuse attenuation coefficient is operationally derived from the derivative of the neperian log of the vertical radiance profile and is usually assumed to be constant over the interval 0^- to z_0; analogous terms can be computed for the various irradiances.

Note that the diffuse attenuation coefficient is derived from radiometric properties that are never constant as a function of depth, and therefore is an AOP. The rapid modification with depth of the radiance distribution in the upper optical depth, even with constant IOPs, implies that the assumption of homogeneity is almost assuredly invalidated. Careful measurements, well-resolved in the vertical and taken near the sea-surface minimize this error, at least in the absence of significant surface roughness. At longer wavelengths (>650 nm), the strong attenuation of water conspires with increased instrument shading, fluorescence and Raman scattering to render this extrapolation extremely tenuous.

It is also necessary to determine the downwelling irradiance just beneath the sea surface, in order to determine the reflectance for the ocean alone. This is even more difficult as the influence of waves is larger on the downwelling irradiance than the upwelling radiance (Zaneveld et al., 2001). Even for small waves horizontal gradients can be much larger than vertical ones. A thorough discussion of light fields beneath waves can be found in Walker (1994).

In addition to providing a means to extrapolate radiances (and irradiances) within the upper ocean, the diffuse attenuation coefficients themselves are of considerable interest. Often viewed as "quasi-inherent" optical properties (e.g., Morel, 1988; Gordon et al., 1988; Morel and Maritorena, 2001), the close correspondence between K and the absorption coefficient places variations in K central to a wide range of applications, including the computation of photosynthesis (regulates the penetration of irradiance available for photosynthesis, as well as light absorbed; e.g., Behrenfeld and Falkowski, 1997), the computation of local heating rates due to absorption of solar radiation (e.g., Zaneveld et al., 1981; Lewis et al., 1990), the photochemical degradation of organic matter (e.g., Johannessen et al., in press), lidar system performance (e.g., Allocca et al., 2002), and underwater visibility (e.g., Zaneveld and Pegau, 2003).

Remote sensing applications often derive K as an output product from measurements of normalized water-leaving radiances through empirical and semi-empirical approaches. For example, K can, with some accuracy, be decomposed into component contributions as with the IOPs in Eqs.(37-41), and can be used as diagnostics for constituents in the ocean, in particular the derivation of chlorophyll concentrations in oceanic Case 1 waters (see Morel, 1988; Gordon et al., 1988; Morel and Maritorena, 2001).

The above sections deal with variations in apparent optical properties in the ocean interior, and their propagation to and through the sea-surface for the estimation of the water-leaving radiances required for calibration and validation of sensors on orbit. As an alternative approach, measurements of upwelling radiance can be made above the sea-surface from ship, buoy or tower platforms. Such measurements are appealing in principle, as they provide a direct measurement of radiance leaving the ocean, and are free from errors in propagation in the upper layer. However, in addition to the desired water-leaving photons, such measurements suffer from the inclusion of photons reflected off the sea-surface. This Fresnel reflectance includes both radiance resulting from the direct reflection of the Sun and from sky reflectance.

For all but the calmest of seas, the contribution from surface reflectance is complex, and can often overwhelm the water-leaving signal (see full discussion in Walker, 1994 and Mobley, 1994). For a flat sea-surface and uniform sky radiance distribution, it is straightforward to compute the Fresnel reflectance over a small subtended solid angle looking down at the sea-surface. It is the downwelling radiance at equivalent relative azimuth and at the complementary zenith angle, multiplied by the Fresnel reflectance, which varies with respect to zenith angle from ~0.02 for normal incidence to ~0.03 at 40°, and then increases strongly with increasing angle in a well-behaved and well-known manner.

In practice, even the lightest of winds ruffle the sea-surface, and uniform sky conditions are rarely encountered, except in heavily overcast days which are not of relevance to remote sensing applications as the sea-surface cannot be viewed from above. The physics are known; the difficulty is in the measurement (or in reality, parameterization) of the convolution of the sea-surface slope spectra (relative to the field of view) with the full sky radiance distribution, and the appropriate time-integration of the resulting at-sensor radiance time-series. With careful attention to detail, and under conditions approaching ideal, correspondence between water-leaving radiances determined from in-water approaches and above water measurements can be as good as 5%; under most realistic conditions, deviations >20% are more common. Current measurement approaches to the estimation of water-leaving radiances from above water platforms and caveats are discussed in Mueller et al. (2003).

7. Conclusions

Radiative transfer as related to optical remote sensing is a complex field that requires physical understanding of the absorption and scattering processes in the atmosphere and oceans, as well as sea surface and bottom reflection characteristics. Given this complexity, as briefly described in this chapter, the success with which IOP and particulate and dissolved properties have been derived from remotely sensed radiance is impressive, but many improvements remain to be made. The remainder of this book describes many of these inversions. In order to advance the field it will be necessary to obtain more detailed descriptions of the IOP and AOP, together with particulate and dissolved properties, particularly in coastal zones.

8. Acknowledgments

Support for this work is gratefully acknowledged from the National Aeronautics and Space Administration, Ocean Biology and Biogeochemistry Program, the Office of Naval Research Optics and Biology Program, and the Natural Sciences and Engineering Research Council.

9. References

Aas, E. 2000. Spectral slope of yellow substance: problems caused by small particles. Proceedings of Ocean Optics XV, 16-20 October, Monaco, Office of Naval Research, USA, CD-ROM.

Allocca, D. M., M.A. London, T.P. Curran, B.M. Concannon, V.M. Contarino, J. Prentice, L. J. Mullen, and T. J. Kane. 2002. Ocean water clarity measurement using shipboard lidar systems. Ocean Optics: Remote Sensing and Underwater Imaging, Robert J. Frouin and Gary D. Gilbert, [Eds.], Proceedings of SPIE 4488:106-114.

Austin, R.W. and G. Halikas. 1976. The index of refraction of seawater. SIO Ref. No. 76-1, Scripps Institution of Oceanography, La Jolla, 121pp.

Barnard,A.H., W.S.Pegau, and J.R.V.Zaneveld. 1998. Global relationships of the inherent optical properties of the ocean. Journal of Geophysical Research, 103:24,955-24,968.

Barth, J.A. and D. Bogucki. 1999. Spectral light absorption and attenuation measurements from a towed undulating vehicle. Deep Sea Research I., 47:323-342.

Behrenfeld, M., and P. Falkowski. 1997. A consumer's guide to phytoplankton primary productivity models. Limnology and Oceanography, 42(7):1479-1491.

Berwald,J., D.Stramski, C.D.Mobley, and D.A.Kiefe. 1995. Influences of absorption and scattering on vertical changes in the average cosine of the underwater light field. Limnology and Oceanography, 40:1347-1357.

Chandrasekhar, S., 1960, Radiative Transfer, Dover, New York, 393 pp.

Colbo, K.M. and M. Li. 1999. Parameterizing particle dispersion in Langmuir circulation. Journal Geophysical Research, In Press.

Dera, J., 1992. Marine physics. Elsevier, New York, NY, 515 pp.

Farmer, D.M. and M. Li. 1995, Patterns of bubble clouds organized by Langmuir circulation. Journal Physical Oceanography, 25:1426-1440.

Garver, S.A. and D.A. Siegel. 1997. Inherent optical property inversion of ocean color spectra and its biogeochemical interpretation: 1. Time series from the Sargasso Sea. Journal of Geophysical Research, 102:18,607-18,625.

Gordon, H.R., O.B. Brown, R.H. Evans, J.W. Brown, R.C. Smith, K.S. Baker, and D.K. Clark. 1988. A semianalytical radiance model of ocean color. Journal of Geophysical Research, 93:10,909-10,924.

Gordon, H.R., O.B.Brown and M.M.Jacobs. 1975, Computed relationships between the inherent and apparent optical properties. Applied Optics, 14: 417- 427.

Hoge, F.E., C.W. Wright, W.B. Krabill, R.R. Buntzen, G.D. Gilbert, R.N. Swift, J.K. Yungel, and R.E. Berry. 1988. Airborne lidar detection of subsurface oceanic scattering layers. Applied Optics, 27:3969-3977.

Hoge, F.E. and P.E. Lyon. 1999. Spectral parameters of inherent optical property models: Methods for satellite retrieval by matrix inversion of an oceanic radiance model. Applied Optics, 38:1657-1662.

Højerslev, N. and J.R.V.Zaneveld. 1977. A theoretical proof of the existence of the submarine asymptotic daylight field. University of Copenhagen Oceanography Series, Report #34, 16 pp.

Jerlov, N.G., 1976. Marine Optics, Elsevier, 231pp.

Johannessen, S.C., W.L. Miller, and J.J. Cullen. 2003. Calculation of CDOM absorbance spectra and UV attenuation from satellite ocean colour data. Journal of Geophysical Research, In Press.

Kirk, J.T.O. 1994. Light and photosynthesis in aquatic ecosystems, 2nd ed. Cambridge, 509 pp.

Lee, Z.P, K.L.Carder, and R. Arnone. 2002. Deriving inherent optical properties from water color: A multi-band quasi-analytical algorithm for optically deep waters. Applied Optics, 41:5755-5772.

Lee, M.E., and M.R. Lewis. 2003. A new method for the measurement of the optical volume scattering function in the upper ocean. Journal of Atmospheric and Oceanic Technology, 20(4):563-571.

Lewis, M.R., M.E. Carr, G. Feldman, W. Esaias, and C. McClain. 1990. The influence of penetrating irradiance on the heat budget of the equatorial Pacific Ocean. Nature, 347:543-545.

Liou, K-N, 1980. An introduction to atmospheric radiation. Academic Press, 392 pp.

Marshall, B.R. and R.C. Smith. 1990. Raman scattering and in-water ocean optical properties. Applied Optics, 29:71-84.

Mobley, C.D. 1994. Light and water: radiative transfer in natural waters. Academic, San Diego, CA, 592 pp.

Morel, A. and L. Prieur. 1977. Analysis of variations in ocean color. Limnology and Oceanography, 22:709-722.

Morel, A. 1988. Optical modeling of the upper ocean in relation to its biogenous matter content (case I waters). Journal of Geophysical Research, 93:10,749-10,768.

Morel, A. and B. Gentili. 1996. Diffuse reflectance of oceanic waters, III: implications of bidirectionality for the remote sensing problem. Applied Optics, 35:4850-4862.

Morel, A., K. J. Voss, and B. Gentilli. 1995. Bi-directional reflectance of oceanic waters: A comparison of model and experimental results. Journal of Geophysical Research, 100:13,143-13,150.

Morel, A. and S. Maritorena. 2001. Bio-optical properties of oceanic waters: a reappraisal. Journal of Geophysical Research, 106(C4):7163-7180.

Mueller, J.L., G.S. Fargion, and C.R. McClain [Eds]. 2003. Ocean optics protocols for satellite ocean color sensor validation. Revision 4, Volume IV. NASA, Goddard Space Flight Center, Greenbelt, MD.

Petzold, T.J., 1972. Volume scattering functions for selected ocean waters. Scripps Institution Oceanography, ref. 72-78.

Preisendorfer, R.W., 1965, Radiative transfer on discrete spaces. Pergamon, Oxford. 462 pp.

Preisendorfer, R.W., 1976, Hydrologic optics. U.S. Department of Commerce, National Oceanic and Atmospheric Administration, Environmental Research laboratories, 6 Volumes.

Roesler, C.S., and E. Boss. 2003. Ocean color inversion yields estimates of the spectral beam attenuation coefficient while removing constraints on particle backscattering spectra. Geophysical Research Letters, 30(9), doi: 10.1029/2002GL016366, 2003.

Roesler, C.S., and M.J. Perry. 1995. In situ phytoplankton absorption, fluorescence emission, and particulate backscattering spectra determined from reflectance. Journal of Geophysical Research, 100:13,279-13,294.

Shifrin, K. 1988. Physical optics of ocean water. American Institute of Physics, New York, 285 pp.

Smith, R.C., R.W. Austin, and J.E. Tyler, 1970. An oceanographic radiance distribution camera system. Applied Optics, 9:2015-2022.

Stramski, D., E. Boss, D. Bogucki, and K. Voss. 2004. The role of seawater constituents in light backscattering in the ocean. Progress in Oceanography, Submitted.

Twardowski, M.S., and P.L. Donaghay. 2001. Separating in situ and terrigenous sources of absorption by dissolved material in coastal waters. Journal of Geophysical Research, 106(C2):2545-2560.

van de Hulst, H.C., 1981. Light scattering by small particles, Dover, 470 pp.

Voss, K.J., 1989. Electro-optic camera system for measurement of the underwater radiance distribution. Optical Engineering, 28:241-247.

Voss, K.J., J.A. Chapin, and H. Zhang. 2000. An instrument to measure the bi-directional reflectance disribution function (BRDF) of surfaces. Applied Optics, 39:6197-6206.

Voss, K.J., C.D. Mobley, L.K. Sundman, J.E. Ivey, and C.H. Mazel. 2003. The spectral upwelling radiance distribution in optically shallow waters. Limnology and Oceanography, 48:364-373.

Walker, R.E. 1994. Marine light field statistics, Wiley, New York, NY, 692 pp.

Weidemann, A.D. R.H.Stavn, J.R.V.Zaneveld, and M.R.Wilcox. 1995. Error in predicting hydrosol backscattering from remotely sensed reflectance. Journal Geophysical Research, 100(C7):13,163-13,177.

Zaneveld, J.R.V., J. Kitchen, and H. Pak. 1981. The influence of optical water type on the heating rate of a constant depth mixed layer. Journal Geophysical Research, 86(C7):6426-6428.

Zaneveld, J.R.V. 1982. Remotely sensed reflectance and its dependence on vertical structure: a theoretical derivation. Applied Optics, 21:4146-4150.

Zaneveld, J.R.V., 1989. An asymptotic closure theory of irradiance in the sea and its inversion to obtain the vertical structure of inherent optical properties. Limnology and Oceanography, 34:1442-1452.

Zaneveld, J.R.V., and W.S. Pegau. 1998. A model for the reflectance of thin layers, fronts, and internal waves and its inversion. Oceanography, 11:44-47.

Zaneveld, J.R.V., 1995. A theoretical derivation of the dependence of the remotely sensed reflectance on the inherent optical properties. Journal Geophysical Research, 100(C7):13,135-13,142.

Zaneveld, J.R.V., E. Boss, and A.H. Barnard. 2001. The influence of surface waves on measured and modeled irradiance profiles. Applied Optics, 40(9):1442-1449.

Zaneveld, J.R.V., and W.S. Pegau. 2003. Robust underwater visibility parameter. Optics Express, 11(23):2997-3009.

Zhang, H., K.J. Voss, D. Mobley, L.K. Sundman, J.E. Ivey, and C.H. Mazel. 2003. Bidirectional reflectance measurements of sediments in the vicinity of Lee Stocking Island, Bahamas. Limnology and Oceanography, 48:380-389.

Chapter 2

AN INTRODUCTION TO SATELLITE SENSORS, OBSERVATIONS AND TECHNIQUES

[1]CHRISTOPHER W. BROWN, [2]LAURENCE N. CONNOR,
[3]JOHN L. LILLIBRIDGE, [4]NICHOLAS R. NALLI AND [3]RICHARD V. LEGECKIS

[1]*Cooperative Research Programs, Office of Research and Applications (ORA), National Environmental Satellite, Data and Information Service (NESDIS), National Oceanic and Atmospheric Administration (NOAA), College Park, MD, 20742USA*
[2]*Oceanic Research and Applications Division, ORA, NESDIS, NOAA, Camp Springs, MD, 20746 USA*
[3]*Oceanic Research and Applications Division, ORA, NESDIS, NOAA, Silver Spring, MD, 20910 USA*
[4]*Cooperative Institute for Research in the Atmosphere, Colorado State University, Camp Springs, MD, 20746 USA*

1. Introduction

Satellite remote sensing is an excellent tool for monitoring coastal waters. The periodic overpass of satellites allows the routine and cost effective collection of a variety of observations over large and often inaccessible expanses of the coast and adjacent waters within a short period of time. This suite of satellite observations satisfies the needs of numerous users. Coastal managers, recreational boaters, commercial fishermen, and environmental scientists are some of the many groups that utilize oceanographic products derived from remotely sensed observations. These products include: estimates of sea surface winds, temperature, sea level height, and chlorophyll concentration. Due to the variety of sensors, techniques and platforms employed, satellite observations differ in their temporal, spatial and spectral characteristics. The sampling domain of current satellites spans from hours to years in time and extends from meters to global in space (Fig. 1). Consequently, different remote sensing systems vary in their ability to meet the demands of a given application. In order to efficiently exploit these systems, the user must consider the capabilities and limitations of each and match an appropriate system to their target application and environment.

Processes and events within the coastal zone, which for our purpose extends from the littoral to the Exclusive Economic Zone (EEZ) or 370 nautical miles offshore, vary on scales that span several orders of magnitude in time and space (Fig. 1). The dominant variability in this dynamic region is on time scales of six hours to days and on spatial scales of meters to hundreds of kilometers. To prove useful, the satellite must be able to observe the phenomena in question at the appropriate scales of interest. Spatial resolution, field of view, and sampling frequency are consequently three key parameters that must be considered when selecting a sensor for an application. Requirements of spatial resolution, for example, vary widely. Imagery with a one-kilometer resolution may be suitable for large features on the continental shelf, but 30-meter resolution is

R.L. Miller et al. (eds.), Remote Sensing of Coastal Aquatic Environments, 21–50.
© 2005 *Springer.*

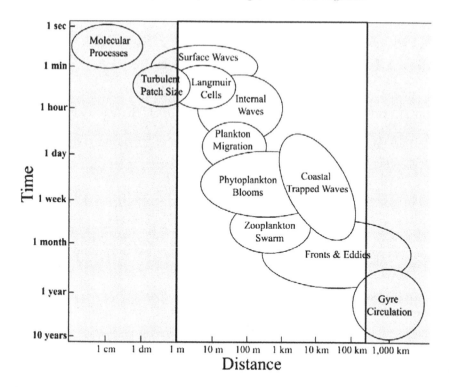

Figure 1. Space and time scales of various oceanic phenomena (adapted from Dickey, 1991). The unshaded region illustrates the appropriate spatial scales for coastal processes and for observations currently available from satellite remote sensing. Reproduced by permission of the author and publisher (American Geophysical Union).

likely too coarse for use in the littoral zone. These and other attributes are worth considering when choosing satellite observations in support of a project.

This chapter presents a brief overview of satellite-based oceanographic remote sensing systems by highlighting their capabilities and limitations for use in the coastal ocean. The reader is urged to consult publications with extended treatment of oceanographic remote sensing (e.g., Maul, 1985; Robinson, 1985; Stewart, 1985), sensor instrumentation and technology (e.g., Pease, 1991), or image processing (e.g. Russ, 1995; Schowengerdt, 1997) for detail discussions on these subjects. Our goal here is to present sufficient information about the diverse array of satellite systems and their observations to enable the reader to select the appropriate sensor(s) for an application. While our discussion focuses on the coastal ocean, the basic approach behind selecting a satellite system to study a particular phenomenon can be applied to the open ocean and inland waters as well.

We begin with the general theory and terminology of satellite remote sensing systems, starting with the nature of light and ending with the characteristics of satellite sensors. Secondly, an overview of general observational categories that each satellite system offers is presented. Finally, the various sensor systems and the techniques employed to retrieve oceanographic parameters are discussed, and the usefulness of each observing system in the coastal ocean is examined.

2. An Overview of Remote Sensing from Space

2.1 INTRODUCTION TO ELECTROMAGNETIC RADIATION

Satellite remote sensing exploits the properties of electromagnetic radiation (EMR), its intensity, frequency spectrum, polarization and time delay of received energy, to infer properties of a target. EMR is characterized interchangeably by its wavelength or frequency with the relationship $v = c/\lambda$, where c is the speed of light in the medium (3.0 x 10^8 m s^{-1}), v is the frequency (cycles s^{-1} or hertz, Hz) and λ is wavelength (m). Satellite oceanography typically employs EMR (Fig. 2), extending over seven orders of magnitude from visible light wavelengths (400-700 nm) to microwave frequencies (3.0 x 10^8 – 1.0 x 10^{11} Hz). Though oceanic sensors are beginning to explore the use of shorter ultraviolet wavelengths, such as 380 nm, radiation at these shorter wavelengths is strongly attenuated by the atmosphere. Typically, wavelength is used when discussing observations within the visible to thermal infrared spectrum, whereas frequency or wavenumber (=$1/\lambda$) is used to describe observations within the microwave.

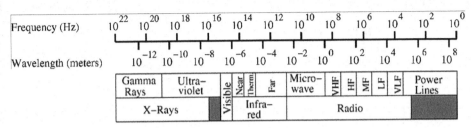

Figure 2. The electromagnetic spectrum.

All matter emits electromagnetic radiation. The properties and behavior of this radiation are described through Planck's quantum theory of blackbody radiation (Planck, 1959). A blackbody is an idealized object that absorbs all radiation striking its surface and re-emits radiation at a maximum rate that is dependent only upon temperature. The spectral characteristics and power of the emitted radiation are determined by the temperature of the blackbody. The power radiated at a given frequency per solid angle per surface area is given by:

$$B_v = \frac{2hv^3}{c^2} \frac{1}{\exp(hv/kT)-1},$$ (1)

where B_v is the spectral brightness of the black body, T is the temperature in degrees Kelvin, v is the frequency, c is the speed of light, h is Planck's constant, and k is Boltzmann's constant. The term brightness, B, is usually used to specify incoming radiant energy while the alternative term radiance, L, indicates outgoing radiant energy. Thus, mathematically, $B = -L$. The degree to which a real object is described by a blackbody radiator is often expressed through the ratio of the object's actual spectral brightness to the spectral brightness of a blackbody, namely it emissivity, e:

$$e = \frac{B_v}{B_{v(blackbody)}},$$ (2)

The emissivity of a blackbody, by definition, is unity. Objects in nature (rocks, vegetation, ice, parcels of air and water) emit radiation at a lower rate than a blackbody. It is this departure from the brightness characteristics of a blackbody that permits a material to be identified and quantified remotely. The emissivity of an object is primarily determined by its physical attributes, such as its chemical composition and surface roughness, and is essentially independent of the objects temperature. A material's emissivity is generally different in different parts of the EMR spectrum permitting the identification of most substances by their unique radiative properties (Thomas, 1990).

In addition to spectral radiances and brightness temperatures, wave polarization is another characteristic utilized in remote sensing. Polarization describes the spatial orientation of the electric and magnetic fields that compose an electromagnetic wave. Vertical and horizontal polarizations are the most common designations and refer to the orientation of the electric field of a wave. For example, a horizontally polarized wave has electric field lines parallel to the Earth's horizon, while a vertically polarized wave has field lines oriented perpendicular to the horizon. A considerable amount of information may be retrieved from measurements of specific polarization intensities and knowledge of how polarizations are affected by surface and atmospheric interactions.

Before a satellite receives an electromagnetic signal from earth, the EMR must pass through the atmosphere. The intervening atmosphere may modify the original signal by scattering or absorbing various spectral portions of the signal. This atmospheric effect presents several challenges to observing and interpreting the observations collected by satellites. Furthermore, some regions of the spectrum are less affected by the atmosphere, i.e. are more transparent, than others.

Earth observing satellites exploit regions of the EMR spectrum for which the atmosphere is relatively transparent. Figure 3 illustrates the atmospheric transmission of EM radiation from visible to microwave wavelengths. Note that the atmosphere is nearly opaque in part of the mid-IR (Infrared) and all of the far-IR regions. As a consequence, few earth observing systems utilize these wavelengths. By contrast, several regions, or "windows" exist, in the visible through thermal infrared wavelengths in which the surface can be viewed, and most of the microwave region is unimpeded by the atmosphere. As a consequence, retrieval of information from these regions is possible.

2.2 BASIC CHARACTERISTICS OF SATELLITE REMOTE SENSING SYSTEMS

The physical attributes of a satellite remote sensing system govern the properties of the observations and hence their application. A satellite system consists of several components, including sensors, a physical structure (i.e., bus), apparatus to maintain the satellite's orientation and orbital position, data storage devices, telemetry equipment to transmit the observations to Earth, and the ground segment that receives, processes and distributes the observations and derived parameters (Stewart, 1985).

2.2.1 *Satellite orbits*

Satellites travel in a roughly elliptical orbit around the earth, dictating the location, time and frequency that observations are acquired. Orbits are designed to meet the requirements of the satellite mission objectives and reduce the complexity of interpreting sensor observations. Knowledge of the satellite's orbital position and a sensor's viewing direction is required to construct a geographically referenced, two-dimensional image from individual observations acquired by the sensor.

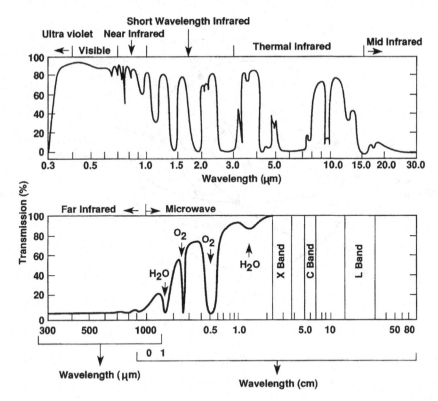

Figure 3. Atmospheric transmission as a function of wavelength within the electromagnetic spectrum employed for satellite oceanography (from Thomas, 1990). Reproduced by permission of the author and publisher (Joint Oceanographic Institutions).

The position of a satellite orbiting the Earth is specified by six Keplerian elements within a geocentric coordinate system: the semi-major axis of the elliptical orbit (a), the orbit's shape or eccentricity (e), the inclination (i) of the satellite's orbital plane relative to the Earth's equatorial plane, the right ascension of the ascending node (Ω), the argument of perigee (ω), and the true anomaly (θ) (Fig. 4). Together, these six orbital elements define the size and shape of the satellite orbit, specify the orientation of the satellite's orbit around the Earth, and describe the satellite's position in orbit.

The orbital elements are described within a coordinate system with Earth as its center. The size and shape of the satellite's orbit are defined by the semi-major axis a and its eccentricity e, the departure from a circular orbit of radius a. The right ascension, Ω, and inclination, i, position the satellite's orbital plane in space. The inclination i is the angle between the Earth's equatorial plane and the satellite's orbital plane, and may be fixed from $0° \leq i \leq 180°$. A satellite with an orbit of $i = 90°$ is in a polar orbit. If $i = 0°$, the orbit is equatorial. The inclination also prescribes the maximum latitudinal extent of the subsatellite or nadir track, the position of the satellite projected perpendicularly onto the Earth's surface (Fig. 4). For example, if $i = 65.0°$, the maximum latitude of the subsatellite track in $\pm 65.0°$ latitude.

Two broad classes of satellite orbits exist. Polar orbiting satellites have an orbital inclination near 90°. These 'low-earth orbit' (LEO) satellites typically fly at an altitude

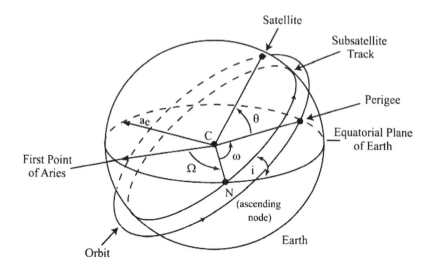

Figure 4. Coordinates and orbital elements describing a satellite's orbit around the Earth (after Stewart, 1985). Reprinted with permission of the author.

of 700 to 800 km above earth and acquire observations at nearly all latitudes while revolving around the earth approximately 14 times a day. A variety of specialized polar orbits exist, such as the sun-synchronous orbit. If the inclination and height of the satellite orbit are carefully selected, a sun-synchronous orbit can be established where the rotation of the satellite's orbital plane will match the rotation of the Earth about the sun, i.e. approximately one degree per day. As a consequence, the satellite crosses the equator (and each latitude) at the same local solar time each day (Stewart, 1985). Satellites with visible ocean color sensors occupy sun-synchronous orbits. Sun-synchronous orbits, however, are not ideal for altimetric missions that measure sea surface height because they alias the tidal signal into low frequencies that cannot be adequately resolved. Instead, orbits with $65° > i > 120°$ are employed to alias the tides into frequencies greater than two cycles per year and permit the satellites to directly measure tides (Stewart, 1985).

In contrast to polar orbiting satellites, geosynchronous or geostationary satellites are in an orbit fixed to the earth's equatorial plane ($i = 0°$), with a significantly higher altitude of 35,800 km. These satellites travel at the same angular rate as the earth's rotation, appearing to remain stationary over a fixed position on Earth. At this high altitude, the sensor can provide observations of a particular region at essentially any desired time interval. For example, the NOAA Geostationary Operational Environmental Satellites (GOES) provide visible and infrared images every 30 minutes with a spatial resolution of 1 and 4 kilometers. Due to their position over the equator, geostationary platforms offer hemispheric coverage, though retrieval of geophysical parameters is questionable beyond ~60° latitude (Stewart, 1985). The high frequency of observations available permits the monitoring of diurnal change which makes geostationary platforms indispensable for weather forecasting and advantageous for examining the dynamic coastal ocean. In addition to observing short-term variability in the ocean and atmosphere, the frequent measurements can be employed to effectively increase the area of cloud-free observations through cloud filtering and compositing of multiple images over a specified period. For an expanded introduction to satellite orbits, see Brooks (1977), Logsdon (1998), Stewart (1985) and references therein.

2.2.2 *Sensor attributes and observational characteristics*

Space-borne remote sensing instruments measure the intensity of received radiation at some polarization and within a prescribed wavelength band and utilize a myriad of physical principles and engineering techniques to acquire the observations for a particular geophysical parameter. The following subsections introduce important concepts and terminology useful in understanding the relationship between sensor attributes and observations.

While all sensors measure energy emitted or reflected from a 'target' on earth, passive sensors rely on an external energy source such as the Sun, whereas active sensors transmit an EMR signal with specific characteristics and measure energy reflected back to it by the target. For example, thermal infrared sensors are passive sensors, while microwave scatterometers and altimeters are active sensors.

Sensors can be classified as imaging or non-imaging. Non-imaging sensors measure a target and record the observation as a single electrical signal. Imaging sensors, by contrast, use the measured electrical signal to photoelectrically drive a 2-D imaging device, such as charge coupled device (CCD) array.

Instruments provide either point or swath measurements. To obtain swath measurements, some form of scanning is employed, such as cross-track (Fig. 5A) or along-track (Fig. 5B), whereby sequential observations are pieced together to form a row or column. Space-borne radiometers usually incorporate a scanning mechanism whereby the observing beam is swept transverse to the orbital motion, providing a wider swath of coverage. Combined with the forward motion of the satellite, a 2-D image is generated. AVHRR, SeaWiFS and Landsat TM all employ cross-track scanning. In along-track scanners, multiple parallel observations of the earth's surface are acquired simultaneously along the path of the satellite using detectors oriented orthogonal to the satellite's motion (Fig. 5B). The SPOT sensor is one example of these "pushbroom scanners". In some cases, conical scanning is employed to view a location on earth at a fixed angle with respect to the vertical.

Resolution is defined as the ability to discriminate between targets and operationally represents the smallest resolvable scale of an observation. The four major categories in which resolution is an important attribute (spatial, temporal, spectral and radiometric) are described briefly below.

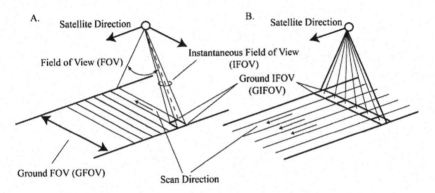

Figure 5. Cross track (A) and along track (B) scanners.

The spatial resolution of a sensor is dictated by its Instantaneous Field of View (IFOV). The IFOV is the angle of view from which a signal is received by a sensor (Robinson, 1985) and is represented by the angle subtended by a single detector element of an optical system (Fig. 5) and the geometry of the antenna. IFOV is independent of sensor altitude. The geometric projection of the IFOV onto the ground is called the Ground-projected Instantaneous Field of View (GIFOV) or Ground Sample Distance (GSD), and is determined by IFOV and sensor height (Schowengerdt, 1997). GIFOV is the spatial resolution of the sensor. In combination with sampling rate, the GIFOV determine the spatial dimensions of the picture element (i.e. pixel) in an image.

Similarly, the angular extent of observations acquired perpendicular to the satellite's path is defined as the Field of View (FOV). The dimensions of the projected FOV onto the ground is the Ground Field of View (GFOV) or swath width (Schowengerdt, 1996; Fig. 5).

Because the acquisition of an observation by a satellite sensor is essentially instantaneous, temporal resolution refers to the frequency with which the same location on Earth will be observed by the sensor. The revisitation rate of a region is dependent upon several factors, including satellite orbital characteristics, the latitude of the observed area, and the field of view of the instrument (swath vs. point measurements). Instruments with a single, fixed viewing direction are dependent upon the satellite's orbital characteristics. The orbital cycle, the period between exactly repeating orbits, ranges from 2 to 35 days. In general, repeat frequency is inversely related to the spatial resolution in polar orbiting sensors; the higher the spatial resolution of the observation, the lower the revisitation rate. Sensors that possess the capability to point and observe the same region over sequential satellite passes effectively increase their revisitation rate for a particular location. Flying numerous sensors in a "constellation" may also increase the repeat frequency of observations. Locations at higher latitudes are more frequently observed than those near the equator with polar orbiters.

Satellite sensors vary in the number and bandwidth of their spectral channels and consequently in their ability to observe discrete spectral regions. Sensors possess a single broad panchromatic spectral band (to provide high spatial resolution observations), multispectral to hyperspectral bands (to acquire images in many spectral bands of intermediate to narrow bandwidth), or a combination of both.

Radiometric resolution refers to the fineness at which the observed measurements can be resolved into discrete radiometric intervals. Radiance within the GIFOV measured by a detector is converted to an electrical signal and a discrete integer value. Only a finite number of bits can be used to code the continuous data measurements as binary numbers onto a recording device. The larger the number of bits recorded, the more closely the quantized data approximates the original data, and the higher the radiometric resolution of the sensor (Schowengerdt, 1997). SeaWiFS (Sea-viewing Wide Field-of-view Sensor) and MODIS (Moderate Resolution Imaging Spectrometer), for example, possess radiometric digitizations of 10 and 12 bits, respectively.

SNR is a measure of signal strength relative to background noise and is one of the key scientific requirements for retrieving geophysical parameters from satellites, especially bio-optical properties and sea surface temperature where the strength of the signal emanating from the ocean is weak.

Comparisons of instrument quality often refer to SNR. In making such comparisons, however, it is important to remember that a measured SNR depends upon the input signal and spectral position (W. Esaias, pers. comm.).

3. Observational Categories and Corresponding Sensors

Various physical and biological properties of the coastal ocean can be retrieved remotely from satellites. Retrieval in this context refers to the quantitative procedure employed for estimating a geophysical parameter from radiation measurements. Table 1 lists the oceanographic parameters that are derivable from present and future satellite remote sensing missions, the electromagnetic spectrum or observational category employed to derive them, and the names of some of the sensors that supply these observations. Note that a single parameter or class of parameters, such as bio-optical properties, may be derivable only from measurements collected using a specific observational category. Conversely, a single geophysical parameter may be derived from observations from different observational categories. Sea ice, for example, can be detected and monitored using VNIR, scatterometry and SAR, though the individual sensors may provide slightly different estimates. Used in combination, the observations complement each other.

Wind speed over the ocean can be derived from passive instruments, such as microwave radiometers, and both active radar altimeters and scatterometers, but only scatterometers can provide wind direction.

Though it is possible to retrieve sea surface salinity (SSS) from microwave observations, i.e. in the low frequency L-band, no such sensor is currently flying. Two future sea surface-salinity missions, Aquarius (Koblinsky et al., 2003) and the Soil Moisture and Ocean Salinity (SMOS) mission, are planned for launch in 2008.

Table 1. A listing of remotely sensed oceanographic parameters, their observational / instrument class and representative sensors.

Parameter	Observational Category	Example Satellite/Sensors
Bio-Optical	Visible – Near Infrared	ENVISAT/MERIS, AQUA/MODIS, OrbView-2/SeaWiFS
Bathymetry	Visible – Near Infrared	Landsat, SPOT, IKONOS
Sea Surface Temperature	Thermal Infrared Microwave Radiometers	POES/AVHRR, GOES/Imager DMSP/SSM/I, TRMM/TMI
Sea Surface Salinity	Microwave Radiometers & Scatterometers	---
Sea Surface Roughness, Wind Velocities, Waves & Tides	Microwave Scatterometers & Altimeters Synthetic Aperture Radar	ERS-1 & -2/AMI QuikSCAT RADARSAT-1
Sea Surface Height, Wind Speeds	Altimeters	Topex/Poseidon, Jason-1
Sea Ice	Visible – Near Infrared Microwave Radiometers, Scatterometers & Altimeters Synthetic Aperture Radar	POES/AVHRR DMSP/SSM/I ERS-1 & -2/AMI RADARSAT-1
Surface Currents, Fronts & Circulation	Visible – Near Infrared, Thermal Infrared Microwave Scatterometers & Altimeters	POES/AVHRR, GOES/Imager Topex/Posiedon, Jason-1
Surface Objects – Ships, Wakes & Flotsam	Synthetic Aperture Radar	RADARSAT-1, Envisat/ASAR

4. Ocean Remote Sensing Systems

4.1 VISIBLE – NEAR INFRARED OCEAN COLOR

Visible - near infrared (VNIR) instruments measure the upwelling radiant flux emanating from the top of the Earth's atmosphere at discrete visible and near-infrared wavelengths. As a consequence, observations are limited to daylight hours. VNIR sensors are grouped into two, somewhat arbitrary categories based on their principal environment of use and function. Those sensors primarily for terrestrial use, with its high reflectance signal and an emphasis on spatial detail, are designed to acquire high spatial resolution observations over broad to panchromatic spectral regions and are commonly called VNIR sensors. Landsat, Ikonos and SPOT belong to this category. Those sensors with aquatic oriented applications that must contend with the low water-leaving signal, sometimes 1% or less of downwelling irradiance, are referred to as "ocean color" sensors. Ocean color sensors require a set of narrow, sensitive spectral channels, to remove atmospheric effects and resolve the spectral signals used to estimate phytoplankton abundance and other radiatively active constituents. They generally offer multispectral observations with a spatial resolution coarser by an order of magnitude than terrestrial VNIR sensors. This classification, however, does not preclude ocean color sensors from supplying information about land surfaces, such as the estimation of Normalized Difference Vegetation Index (NDVI), or terrestrial sensors from operating within the aquatic environment, such as mapping the location of coral reefs.

Visible radiation is the only portion of the electromagnetic spectrum that appreciably penetrates into the water column, reaching depths of up to 30 m under clear conditions. This capability permits the retrieval of particulate and dissolved substance concentrations and inherent optical properties within the ocean's surface layer and the mapping and monitoring of subsurface, such as shallow-water ($\leq$ 10 m) bathymetry, coral reefs and aquatic vegetation. In addition, its short wavelength allows it to resolve targets at fine resolution.

4.1.1 *Basic theory of observations*

Ocean color is defined as the spectrum of water-leaving radiances (L_w) or reflectance ($\rho = \pi L_w / F_o \cos \theta_o$, where F_o is the extraterrestrial solar irradiance and θ_o is the solar zenith angle) exiting the water column. All bio-optical quantities of interest, such as chlorophyll concentration, are derived from these values. Ocean color observations are acquired by measuring the radiances of different visible wavelengths at the sensor, computing and subtracting components of this total radiance due to specular reflection and the atmospheric contributions, and converting the resulting water-leaving radiances into meaningful geophysical parameters, such as chlorophyll and suspended sediment concentrations. The spectral radiances measured by the satellite, L_t (λ), can be partitioned as:

$$L_t(\lambda) = L_r(\lambda) + L_a(\lambda) + L_{ra}(\lambda) + L_g(\lambda) + t(\lambda)L_w(\lambda), \tag{3}$$

where L_r is the radiance resulting from multiple (Rayleigh) scattering by air molecules in the absence of aerosols, $L_a(\lambda)$ is the radiance resulting from multiple scattering by aerosols, L_{ra} is the interaction term between molecular and aerosol scattering, L_g is due to the reflectance of the solar beam from the ocean surface, and L_w is the water-leaving radiance (Gordon and Wang, 1994). The water-leaving radiances, the property of interest, are reduced by the atmospheric diffuse transmittance, $t(\lambda)$, from the water's surface to the satellite.

Given the measurement of $L_t(\lambda)$ by the sensor, retrieval of L_w (λ) requires estimation of the remaining terms on the right-hand side of Eq. (3). Specular reflection L_g can be ignored or reduced if the sensor can tilt the scan plane away from the specular image of the sun, such as is the case for SeaWiFS, and affected areas may be excluded during image post-processing. Removal of the atmospheric contribution to $L_t(\lambda)$, which represents 90% or more of the total signal, is critical in accurately retrieving any bio-optical or bathymetric properties. This process, called atmospheric correction, traditionally employs ancillary meteorological data and makes several assumptions to estimate the individual components, with the aerosol terms being most problematic (Gordon and Wang, 1994; Siegel et al., 2000; Stumpf et al., 2003; Wang, 1999; Wang and Gordon, 1994).

The resulting water-leaving radiances (L_w) or equivalent reflectance (ρ_w) are used to estimate the desired bio-optical property through empirical or semi-empirical methods (Carder et al., 1999; D'Sa et al., 2002; Maritorena et al., 2002; O'Reilly et al., 1998). Light with known spectral properties enters a body of water body and its spectral character is altered by suspended particulates and dissolved substances within the water through scattering and absorption. Chlorophyll, for example, absorbs primarily within the blue, with a secondary peak in the red blue, and backscatter in the green. Changes in the blue/green ratio are used to estimate phytoplankton abundance. A portion of the light containing this spectral information exits the water body, i.e. water-leaving radiances, and is related to the geophysical property.

Traditionally, bio-optical algorithms, such as the one for chlorophyll concentration, C, have taken the general form:

$$C = A\left[\frac{L_w(\lambda_i)}{L_w(\lambda_j)}\right]^B , \tag{4}$$

where A and B are empirical constants (Gordon and Morel, 1983).

A recent and ongoing experimental development in satellite ocean color processing is the simultaneous estimation of atmospheric and oceanic properties in a radiatively coupled atmosphere-ocean model, (e.g., Stamnes et al., 2003). Details about bio-optical algorithms and atmospheric correction will be described in the chapter on ocean color analysis by Mueller-Karger et al. (Chapter 5).

4.1.2 Sensors – past, present and future

The Coastal Zone Color Scanner (CZCS) (Hovis et al., 1980) onboard Nimbus-7 was the first satellite "ocean color" sensor and demonstrated that bio-optical properties could be extracted from space-borne measurements within the visible - near infrared spectrum. A scanning radiometer that observed the ocean in six coregistered spectral bands, CZCS was a proof-of-concept mission with a limited duty schedule. Image coverage was often poor, varying temporally and spatially during the mission. With the limited number and placement of the spectral bands of the CZCS, atmospheric correction and bio-optical algorithms were derived using assumptions valid only for the open ocean.

Spatial, spectral, and temporal resolution is increasing in satellite ocean color sensors, based on a demonstrated need for higher precision and accuracy at very low chlorophyll levels to address vast open ocean provinces, and to accommodate improvements in atmospheric and bio-optical algorithms.

The new generation of ocean color sensors significantly improves our ability to monitor the coastal environment. The coastal environment presents several obstacles for the use of satellite ocean color data, both in terms of the frequency of data necessary to monitor this highly variable environment, and in the complexities of deriving useful geophysical parameters.

Over the next decade, polar orbiting platforms will carry several visible - near infrared sensors, such as the SeaWiFS, MODIS, and the planned Visible/Infrared Imager Radiometer Suite (VIIRS), capable of furnishing highly accurate radiances with high spectral and one kilometer spatial resolution at a global revisit period of approximately two days (Table 2). These rates of coverage adequately image the larger features on the continental shelf and in the open ocean. Phytoplankton chlorophyll and productivity, water clarity and other products are routinely produced from this data for a variety of applications including Naval operations, fisheries, and global carbon modeling.

Closer to shore and in bays and estuaries, features are smaller and change more rapidly. Sensors with a spatial resolution on the order of 100 to 300 m are required. These intermediate resolution systems are sufficient for many coastal ocean applications that require the imaging of dynamic water features, such as fronts and red tides. The high-resolution mode of Medium Resolution Imaging Spectrometer (MERIS, 300 m) may be particularly useful for studies of coastal features that require better than 1-km resolution, but also large area coverage.

Applications that require the retrieval of coastal or bottom features or manmade objects, such as fish farms, require 30 m or better resolution. In addition, the system must have many narrow spectral channels and a high Signal-to-Noise Ratio (SNR) to distinguish and separate the various constituents in coastal water applications.

Existing sensors, such as MERIS and MODIS, have the appropriate signal-to-noise ratio and a suite of spectral bands for open ocean waters plus additional bands suitable for measuring chlorophyll fluorescence, suspended sediments and for performing atmospheric correction over coastal waters. Imaging at 1-km resolution allows for imaging an entire continental shelf and is particularly useful for addressing large-scale ocean forcing, such as El Niño effects, on coastal ocean dynamics.

Terrestrial oriented satellite instruments, such as Landsat Thematic Mapper (TM), the Enhanced Thematic Mapper Plus (ETM+), and Satellite Pour l'Observation de la Terre (SPOT), with their high spatial resolution, have been used to map shallow bottom features and adjacent land areas. These land sensors, designed with moderate SNR and broad spectral bands for the brighter land signals, are not suited for coastal ocean applications because they lack the attributes required for accurately retrieving bio-optical properties. For instance, both the blue and green wavelengths are measured in the first band of Landsat and SPOT and the reflectance in these two important spectral regions cannot be used independently to estimate chlorophyll concentration.

The Enhanced Thematic Mapper Plus (ETM+), aboard Landsat-7, provides Landsat data continuity and improved land remote-sensing capabilities. ETM+ replicates the capabilities of the successful Thematic Mapper instruments on Landsats 4 and 5 and enhances its capabilities by including a 15-meter resolution panchromatic band, a Long Wave Infrared 60-meter resolution band, and six 30-meter resolution multispectral bands (Table 2). The ETM+ also includes new features that make it a more versatile and efficient. The ETM+ instrument is a fixed whisk-broom, eight-band, multispectral scanning radiometer capable of providing high-resolution imaging information of the Earth's surface. An ETM+ scene has a GIFOV of 15, 30, and 60 meters in bands 8, 6, and 1-5 and 7, respectively.

Individual polar orbiting ocean color sensors provide an inadequate number of observations to resolve processes operating at hourly to daily time scales, such as those characteristics of the coastal ocean. More frequent observations are required to observe these ephemeral events and processes. To study short-term events, data from polar orbiting sensors need to be augmented by comparable data with a higher revisitation rate. This can be accomplished in part by operating and combining the observations of several polar orbiting ocean color sensors with different equatorial crossing times, or more adequately by flying an ocean color sensor on a geostationary platform.

Geostationary platforms offer the opportunity to revisit a location as frequently as desired and remedy the coverage constraints imposed by polar orbiting platforms. Coastal features that vary in time and space in response to weather and tides may be best observed from a geostationary satellite instrument with 300 m or finer spatial resolution. These geostationary ocean color observations could acquire relatively high SNR by "staring" at a location using a 2-D imaging system and integrating the signal in each pixel and spectral channel over time. Applications could include, but are not limited to: 1) quantifying the response of marine ecosystems to short-term physical events, such as passage of storms and tidal mixing; 2) monitoring biotic and abiotic material in transient surface features, such as river plumes and tidal fronts; 3) detecting, tracking, and predicting the location of hazardous materials, such as oil spills, ocean waste disposal, and noxious algal blooms; 4) initializing, validating, and running coupled biological-physical coastal ecosystem models; and, 5) assessing the effect of tidal aliasing and sub-pixel variability on global estimates from polar orbiting observations. Although no sensors of this type are currently in operation, the High Resolution Visible and IR Sensor (HiRVIS) of the Korea Aerospace Research Institute and the GOES-R Hyperspectral Environmental Suite (HES-CW; Table 2) of the National Oceanic and Atmospheric Administration (NOAA), are being considered for launch in 2008 and 2012, respectively.

Only a few of the many available ocean color sensors are mentioned here. The International Ocean Colour Coordinating Group (IOCCG) maintains a web site with information on these and other past, present and future sensors.

4.1.3 *Coastal capabilities and limitations*

As noted in the preceding section, the variety of VNIR sensors available provide observations with spatial resolutions ranging from 10's of meters to kilometers, making them useful in resolving the fine to mesoscale features of the coastal zone. These observations, which are currently collected from polar orbiting satellites with a revisit rate of approximately two or more, do not adequately resolve the short-term events and processes characteristic of the coastal environment. More frequent observations are required to observe these ephemeral events and processes.

The coastal environment is optically complex and retrieval of in-water constituents is difficult. The inherent difficulty associated with atmospheric correction of satellite ocean color imager is made more difficult by the potential influence of the bottom and the highly scattering optical environment in coastal waters. Many of the assumptions employed during atmospheric correction are invalid. Also, the bio-optical complexity of these waters makes it very difficult to separate the optical signal into its radiatively active constituents. In contrast to open ocean waters where the in-water optical environment is a function primarily of phytoplankton and their co-varying by-products (Case 1 waters), the optical properties of coastal waters are attributable to a mixture of phytoplankton, particulate organic detritus, suspended sediments, and chromophoric dissolved organic matter that may vary independently of one another (Case 2 waters). Furthermore, in optically shallow waters where the bottom influences the water-leaving radiances, the bottom signal must be adequately removed to ascertain the water column

Table 2: Characteristics of several modern visible – infrared satellite sensors. Abbreviations: HS: hyperspectral, MS: multispectral, and Pan: panchromatic.

Sensor / Mission	Satellite	Dates of Operation	Swath (km)	Resolution (m)	Number of Vis-NIR Bands	Spectral Coverage (nm)	Repeat Viewing Frequency (days)
AVHRR	NOAA			1100			1 – 2
CZCS	Nimbus-7	24/10/78 - 22/06/86	1556	825	5	433 – 800	1 – 2
OCTS	ADEOS	17/08/96 - 1/07/97	1400	700	8	402-12500	3
GLI	ADEOS-II	14/12/02 - 25/10/03	1600	250/1000	6/23	460-2210/375-1380	4
MSS	Landsat 4-5	16/6/82- 8/3/94-	185	80	4	500-1100	16
TM	Landsat 4-5	16/6/82- 8/3/94-	170	30	6	450-2350	16
ETM+ (MS/Pan)	Landsat 7	15/4/99 -	183	30/15	6/1	450-2350/520-900	16
MultiSpectral / Panchromatic	IKONOS	24/09/99 -	11/5	4/1	4/1	445 – 900	3
HRG (MS / Pan)	SPOT 5	05/04/2002 -	60	10/5, 2.5	4/2	500-1750/480-710	1 - 4
VEGETATION 2	SPOT 5	05/04/2002 -	2,250	1000	4	450 - 1750	1
SeaWiFS	OrbView 2	01/08/97 -	2806	1100	8	402 - 885	1 – 2
MODIS	Terra / Aqua	18/12/99 / 04/05/02 -	2330	250/500/1000	2/5/12	620-876/459-2155/405-965	1 – 2
OCM	IRS-P4	26/05/99 -	1420	350	8	402-885	1 – 2
MOS	IRS-P3	21/03/96	200	520	18	400-1600	
MERIS	ENVISAT-1	01/03/02 -	1150	300/1200	15	390-1040	1 - 3
S-GLI	GCOM	~ 2007	1600	750	11	412 – 865	N/A
VIIRS	NPP / NPOESS	~ 2006 / ~2009	3000	750	22	402-11800	1 - 2
Imager	GOES I-M	04/94 -	N/A	1		550-750	30 minutes
HES CWI	GOES-R	~ 2012	400	250	8 to HS	440 – 1000	30 minutes

contribution and retrieve estimates of its constituents, or vice versa to estimate bathymetry and surface features. Fortunately, various techniques are being developed to overcome these problems.

4.2 THERMAL INFRARED

Thermal infrared sensors provide estimates of the sea surface temperature. Strictly speaking, the radiometric skin temperature of the water, not the bulk temperature of the underlying water, is detected by the sensor.

4.2.1 Basic theory of observations

IR wavelengths are well suited for remote sensing of sea surface temperature (SST) from satellites because the earth's thermal emission peaks in the infrared (IR) spectrum, and the emissivity of water in the IR is very close to unity. In spite of this, there are two physical problems that must be surmounted to derive accurate satellite SST (Nalli and Smith, 2003). The first problem is to account for the surface-leaving radiance, which varies with surface roughness (i.e., sea state) and atmospheric downwelling radiance. These variations are relatively small and only become significant for applications requiring a high degree of absolute accuracy and precision (e.g., ~0.1°C rms required for climate change detection). The second problem is to account for the attenuation of this source term through the intervening atmosphere, namely clouds, gases (primarily water vapor), and aerosol. The atmosphere can impact IR window brightness temperatures by as more than 1 °C.

Retrieval algorithms assume some degree of cloud-free radiative transfer, even in partly cloudy situations because clouds are typically opaque in the IR,. Therefore, the problem of cloud detection, a difficult undertaking in its own right, is usually treated as a problem distinct from retrievals. Atmospheric water vapor, aerosol and temperature, on the other hand, must be corrected for. To mitigate the impact on satellite observations, sensors are designed to operate in spectral windows where the atmosphere is semi-transparent. Two window regions commonly used for SST are the long-wave IR (LWIR; 10.0–12.5 μm) and the short-wave IR (SWIR; 3.7–4.2 μm) regions. Unfortunately, even within these windows, thermal IR radiation emitted from the ocean surface is partially absorbed by the atmosphere. The amount of absorption is different for each IR window and these differences can be used to correct for atmospheric effects and estimate the temperature at the ocean surface. Algorithms are thus used to retrieve SST from cloud-free, multispectral IR window-channel radiance observations (e.g., Anding and Kauth, 1970; McClain et al., 1985; McMillin, 1975; Merchant et al., 1999; Nalli and Smith, 2003).

The typical retrieval of SST from satellites relies on empirical relationships between SST and observed brightness temperatures (predictors) of the form:

$$T_{b_i} = b_0 + \sum_i b_i T_i, \tag{5}$$

where T_b is the brightness temperature for observation i. This methodology is known as a statistical retrieval (e.g., Kidder and Vonder Haar, 1995). The coefficients b_i are determined from statistical regression of T_b against T_i given a representative training data sample. In practice, there are two ways of performing the training regression analyses: (1) empirically against coincident *in situ* buoy-measured SST (e.g., McClain et al., 1985; Walton et al., 1998); or, (2) synthetically from radiative transfer model calculations (e.g., Merchant et al., 1999). Figure 6 illustrates the observed linear relationship between channel brightness temperatures and SST measured by *in situ* buoys (~ 1 meter depth) that are coincident in space and time. The strong correlation (e.g, $R^2=0.949$), and statistical significance (e.g., $F=2296.1$ and $p=0$) of such observed data justifies the statistical retrieval algorithms based upon Eq. 5. One potential means for improving radiative transfer based retrievals is to calculate a direct solution of the IR radiative transfer equation. Algorithms of this type are based upon radiative transfer physics instead of inferential statistics and are thus referred to as physical retrieval methods (see Kidder and Vonder Haar, 1995)

Apart from their simplicity, the primary advantage of empirical approaches is that they implicitly account for potential sources of uncertainty, including surface and atmospheric effects, instrument noise, and calibration. Furthermore, non-IR parameters may be used as predictors. A recent example of this has been the implementation of aerosol bias-correction algorithms derived using slant-path aerosol optical depth retrievals from solar-reflectance channels (Nalli and Stowe, 2002). The disadvantages of empirical retrievals include the difficulty in acquiring high-quality, globally representative buoy observations and accounting for uncertainties caused by the differences between the radiometric skin SST (detected by the satellite) and *in situ* SST (measured by the buoy at ~1 m depth). Radiative transfer based statistical retrievals have been devised to circumvent these disadvantages (e.g., Merchant et al. 1999), but nevertheless rely upon a parametric linear equation such as Eq. (5).

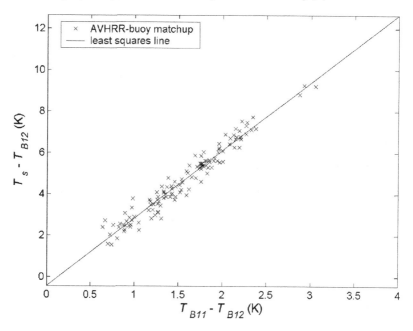

Figure 6. Observed linear relationship between cloud-free window channel brightness temperatures (11 and 12 µm) and buoy-measured SST matched up in space and time. The data shown here are taken from the Pathfinder Oceans Matchup Database (PFMDB; Kilpatrick et al., 2001) for daytime drifting buoy matchups during January 1999.

The accuracy of SST in cloud free regions derived from thermal IR data, assessed by comparing them against *in situ* buoy measurements at 1-meter depth, is approximately 0.5°C. Some of the error may be attributed to the diurnal SST cycle between the bulk and skin temperatures that can differ by several degrees Celsius during summer months under low wind speed conditions, especially at lower latitudes. Mutichannel SST accuracy can be degraded by intermittent atmospheric aerosols from volcanoes, smoke and desert sands.

4.2.2 Sensors – past, present and future

The first Advanced Very High Resolution Radiometer (AVHRR) went operational in 1978 (Table 3). It had three IR channels and a spatial resolution of 1-km for coastal oceans and 4-km for the entire globe. In 1982, the first multi-channel sea surface temperature (MCSST) product was introduced and has provided a continuous global data set. However, large volcanic eruptions in 1982 (El Chichón) and in 1991 (Pinatubo) resulted in a prolonged increase of atmospheric aerosols that introduced a bias in the MCSST that was not corrected in real time. Another bias emerged when the satellite equatorial crossing time increased due to a slow orbital drift and diurnal effects became apparent. Consequently, a Pathfinder project was initiated to remove the bias. The resulting retrospective Pathfinder MCSST Data Set is distributed through the Jet Propulsion Laboratory.

To improve SST accuracy, the Along Track Scanning Radiometer (ATSR) was launched on ERS-1 in 1991. The ATSR possesses the same IR channels as the AVHRR, but exhibits lower instrument noise. While the AVHRR scanned across its flight track, the ATSR had a conical scan design that provided a nadir and forward look at the same area of the ocean. In essence, the same ocean target was observed through different atmospheric path lengths, allowing atmospheric correction to be more accurately estimated for the SST. The conical scan, however, covered a swath of only 500 km and the satellite was in a 35-day repeat cycle, so it took much longer to create global SST maps. The Advanced ATSR on Envisat was launched in 2002 with improved cloud detection and calibration options.

In 1994, GOES-8 was launched. It had three IR channels at a spatial resolution of 4 km and a repeat cycle of 30 minutes. This frequency of observations allowed a new test for clouds since in one hour (using 3 IR images) it was possible to distinguish between nearly stationary ocean SST features and rapidly moving clouds. With two GOES satellites positioned above the equator at 75°W and 135°W, an operational GOES SST product was introduced at hourly intervals and provides SST in the Atlantic and Pacific oceans between longitudes 180°W and 30°W. The GOES SST product incorporated radiative transfer model calculations to estimate the SST algorithms and to detect clouds. The GOES SST product is readily available in near real-time from NOAA's Satellite Active Archive (SAA). In 2003, the new GOES-12 (M) was launched but it has only two IR channels since the 12 μm channel was replaced by a 13 μm channel to facilitate the detection of clouds for atmospheric applications. Since the 3.9 channel is sensitive to solar radiation, SST accuracy is affected adversely during the day.

The MODIS radiometers on the NASA polar orbiting Terra and Aqua satellites provide a test of new multi-channel (5 wavelengths) SST estimates using multi-detector arrays instead of a single detector. MODIS obtains global measurements at a spatial resolution of 1-km and therefore provides better cloud discrimination. The MODIS also provides a test of the new technologies that will be used in the future VIIRS instrument that is being designed for the National Polar-orbiting Operational Environmental Satellite System (NPOESS) project in the U.S. The VIIRS should be ready for launch in 2007 on the NPOESS Preparatory Project (NPP) satellite and it will be the first satellite system used for both operational and research applications.

The European contribution to the satellite systems will include AVHRR/3 which will be launched on METOP and provide a global polar-orbiting sensor in a morning equatorial crossing time while NOAA will provide the afternoon polar orbiter. In addition, the SEVERI radiometer on the new geostationary Meteosat Second Generation (MSG) satellite was launched in 2002 and positioned at longitude 0°W, providing a view of Europe, Africa and adjacent seas. With three IR channels, the MSG will provide hourly SST estimates during the next decade. In 2012, the Advanced Baseline instrument (ABI) on the NOAA GOES-R satellites will provide 2 km spatial resolution, 15-minute full disk images, and three IR channels suitable for SST. In addition, there are a series of atmospheric sounders with spectrometers that provide nearly continuous IR spectra such as HES on GOES-R, AIRS on Aqua, CrIS on NPP and IASI on METOP. The spatial resolution of the sounders is coarse (~15 km), but it will be possible to investigate the advantages of continuous multi-spectral SST observations.

4.2.3 *Coastal capabilities and limitations*

Two basic types of satellite SST products exist for coastal applications: 1) individual real-time SST images at full spatial resolution that include ocean, land, and clouds; and, 2) cloud cleared, composite SST maps at weekly or longer intervals with data gaps filled by interpolation. The real-time SST images are useful for monitoring the SST patterns of ocean fronts, eddies and offshore currents such as the Gulf Stream and the Loop Current in the Gulf of Mexico. The variability of these currents and eddies influences the coastal circulation. Pairs of IR images, separated by hours or days, can be used to estimate the direction and magnitude of surface currents. This information is useful to commercial and sport fishermen, scuba divers, sailors, search and rescue activities, oil spill management and offshore oilrig operators. Winds along the coast, when combined with the Earth's rotation, can produce upwelling that brings cool subsurface water to the surface along the coast. Upwelling tends to occur seasonally during the spring and summer months off the western coastlines of Africa and North and South America. Such upwelling events are especially noticeable in SST images during summer months along the eastern coast of the United States and can last for as long as a week. Tidal mixing of waters in shallow areas of the Gulf of Maine also produces a significant SST pattern. For example, during the summer, cool subsurface water appears over the shallow waters of the Georges Bank. The health and abundance of aquatic life near the ocean surface is dependent upon the environmental temperature. Because ocean surface biota depends upon nutrients, which in turn depend upon SST, fisheries benefit from real-time knowledge of the local SST distribution. NOAA Marine Fisheries rebuild and maintain sustainable fisheries, promote the recovery of protected species, and protect and maintain the health of coastal marine habitats.

The spatial and temporal scales of SST patterns detected by the satellite are limited by the IR sensor design, e.g. 1-km and 12-hours for polar orbiters, and 4-km and 30-minutes for geostationary orbiters. These spatial scales limit satellite coastal observations such that temperatures of most rivers and small lakes are not resolved, but large bays, such as the Chesapeake Bay, and the Great Lakes can be monitored. For SST pattern recognition, the coastal patterns must have a spatial scale about 5 times larger than the satellite sensor sample size. Therefore, polar satellites can resolve features greater than 5 km while geostationary satellites can resolve features greater than 20-km. The temporal sampling of SST sensors and the presence of clouds produce large gaps in SST images with daily, seasonal and regional cycles. Because clouds change and move rapidly during the day, the geostationary satellite provides more opportunities for accumulating cloud free ocean areas. For example, 2 to 4 day composites of hourly

Table 3. Characteristics of several modern thermal infrared sensors. Legend: LEO = Low Earth Orbit; GEO = Geostationary Orbit; NBR = Narrowband Radiometer; GS = Grating Spectrometer; FTS = Fourier Transform Spectrometer; CS = Conical Scan.

Sensor	Sensor Type	Satellite (Orbit)	Dates of Operation	Swath (km)	Resolution (m)	IR Spectral Channels	IR Spectral Coverage (μm^{-1}) *	Repeat Frequency (days)
AVHRR/1-3	NBR	NOAA/6-17 (LEO)	1978--present	2800	1100	3	3.7, 10.8, 12	1
AVHRR/3	NBR	Metop, NOAA/N-N'(LEO)	2005--?	2800	1100	3	3.7, 10.8, 12	1
ATSR/1-2	NBR-CS	ERS (LEO)	1991--present	500	1000	3	3.7, 10.8, 12	35
AATSR	NBR-CS	Envisat (LEO)	2002--present	500	1000	3	3.7, 10.8, 12	35
AIRS	GS	Aqua (LEO)	2002--present	1650	13500	2378	3.74 – 15.4	16
CrIS	FTS	NPP/NPOESS (LEO)	2006--?	2200	14000	1300	3.9 – 15.4	16
Imager	NBR	GOES/8-11 (GEO)	1994--present	+/- 60 deg from nadir	4000	3	3.9, 10.8, 12	30 min
Imager	NBR	GOES/M-Q (GEO)	2003--present	+/- 60 deg from nadir	4000	2	3.9, 12	30 min
ABI	NBR	GOES-R (GEO)	2012--?	+/- 60 deg *	~2000	3	4.0, 11, 12	15 min
HES	FTS	GOES-R (GEO)	2012--?	+/- 60 deg *	~4000	3	4.0, 11, 12	15 min
IASI	FTS	Metop (LEO)	2006--?	2200	25000	8461	3.6 – 15.5	2
MODIS	NBR	Terra	1999--present	2330	1000	5	3.78, 3.97, 4.05, 11, 12	16
MODIS	NBR	Aqua	2002--present	2330	1000	5	.78, 3.97, 4.05, 11, 12	16
SEVIRI	NBR	MSG (GEO)	2002--present	+/- 60 deg *	3000	3	3.9, 11, 12	NA
VIIRS	NBR	NPP/NPOESS (LEO)	2007 --?	3000	750	4	3.7, 4.0, 10.8, 12	1

*Spectral coverage refers central wavelengths of bands used to derive sea-surface temperature. Spectral selections for future satellites are subject to change.
** Longitudinal and latitudinal limits from GEO nadir for SST estimates.

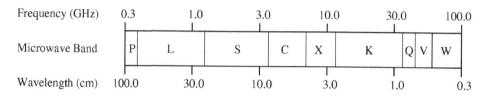

Figure 7. Microwave spectrum.

GOES SST can provide nearly continuous SST patterns in some ocean areas, such as mid-latitudes. However, high latitude areas (Gulf of Alaska, Labrador Sea) tend to be very cloudy except during summer months and SST coverage is limited.

4.3 PASSIVE MICROWAVE RADIOMETERS

The microwave portion of the electromagnetic spectrum lies between 0.3 cm and 100 cm (Figs. 2, 7) and offers some unique qualities for geophysical remote sensing. One important attribute is that the atmosphere is essentially transparent at microwave frequencies (Ulaby et al., 1981; Fig. 3), permitting observations of the surface through clouds. As with thermal IR, the observations can also be made at night. The major tradeoff of using microwaves is the relatively course spatial resolution of its observations. The surface "footprint" of a space-borne microwave antenna is proportional to $h\lambda/L$, where λ is the microwave wavelength, h is the altitude of the spacecraft, and L is the physical dimension of the antenna. With a typical polar orbit configuration (h = 750 km) and the inherent limitations on antenna size, the surface "footprints" are on the scale of several kilometers. Sophisticated processing techniques may be employed to effectively create a synthetic aperture of much larger dimensions enabling remarkably fine spatial resolution. Inevitably, however, this is accomplished through the sacrifice of near-real time operational capabilities and global coverage.

4.3.1 *Basic theory of observations*

Microwave radiometers are devices that measure radiation in the microwave spectrum within a controlled field of view. The intensity of the microwave radiation measured by a radiometer is conveyed as brightness temperature in degrees Kelvin (K). The antenna of a satellite based radiometer intercepts electromagnetic energy from three fundamental sources: 1) the radiation emitted from the earth's surface; 2) the upwelling radiation emitted by the atmosphere; and, 3) the downwelling emitted by the atmosphere and reflected upward by the ground. These three sources encompass a plethora of geophysical processes including the ocean surface dynamics, clouds, rain and other precipitation, and molecular emissions by oxygen and other atmospheric constituents (Wallace and Hobbs, 1977). These factors contribute not only to the basic emission sources, but also to energy absorption and reflection as radiation propagates towards the instrument's antenna. These instruments often possess channels that are sensitive to the intervening atmosphere. Information from these channels is used to quantify and remove the interference caused by the radiation absorption, reflection and emission between the satellite and the ground.

Modern microwave radiometers routinely retrieve ocean surface wind speeds, sea ice coverage and type, sea surface temperature, sea state, and a variety of meteorological properties. Sea surface salinity (SSS) can also be retrieved from low frequency microwave observations (L-band), where the sensitivity of brightness temperature to ocean salinity is maximal, due to its measurable effect on the dielectric constant of seawater. Though SSS has been estimated using observations from aircraft studies, no comparable satellite sensor exists. As noted earlier, two future sea surface-salinity missions – NASA's Aquarius (Koblinsky et al., 2003) and the European Space Agency's Soil Moisture and Ocean Salinity – are planned for launch in 2008 to provide global estimates of sea surface salinity with an accuracy of 0.2–0.3 practical salinity units. Due to the coarse spatial resolution (100 – 200 km) of their observations, data from these space-borne salinity sensors will be of limited use in coastal waters.

4.3.2 *Radiometers – past, present and future*

Table 4 highlights the past, present and planned future of earth orbiting microwave radiometers used for geophysical remote sensing, including their fundamental capabilities as remote sensing tools. Perhaps the best-quantified space-borne microwave radiometer is the Special Sensor Microwave Imager (SSM/I). Since 1987, it has been flown on a series of satellites launched and operated by the Defense Meteorological Satellite Program (DMSP). Presently, SSM/I flies on three operational satellites and its observations provide a variety of oceanographic and atmospheric parameters with essentially global coverage in less than 24 hours (Goodberlet et al., 1989; Weng and Grody, 1994; Wentz, 1997). The next generation of SSM/I, the Special Sensor Microwave Imager Sounder (SSMIS), was launched in 2003. It possesses all of the SSM/I channels as well as microwave channels capable of measuring vertical profiles of temperature, humidity, and water vapor in the troposphere. The Advanced Microwave Scanning Radiometer (AMSR-E) aboard the EOS AQUA spacecraft and the TRMM (Tropical Rainfall Measuring Mission) Microwave Imager (TMI) represent more advanced passive microwave instruments. For example, TMI, which possesses the same channels as SSM/I plus two that are relatively insensitive to precipitation, has a smaller footprint than SSM/I (Table 4). TMI's low altitude tropical orbit ($i = 40°$), however, limits it coverage to $±40°$ (Connor and Chang, 2000).

A new generation of satellite remote sensing instruments made its debut with the launch of WindSat on the Coriolis satellite in January of 2003. WindSat is a five-frequency microwave radiometer capable of resolving the full wind vector, i.e. both its speed and direction, at the ocean surface. In addition, a large rotating dish reflector antenna improves the resolution of WindSat's ground footprints over SSM/I, increasing its utility between the coastline and shelf break in many areas.

4.3.3 *Coastal capabilities and limitations*

Microwaves are several orders of magnitude larger in wavelength than the visible portions of the spectrum and correspondingly view the earth's surface at much coarser spatial resolution. The spatial resolution of typical microwave radiometer observations is ~25 km, far too coarse for all but the largest scale processes on the continental shelf. Microwave instruments may also suffer from land side-lobe contamination, limiting the accurate retrieval of oceanographic parameters from within a couple of beam footprints of the coast (~50km). While clouds are commonly considered "invisible" to microwaves, precipitation in the atmosphere can severely limit the retrieval of certain geophysical parameters.

4.4 SCATTEROMETERS

From the perspective of modern satellite geosciences, active microwave remote sensing implies the use of radars, including scatterometers, altimeters, profiling radars, and synthetic aperture radars. All of these instruments employ a controlled transmission of an electromagnetic wave to a specified target and measurement of the portion of the wave that is reflected back to a receiver. The amplitude and phase information of the return wave can be used to determine characteristics of the target and often to infer other geophysical properties associated with the target.

4.4.1 *Basic theory of observations*

Scatterometers are radars capable of inferring the ocean surface wind vector by accurately measuring the radar backscatter from wind-induced capillary waves on the ocean's surface. The frequencies of scatterometers are selected so that the wavelength of the transmitted pulse geometrically matches the scale of the ocean surface waves to produce a constructive reflection condition called Bragg scattering and enhancing the returned signal. It is possible to retrieve ocean surface wind velocity, which is both speed and direction, from scatterometer measurements because the surface wave characteristics responsible for the strong reflection condition are produced and modulated by the atmospheric winds at the surface. Similarly, sea surface roughness and associated properties can be estimated by scatterometers.

4.4.2 *Sensors – past, present and future*

Table 5 lists the major satellite based scatterometers that are currently employed or planned for the retrieval of ocean surface wind vectors. The first scatterometer to be operated in space, the S-193, was carried aboard SKYLAB during 1973 and 1974 and demonstrated the feasibility of using scatterometers to measure the ocean surface winds (Moore and Young, 1977). This led to the development and launch of the Seasat-A Satellite Scatterometer (SASS) onboard the Seasat oceanographic satellite in 1978. Though the instrument operated for only 100 days, SASS data was successfully used to generate accurate ocean surface wind velocity measurements (Grantham et al., 1977). The European Space Agency (ESA) followed with scatterometers aboard their ERS-1 and ERS-2 satellites, launched in 1991 and 1995 respectively (Table 5) and validated the operational feasibility of space borne scatterometers.

In 1996, NASA launched the NASA Scatterometer (NSCAT) aboard the ADEOS satellite. NSCAT operated successfully for 9 months, producing ocean surface wind vectors with a 25 km spatial resolution, until a catastrophic power failure occurred on May 9, 1997. After this failure, NASA quickly developed a stop-gap mission to keep a scatterometer in orbit until the launch of ADEOS II. The mission used the Seawinds scatterometer on the QuikSCAT satellite, which was launched in July of 1999 and has been operational since (Lungu, 2001). QuikSCAT proved to be a particularly valuable mission because ADEOS II failed roughly 9 months after it became operational in October 2003. Though Seawinds on ADEOS II provided over 9 months of valuable data, QuikSCAT presently carries the only microwave scatterometer capable of global ocean surface wind measurements. The most immediate plan for another scatterometer is the ASCAT mission of ESA. This C-band instrument is scheduled for launch in 2005.

4.4.3 *Coastal capabilities and limitations*

As with microwave radiometers, the use of scatterometers is limited in the coastal ocean due to their coarse spatial resolution. The standard QuikSCAT wind product (Table 5) has a 25 km spatial resolution, though an improved 12.5 km product is now operationally available. Other radar instruments, such as the TRMM Precipitation Radar, are being exploited to derive ocean surface winds with 5 km surface footprints using scatterometry techniques and geophysical assumptions about the continuity of the wind field, though this technique has yet to reach a level of practical implementation. Although new techniques will certainly serve to reduce the spatial constraint in the future, the limitations of microwave instruments should be kept in mind, particularly for coastal regions because scatterometers are susceptible to land contamination in the same manner as microwave radiometers.

Table 4. Characteristics of modern passive microwave missions.

Satellite Mission	Sponsoring Agency	Geophysical Data Record Duration	Swath Width (km)	Resolution (km)	Frequencies (GHz)	Environmental Data Products
SSM/I	DMSP	07/1987 –	1400	15-70	19.4, 22.2, 37.0, 85.5	Cloud Water, Precipitable Water, Ice, Snow, Wind Speed
TMI	NASA/NASDA	12/1997 –	759	7-63	10.7, 19.4, 21.3, 37.0, 85.5	Same As SSM/I
WindSat	NRL	01/2003 –	950	13-60	6.8, 10.7, 18.7, 23.8, 37.0	Surface Wind Vector, SST, Precipitation parameters
AMSR-E	NASA/NASDA	06/2002 –	1445	6-75	6.9, 10.7, 18.7, 23.8, 36.5, 89.0	Same As SSM/I
SSMIS	DMSP	In Calibration	1700	12.5-75	24 Channels (19.4 – 183.3)	Same as SSM/I plus Soundings

Table 5. Characteristics of modern active microwave missions.

Satellite Mission	Sponsoring Agency	Geophysical Data Record Duration	Swath Width (km)	Resolution (km)	Frequencies (GHz)	Environmental Data Products
ERS-1,-2	ESA	08/91 – 06/96; 04/95 –	500	50	5.3	Surface Wind Vector
ADEOS	NASA/NASDA	10/1996 – 06/1997	2×600	25	13.995	Surface Wind Vector
QuikSCAT	NASA	07/1999 –	1800	25/12.5	13.4	Surface Wind Vector, Sea Ice
ADEOS-II	NASA/NASDA	12/2002 – 10/2003	1800	25	13.4	Surface Wind Vector

Table 6. Characteristics of modern altimetry missions.

Satellite Mission	Sponsoring Agency	Geophysical Data Record Duration	Maximum Latitude (Degrees)	Groundtrack Separation (km)*	Main Altimeter Frequency (GHz)	Secondary Altimeter Frequency (GHz)	Radiometer Frequencies (GHz)	Repeat Cycle Period (Days)
ERS-1,-2	ESA	08/91 – 06/96; 04/95 –	81.46	74.2 (86) / 6.4 (1002) / 1.3 (4822)	13.8	N/A	36.5/23.8	3.0 / 35.0 / 168.0
Topex/Poseidon	NASA/CNES	09/1992 –	66.04	25.1 (254)	13.6/13.65	5.3	37.0/21.0/18.0	9.95
GFO	U.S. Navy	01/2000 –	71.95	13.1 (488)	13.495	N/A	37.0/22.0	17.05
Jason-1	NASA/CNES	12/2001 –	66.04	25.1 (254)	13.575	5.3	34.0/23.8/18.7	9.95
Envisat	ESA	04/2003 –	81.46	6.4 (1002)	13.575	3.2	36.5/23.8	35.0

*The maximum (equatorial) groundtrack separation is based on the number of half-revolutions (in parentheses) during a repeat-period cycle, based on an earth radius of 6378 km.

4.5 ALTIMETERS

4.5.1 *Basic theory of observations*

Radar altimeters transmit microwave frequency pulses to the sea surface and receive the reflected echoes, measuring range from the round-trip travel time. The primary frequency is in the Ku-band, at around 13.5 GHz, which is highly reflective from water surfaces and not strongly attenuated by atmospheric constituents (see Fig. 3). Radar pulses are both scattered and reflected from the sea surface within a 'footprint' of a few kilometers at nadir. The shape of the return radar echo, as a function of time, provides estimates of range, significant wave height and wind speed. The uncorrected range or distance between the altimeter and the earth's surface is simply $R = \tau * c/2$, where c is the speed of light and τ is the two-way travel time. The height of the sea surface is estimated, by subtracting the altimetric range from the orbital altitude of the satellite (Fig. 8). Sea surface heights are on the order of 100's of meters, which is about four orders of magnitude less than the satellite's altitude and range.

The sea surface height, relative to the reference ellipsoid, is comprised of: 1) the static 'geoid', which represents the equi-potential surface that the ocean would assume in the absence of currents, wind, and tides; 2) height variations due to tides; 3) surface effects from the oceanic response to atmospheric pressure; and, 4) dynamic topography associated with geostrophic currents. The dynamic topography is the small difference, on the order of 1-2 m, between the measured sea surface height and the marine geoid, which has amplitudes of 100 m.

To achieve high accuracy in sea surface height, both the range and orbit heights must be carefully estimated. Precise orbit determination is based on satellite laser ranging, GPS, or Doppler tracking of the spacecraft and is performed independently from the range measurements. The ranges must be corrected for both instrumental and geophysical effects. The geophysical corrections account for refractive path delays due to the atmosphere as well as effects from the interaction of radar pulses with ocean waves. The time-varying component of the corrected sea surface height is a mixture of tidal, geostrophic current, and surface pressure-driven signals. The large static component, due to geographical variations in the earth's gravity field (the marine geoid), can be removed to retain these oceanographic signals. The tidal variations can be estimated using empirical ocean tide models. The oceanic response to changes in atmospheric pressure can be modeled as an 'inverse barometer': sea level rises (falls) approximately 1 cm for every 1 mbar fall (rise) in pressure. Large- and meso-scale surface currents are in geostrophic balance with the slope of the sea surface. Altimetry provides the along-track component of sea surface slope, yielding the cross-track component of velocity. The cross-track separation (in both space and time) is generally too large to provide the along-track component of velocity. The proposed Wide-Swath-Ocean-Altimeter (WSOA), to be flown on JASON-2 in early 2008, will supply both velocity components over the swath area mapped by this new instrument.

Scattering of the altimeter's radar pulse by ocean surface waves causes the received echo to have a broader leading edge as wave height increases. The slope of the waveform's leading edge is used to calculate significant wave height at nadir along the satellite's ground track. No information on wave direction, period, or wavelength is normally provided by altimetry.

Small-scale roughness from capillary waves reduces the amount of backscattered energy returned to the altimeter. This roughness can be related empirically to wind speed above the ocean surface. As the wind speed increases, the radar backscatter decreases. Because the altimeter provides a nadir point measurement, the wind direction cannot be determined. By contrast, scatterometers, which can also operate at Ku-band frequencies, utilize multiple antennae and combinations of horizontal and vertical polarization to obtain a swath of backscatter measurements that provide both wind speed and direction. In coastal regions wind and wave measurements can only be made when the radar's footprint is not affected by land or ice surfaces. An excellent source for more in-depth information on satellite altimetry can be found in Fu and Cazenave (2000).

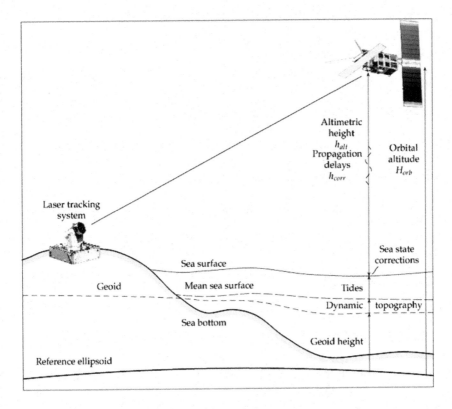

Figure 8. The principal of satellite radar altimetry (Scharroo, 2002). Reprinted with permission of the author.

4.5.2 *Sensors – past, present and future*

The primary characteristics of recent altimetry missions are shown in Table 6. The basic sensor parameters are very similar, based on a primary Ku-band measurement. Most of the older missions, other than Topex, were based on single-frequency altimeters, while all future missions are expected to utilize dual-frequency instruments that allow for correction of ionospheric path delays. All missions after Geosat, a Navy mission that operated from April 1995 to December 1989, have included a dual or three-frequency radiometer to provide tropospheric water vapor corrections.

Future altimetry missions have the potential to solve some of the current limitations for coastal remote sensing. By utilizing Doppler shift information from correlated radar echoes (at higher pulse repetition frequencies) one can effectively increase the along-track spatial resolution to 250 m (Raney, 1998). This will allow sea surface height information to reach much closer to shore. Similarly, a small single-frequency Ka-band (~35 GHz) altimeter 'Altika' will have increased along-track spatial resolution, perhaps less than 5 km, and is more insensitive to ionospheric effects than traditional Ku-band altimeters (Vincent and Thouvenot, 2001). The under-sampling problem inherent in a nadir (vs. swath) instrument can be addressed by constellations of micro-satellites, e.g. Wittex (Raney and Porter, 2001). Finally, the Wide Swath Ocean Altimeter (WSOA), proposed to fly on Jason-2, will utilize a completely new technique based on radar interferometry, mapping 200 km wide swaths of sea surface height (Rodriguez et al.,

2001). The WSOA will yield much better temporal and cross-track resolution than current altimeters, with ~15 km resolution within the swath.

4.5.3 *Coastal capabilities and limitations*

The current state of altimetric technology is not optimal for coastal oceanographic applications. The altimetric footprint 'feels' land as the satellite approaches the coast; the return echoes include the land effects and give inaccurate height measurements within 10 km or so of shore. In addition, although altimetric tide models perform well for the open ocean, they breakdown in coastal regimes where tidal amplitudes increase and spatial scales of variability are too fine.

The biggest drawback is the spatio-temporal sampling pattern of modern altimetry missions (Table 6). With repeat-period orbits of 10 to 35-days, there is a tradeoff between the frequency of revisiting a location in time vs. the separation of ground tracks spatially. Given the short space and time scales of coastal jets and other near shore processes, single-mission sampling cannot resolve the mesoscale motions of interest. Combining sea surface height data from multiple missions provides better sampling, but data 'homogenization' becomes a critical issue.

One strategy is to use dynamical ocean models or coupled ocean-atmosphere models that assimilate altimetry data to provide smaller scale or more synoptic sampling of the coastal regime. These models provide intelligent interpolation of the altimetry, consistent with the underlying physical processes.

4.6 SYNTHETIC APERTURE RADAR

Among the more common applications of Synthetic Aperture Radar (SAR) are measurement of surface ice, near-surface winds, sea state and the detection of algal blooms, oil slicks, and other marine surface pollutants. SAR is also used to monitor fishing vessels (Clemente-Colon, 2002).

4.6.1 *Basic theory of observations*

This active instrument makes use of its own motion relative to the desired target and some sophisticated signal processing to effectively create a very large viewing aperture. By carefully tracking and recording the transmitted and reflected signal of an orbiting radar over a specific arc of the orbit, it is possible to mathematically exploit the effect of the target's (usually the earth's surface) relative velocity to create a radar observation footprint of very fine resolution. The size of the footprint is determined by an equivalent antenna aperture dimension, defined by the length of arc the satellite has traversed. This technique can retrieve spatial resolutions measured in meters rather than the typical several kilometer size associated with most orbiting microwave instruments.

4.6.2 *Sensors – past, present and future*

The first spaceborne SAR was placed in Earth orbit aboard the NASA Seasat-A satellite in 1978. Operating in the microwave L-Band GHz with horizontal polarization, the instrument collected ocean surface radar observations for only 106 days before an electrical failure cut the mission short. Nonetheless, the system quantitatively demonstrated the use of SAR imaging for measuring ocean surface roughness. Several SAR experiments were carried out in the early 1980s as part of Space Shuttle missions, some including more advanced SAR instruments with multi-band and polarization capabilities. ESA launched its ERS-1 and ERS-2 satellites in 1991 and 1995, each possessing the Advanced Microwave Instrument (AMI) capable of operating in either a SAR or conventional scatterometer mode. The AMI operates in the C-Band (5.3 GHz,

5.6 cm) using vertical polarization and, in SAR mode, can generate images of surface dimension 100km x 100km with a pixel-to-pixel distance of 12.5m.

Presently there are two SAR instruments operating in orbit over the Earth that are accessible and of use to the geoscientist; RADARSAT-1, a Canadian satellite that has been flying since 1995, and the Advanced Synthetic Aperture Radar (ASAR) aboard ENVISAT, a multi-instrument orbiting platform launched by ESA in March of 2002. RADARSAT-1 operates in the C-Band using horizontal polarization and has a similar surface resolving capability to the ERS/AMI instrument, but can also operate in a variety of other modes with wider swaths and larger scale surface resolutions (Fig. 8). These different modes can produce images ranging from 500km x 500km with 100m resolution down to 50km x 50km with 8m resolution. ASAR also operates in the C-band of the microwave spectrum and employs both vertical and horizontal polarization to expand the geophysical response information available. ASAR offers a variety of image products encompassing different polarization combinations and spatial coverage and resolution. These images range from the 400km x 400 km Global Monitoring images with 1 km resolution, down to the "precision" images covering 56 km x 100 km with 12.5 m resolution.

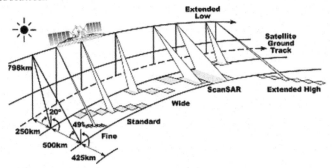

Figure 8. Various observation modes of RADARSAT-1. Reproduced by permission of the publisher (RADARSAT Inernational).

Because of the tremendous quantity of data and the time consuming processing associated with their imaging, SAR instruments presently in Earth orbit cannot provide global and near-real time products typical of other satellite-borne sensors. RADARSAT-1, for example, has an orbital swath between 35 km and 500 km, depending on the imaging mode it is using, and has an orbit that repeats only every 24 days.

Future SAR satellite missions include RADARSAT-2, the follow-on mission to RADARSAT-1 that will have surface resolution capabilities from 100m meters with a 500 km swath, down to 3 meters with a 20 km swath. RADARSAT-2 will be launched in early 2005. The Japanese Aerospace Exploration Agency also plans to launch its Advanced Land Observing Satellite (ALOS) that will carry the Phased Array type L-band Synthetic Aperture Radar (PALSAR). This instrument will produce images with resolutions of either 10 meters or 100 meters swath widths between 70 km and 350 km. ALOS is scheduled for launch in 2004.

4.6.3 *Coastal capabilities and limitations*

The ultra-high spatial resolution abilities of SAR observations make them useful in detecting and examining appropriate phenomena in coastal waters, e.g. oil slicks, sea ice and surface manifestations of internal waves. The large quantity of data and computationally intensive processing of SAR imagery restrict their near-real time use and geographic coverage, and they are currently used primarily in retrospective case studies of specific geographic regions. The low revisitation rate of current SAR imagery also limits their use in monitoring rapidly changing features.

5. Summary

The coastal ocean is a dynamic region with processes operating on time scales of hours to days and on spatial scales of meters to hundreds of kilometers. In order to monitor this complex region, one must be able to spatially and temporally resolve the processes and events occurring within it. Satellites offer a platform from which synoptic observations can be routinely acquired to derive a host of surface ocean properties that cover much of the tempo-spatial domain of the coastal environment.

The spatial and temporal attributes of the various oceanographic properties, imposed by the characteristics of the satellite system employed to derive them, dictate their applicability in the coastal zone. Visible – near infrared observations, which can penetrate water to a greater extent than measurements collected at higher wavelengths, provide information at high to moderate spatial and temporal resolution about the benthos within optically shallow waters and constituents within the upper portion of the water column. Their spatial resolution permits their use in near-shore waters. The retrieval of this information, however, is difficult due to the complexities of the coastal environment and limited to clear weather during daylight hours. Thermal infrared imagery is used to derive sea surface's skin temperatures during both day and night at moderate spatial resolution, yet it is still hampered by clouds. The placement of thermal IR sensors on both polar orbiting and geostationary satellites yields SST observations with a variety of temporal resolutions for large portions of the ocean. Though microwave observations, both from passive and active sensors, permit the retrieval of ocean surface properties through clouds, their use is generally limited to relatively large-scale features located 10 kilometers or more offshore because of their relatively coarse spatial resolution and the contamination of their signal by land. This is less of a problem for SAR imagery, which can furnish very high-resolution imagery. The placement of microwave sensors on polar orbiting satellites, with their infrequent revisitation rate, limits the use of microwave observations to monitoring longer-term processes occurring within the coastal ocean.

Awareness of the capabilities and limitations of the various systems is the first step in selecting an appropriate satellite system from the diverse array currently available to meet the needs of a specific application. Additional detail will be necessary to understand the processing and interpretation of the satellite derived properties, but it is hoped that the information presented in this chapter, though brief and introductory, offers insight into the factors that must be considered when selecting a satellite system for a specific purpose. This understanding will allow the user to avoid potential pitfalls encountered by first time users and efficiently exploit satellite systems and their observations to monitor coastal waters.

6. Acknowledgements

We thank our colleagues who assisted in the preparation of this book, either by providing information or reviewing previous drafts, especially Pablo Clemente-Colon, Tim Mavor, and Stephanie Schollaert Uz. The views, opinions, and findings contained in this chapter are those of the authors and should not be construed as an official

National Oceanic and Atmospheric Administration or U.S. Government position, policy, or decision.

7. References

Anding, D. and R. Kauth. 1970. Estimation of sea-surface temperature from space. Remote Sensing of Environment, 1:217-220.

Brooks, D. R. 1977. An Introduction to Orbit Dynamics and Its Application to Satellite-Based Earth Monitoring Missions. NASA Reference Publication, Report Number 1009, National Aeronautics and Space Administration, Hampton. 80 pp.

Carder, K. L., F. R. Chen, Z. P. Lee, S. K. Hawes and D. Kamykowski. 1999. Semianalytic Moderate-Resolution Imaging Spectrometer algorithms for chlorophyll a and absorption with bio-optical domains based on nitrate-depletion temperatures. Journal of Geophysical Research, 104(C3):5403-5421.

Clemente-Colon, P. 2002. The use of spaceborne synthetic aperture radar in the monitoring and management of Bering Sea fisheries: UNESCO CSI Info Paper 14. pg 117-141.

Connor, L. N. and P. S. Chang. 2000. Ocean surface wind retrievals using the TRMM micrwave imager. IEEE Transactions on Geoscience and Remote Sensing, 38:2009-2016.

Dickey, T. D. 1991. The emergence of concurrent high-resolution physical and biooptical measurements in the upper ocean and their applications. Reviews of Geophysics, 29(3):383-413.

D'Sa, E. J., C. Hu, F. E. Muller-Karger and K. L. Carder. 2002. Estimation of colored dissolved organic matter and salinity fields in case 2 waters using SeaWiFS: Examples from Florida Bay and Florida Shelf. Proceedings of the Indian Academy of Sciences-Earth and Planetary Sciences, 111(3):197-207.

Fu, L.-L. and A. Cazenave. 2000. Satellite Altimetry and Earth Sciences: A Handbook of Techniques and Applications. Vol 69: Academic Press, San Diego. pg 463.

Goodberlet, M. A., C. T. Swift and J. C. Wilkerson. 1989. Remote sensing of ocean surface winds with the Special Sensor Microwave/Imager. Journal of Geophysical Research, 94:14,547-14,555.

Gordon, H. R. and M. Wang. 1994. Retrieval of water-leaving radiance and aerosol optical thickness over the oceans with SeaWiFS: A preliminary algorithm. Applied Optics, 33(3):443-452.

Gordon, H.R. and A.Y. Morel. 1983. Remote assessment of ocean color for interpretation of satellite visible imagery: A Review, Springer-Verlag, New York, 114 pp.

Grantham, W. L., E. M. Bracalente, W. L. Jones and J. W. Johnson. 1977. The SEASAT-A satellite scatterometer. IEEE Journal of Oceanic Engineering, OE-2:200-206.

Hovis, W. A., D. K. Clark, F. Anderson, R. W. Austin, W. H. Wilson, E. T. Baker, D. Ball, H. R. Gordon, J. L. Mueller, S. Z. El-Sayed, B. Sturm, R. C. Wrigley and C. S. Yentsch. 1980. Nimbus-7 Coastal Zone Color Scanner: system description and initial imagery. Science, 210:60-63.

Kidder, S. Q. and T. H. Vonder Haar. 1995. Satellite Meteorology: An Introduction. Academic Press, San Diego. 466 pp.

Kilpatrick, K. A., G. P. Podesta and R. H. Evans. 2001. Overview of the NOAA/NASA Pathfinder algorithm for sea surface temperature and associated matchup database. Journal of Geophysical Research, 106:9179-9198.

Koblinsky, C. J., P. Hildebrand, D. LeVine, F. Pellerano, Y. Chao, W. Wilson, S. Yueh and G. Lagerloef. 2003. Sea surface salinity from space: Science goals and measurement approach. Radio Science, 38:8064, doi:10.1029/2001RS002584.

Logsdon, T. 1998. Orbital Mechanics. Theory and Applications. John Wiley and Sons, Inc., New York. 268 pp.

Lungu, T. 2001. QuikSCAT Science Data Products User's Manual. California Institute of Technology, Pasadena. pg 86.

Maritorena, S., D. A. Siegel and A. R. Peterson. 2002. Optimization of a semianalytical ocean color model for global-scale applications. Applied Optics, 41(15):2705-2714.

Maul, G. A. 1985. Introduction to Satellite Oceanography. Martinus Nijhoff Publishers, Dordrecht. 606 pp.

McClain, E. P., W. G. Pichel and C. C. Walton. 1985. Comparative performance of AVHRR-based multichannel sea surface temperatures. Journal of Geophysical Research, 90(C6):11,587-11,601.

McMillin, L. M. 1975. Estimation of sea surface temperatures from two infrared window measurements with different absorption. Journal of Geophysical Research, 80:5113-5117.

Merchant, C. J., A. R. Harris, M. J. Murray and A. M. Zavody. 1999. Toward the elimination of bias in satellite retrievals of sea surface temperature: 1. Theory, modeling and interalgorithm comparison. Journal of Geophysical Research, 104(C10):23,565-23,578.

Moore, R. K. and J. D. Young. 1977. Active measurement from space of sea surface winds. IEEE Journal of Oceanic Engineering, OE-2:309-317.

Nalli, N. R. and W. L. Smith. 2003. Retrieval of ocean and lake surface temperatures from hyperspectral radiance observations. Journal of Atmospheric and Oceanic Technology, 20(12):1810-1825.

Nalli, N. R. and L. L. Stowe. 2002. Aerosol correction for remotely sensed sea surface temperatures from the National Oceanic and Atmospheric Administration advanced very high resolution radiometer. Journal of Geophysical Research, 107(C10): 3172, doi:10.1029/2001JC001162.

O'Reilly, J. E., S. Maritorena, B. G. Mitchell, D. A. Siegel, K. L. Carder, S. A. Garver, M. Kahru and C. McClain. 1998. Ocean color chlorophyll algorithms for SeaWiFS. Journal of Geophysical Research-Oceans, 103(C11):24937-24953.

Pease, C. B. 1991. Satellite Imaging Instruments: Principles, Technologies, and Operational Systems. Ellis Horwood Limited, Chichester. 336 pp.

Planck, M. 1959. The Theory of Heat Radiation. Dover Publications, Inc., New York. 224 pp.

Raney, R. K. 1998. The delay Doppler radar altimeter. IEEE Transactions on Geoscience and Remote Sensing, 36(5):1578-1588.

Raney, R. K. and R. Porter. 2001. An innovative multi-satellite radar altimeter constellation. In: Report of the High-Resolution Ocean Topography Science Working Group Meeting, D. Chelton (Ed.). Oregon State University, Ref. 2001-4. pg 170-176.

Robinson, I. S. 1985. Satellite Oceanography. An Introduction for Oceanographers and Remote-Sensing Scientists. Ellis Horwood Limited, Chichester. 455 pp.

Rodriguez, E. and B. D. Pollard. 2003. Centimetric sea surface height accuracy using the Wide-Swath Ocean altimeter. Proceedings of the 2003 IGARSS International Geoscience and Remote Sensing Symposium, Vol. 5. pg 3011-3013

Russ, J. C. 1995. The Image Processing Handbook. CRC Press, Boca Raton. 674 pp.

Scharroo, R. 2002. A decade of ERS satellite orbits and altimetry. Ph.D. Dissertation. Delft University Press, 195 pp.

Schowengerdt, R. A. 1997. Remote Sensing. Models and Methods for Image Processing. Academic Press, San Diego. 522 pp.

Siegel, D. A., M. Wang, S. Maritorena and W. Robinson. 2000. Atmospheric correction of satellite ocean color imagery: the black pixel assumption. Applied Optics, 39(21):3582-3591.

Stamnes, K., B. Yan, H. Eide, A. Barnard, W. S. Pegau and J. J. Stamnes. 2003. Accurate and self-consistent ocean color algorithm: simultaneous retrieval of aerosol optical properties and chlorophyll concentrations. Applied Optics, 42(6):939-951.

Stewart, R. H. 1985. Methods of Satellite Oceanography. University of California Press, Berkeley. 360 pp.

Stumpf, R. P., R. A. Arnone, R. W. Gould, P. M. Martinolich and V. Ransibrahmanakul. 2003. A partially-coupled ocean-atmosphere model for retrieval of water-leaving radiance from SeaWiFS in coastal waters. In: Hooker, S. B. and E. R. Firestone (Eds.). National Aeronautics and Space Administration, Greenbelt, MD. 51-59.

Thomas, R. H. 1990. Polar Research From Satellites. Joint Oceanographic Institutions, Inc., Washington, DC. 91 pp.

Ulaby, F. T., R. K. Moore and A. K. Fung. 1981. Microwave Remote Sensing: Active and Passive. Artech House, Norwood, MA. 456 pp.

Vincent, P. and E. Thouvenot. 2001. Status of Ka-band Altimetry Studies. In: Report of the High-Resolution Ocean Topograph Science Working Group Meeting. D. Chelton (Ed.). Oregon State University, Ref. 2001-4. pg 179-187.

Wallace, J. M. and P. V. Hobbs. 1977. Atmospheric Science: An Introductory Survey. Academic Press, Inc., Orlando. 467 pp.

Walton, C. C., W. G. Pichel, J. F. Sapper and D. A. May. 1998. The development and operational application of nonlinear algorithms for the measurement of sea surface temperatures with the NOAA polar-orbiting environmental satellites. Journal of Geophysical Research, 103(C12):27,999–28,012.

Wang, M. 1999. Atmospheric correction of ocean color sensors: Computing atmospheric diffuse transmittance. Applied Optics, 38(3):451-455.

Wang, M. and H. R. Gordon. 1994. A simple, moderately accurate, atmospheric correction algorithm for SeaWiFS. Remote Sensing of Environment, 50:231-239.

Weng, F. and N. Grody. 1994. Retrieval of cloud liquid water using the Special Sensor Microwave/Imager (SSM/I). Journal of Geophysical Research 99:25535-25,551.

Wentz, F. J. 1997. A well-calibrated ocean algorithm for Special Sensor Microwave/Imager. Journal of Geophysical Research, 102:8703-8718.

Chapter 3

OPTICAL AIRBORNE REMOTE SENSING

[1]JEFFREY S. MYERS AND [2]RICHARD L. MILLER

[1]SAIC, *Airborne Sensor Facility, Ames Research Center, Moffett Field, CA, 94035 USA*
[2]*National Aeronautics and Space Administration, Earth Science Applications*
Directorate, Stennis Space Center, MS, 39529 USA

1. Introduction

The technology of airborne remote sensing encompasses a broad class of airborne platforms and a wide suite of remote sensing instruments. Usable platforms range from airships (Hamley, 1994; Windischbauer and Hans, 1994), UAVs (Unmanned Airborne Vehicles, Miller et al., 1999; Petrie, 2001), and helicopters (Gade et al., 1998; Matthews et al., 1997; Schaepman et al., 1997), to standard and high-altitude aircraft (see for example, Carder et al., 1993; Harding et al., 1994, Miller and Cruise, 1995; Mumby et al., 1998). Similarly, the airborne instruments deployed on aircraft vary from analog cameras (Sheppard et al., 1995), digital CCD (Charged Coupled Device) cameras (Howland, 1980; Miller et al., 1997), and analog and digital video cameras, to digital optical systems. In principle, any air-based platform that contains a measurement and recording device can be a remote sensing system for Earth observations. As a consequence, there are numerous, almost countless, airborne remote sensing systems available from commercial, state, and federal organizations world wide. Hence, a complete treatise on airborne remote sensing is far beyond the scope of this chapter. Here, we provide a brief overview of key elements that potential users of airborne remote sensing of coastal environments should consider. We limit our discussion to aircraft platforms that employ digital optical imaging systems and focus on issues that compare and contrast the use of airborne systems with the optical satellite systems described in Brown et al. (Chapter 2).

2. Elements of Airborne Remote Sensing

The wide spread use of airborne remote sensing is due in large-part to the availability of low cost aircraft and imaging systems and the operational characteristics of the aircraft. The major benefit of airborne remote sensing, compared to satellite-based systems, is that the user can define the deployment and operational characteristics of the remote sensing system. In general, an airborne system can provide considerably higher spatial resolution data (e.g., less than a meter to tens of meters). The system can be deployed when atmospheric (i.e., cloud-free), environmental, and solar conditions are acceptable to study specific phenomenon. For example, in some areas, such as the Pacific Northwest of the U.S., airborne remote sensing is the only way to reliably obtain coastal zone imagery. The deployment can also be coordinated with a field program to acquire *in situ* measurements for instrument calibration or algorithm development or validation. These advantages are particularly important in coastal aquatic environments where many processes occur over time and space scales that cannot be adequately sampled by most satellite instruments (Miller et al., 2003). Airborne remote sensing

R.L. Miller et al. (eds.), Remote Sensing of Coastal Aquatic Environments, 51–67.
© 2005 *Springer.*

enables specific coastal events to be studied, for example, such as a particular phase of a tidal cycle or an extreme low-water event in Puget Sound, WA (Fig. 1). In this image obtained by the Airborne Thematic Mapper (5 m resolution) a normally submerged sea grass bed on a tidal flat can be examined. This grass bed is only exposed several days each year at Mean Lower Low Water. Clouds are a persistent feature in this region and often obscure the view of satellite instruments.   The flexibility of user-defined deployments provides a capability to study ephemeral coastal events such as algal blooms or human-induced spills. These events can be imaged over a period of hours or days as they evolve in a manner impossible for most satellite-based systems.

There are however, several additional factors that must be considered to evaluate the utility of airborne remote sensing for a given application. These factors include geometric accuracy, radiometric calibration, and total spatial coverage. As discussed below, these factors can reduce the effectiveness of airborne remote sensing for certain coastal applications.

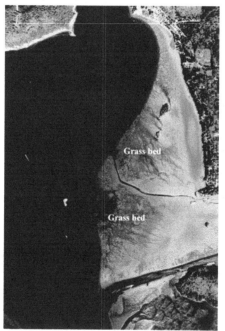

Figure 1. Near-infrared (830 nm) image acquired 30 July, 2000 using the Airborne Thematic Mapper over exposed sea grass beds of Puget Sound, WA. The flexibility of an aircraft system to respond to clear sky conditions enabled the capture of this unique image.

2.1 AIRCRAFT AS REMOTE SENSING PLATFORMS

In the early 19th century, as commercial aircraft came into common usage, it was obvious that they were an ideal platform to observe a variety of geographic features. The introduction of aerial mapping cameras in the 1920's, pioneered by Sherman Fairchild, led quickly to the use of airborne mapping by the U.S. Coast and Geodetic Survey for charting coastlines. Further development of aerial photography was spawned by World War II and the high altitude reconnaissance of the subsequent Cold War period. These developments clearly demonstrated that Earth observations could be acquired from a wide range of aircraft and that thematic information could be extracted from aerial photographs. Aerial photography remains an important aspect of airborne remote sensing. As remote sensing technologies evolved into the electronic age, a wide range of instruments from early spectrophotometers, radiometers, and imaging devices were used to observe oceanographic features. These advances were paralleled by advancements in aircraft platform systems.

The use of aircraft as remote sensing platforms has proliferated because most aircraft can be modified to accommodate a variety remote sensing instruments such as cameras, video systems, and imaging radiometers. However, the aircraft must be reengineered so that the instruments can be mounted without weakening the aircraft structure or disrupting flight systems. For example, optical view ports or windows are frequently constructed in the belly of the fuselage or instrument pods are mounted on the wings or wing struts. Additional considerations must be given to accommodate the space and power requirements of instrument mounting hardware (i.e., stabilization structures) and data acquisition/recording systems. Another factor that has lead to the rapid commercialization of airborne remote sensing is that many aircraft systems can accommodate several instrument packages by simply swapping out different systems.

Airborne remote sensing has been conducted on virtually every class of aircraft, ranging from small single-engine propeller planes to large multi-engine commercial and specialized military platforms. If remote sensing instruments are relatively light weight, and the area to be sampled is relatively small and close to an airfield, a light one- or two-engine private plane is by far the most economical. Typically un-pressurized, these aircraft usually operate at altitudes below 25,000 ft., with a range less than 1,000 miles. As the areas to be imaged and their distance from the operating base increase, larger and more capable platforms are required. Turbine or turbo-charged twin engine propeller aircraft offer the next level of performance, with a significant increase in payload and range, and operating altitudes up to 35,000 ft. This class of aircraft often offers the best compromise between performance and cost for coastal work. Corporate jets, flying above 40,000 ft, and at much higher speeds, can map large areas (at lower resolutions) in a single day, but are relatively expensive to operate. Special-purpose high altitude platforms, such as the NASA ER-2 flying at 65,000 ft, map even larger areas, yet at lower resolutions, but are even more expensive to operate, and take a significant amount of time to climb to their operating altitude.

There are many aircraft-related parameters to consider when planning a field deployment such as operating altitudes, range, payload capacity, endurance, and cost of operation. Of these, the altitude of a deployment, along with the instrument's Instantaneous Field of View (IFOV), determines the spatial resolution of the area (i.e., pixel size) sampled (Fig. 2). For example, the nominal pixel size produced from a sensor with a typical 2.5 milliradian IFOV will vary from three to 50 m when operated at an altitude of 3,900 and 65,000 ft, respectively. Hence, a user can define, within the limitations of the aircraft's operating altitude, the spatial resolution of the data to match the spatial scale of coastal processes. In contrast, most satellite systems have fixed spatial resolutions that are typically too coarse for coastal or estuarine applications. Figure 3 provides a comparison between 50 m resolution airborne data and the same data set re-sampled to 1 Km which is typical of many satellite instruments.

Another important parameter determined by the altitude of the aircraft and instrument characteristics is the total swath width or the area on the ground observed in one line of pixels measured by the sensor. Simply, for a given sensor, the swath or area covered perpendicular to the line of flight, increases with increasing altitude. This distance is determined by the total angle imaged from nadir (directly below the aircraft) and can be approximated by spatial resolution and the number of pixel elements per line imaged (Fig. 2).

An important factor that increases with increasing altitude is the optical thickness of the atmosphere that exists between the Earth's surface and the remote sensing instrument. The increase in atmospheric path length contributes a significant component of the reflected energy (especially in the visible portion of the electromagnetic spectrum) detected by the instrument. The atmospheric component, or path radiance, is

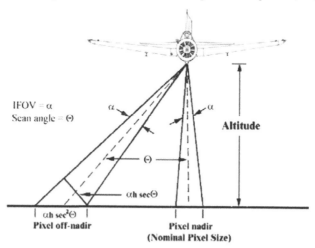

Figure 2. Viewing geometry of an airborne scanner system. Pixel size increases from a minimum directly below the aircraft (nadir) to a maximum at the system's scan angle (Θ). The scan geometry is symmetrical about nadir. The total scan angle is 2Θ.

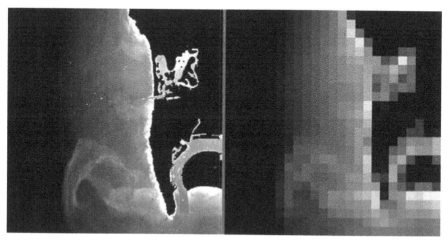

Figure 3. A comparison of data acquired at 50 m (left) and then resampled to a spatial resolution of 1 km that is typical of many satellite-based instruments (right). The details of a sewage spill in the lower portion of the 50 m data is largely obscured in the 1 km image.

primarily due to the back scatter of light into the field of view of the instrument from particles suspended in the atmospheric (i.e., aerosols). This component of the measured signal must be removed to derive the surface radiance or reflectance value. Atmospheric correction algorithms are employed to remove this contamination to obtain the desired surface signal (Richter and Schläpfer, 2002; Miller et al., 1994; Lavender and Nagur, 2002). Many deployments acquire atmospheric profiles of temperature, humidity, and atmospheric pressure using radiosonde balloons during an instrument over flight. These profiles are then input to atmospheric models such as MODTRAN 4.0 (MODerate resolution TRANSsmittance) available from the Air Force Research Lab, Space Vehicles Directorate, to effect an atmospheric correction. There are additional strategies to apply an atmospheric correction scheme that do not require *in situ* measurements of the atmosphere. A common approach is the "clear water pixel" or dark pixel subtraction technique (see for example, Gordon and Morel, 1983; Siegel et al., 2000). This technique is based on the assumption that the remote sensing instrument has a spectral band for which clear water is essentially a black body (i.e., no reflectance). Therefore, any radiance measured by the instrument in this band is due to atmospheric backscatter. This value is then subtracted from all pixels in the corresponding image assuming a homogenous atmospheric particle size distribution over the extent of the scene. An instrument band in the near-infrared region of the spectrum is most commonly used although bands in the red region have been used with success (see for example, Miller et al., Chapter 11). The topic of atmospheric correction is fairly complex and there are numerous papers written on the subject. The reader is directed to Gordon and Morel (1983) and Muller-Karger et al. (Chapter 5) for a detailed discussion.

Airborne platforms, particularly high altitude aircraft, are also used as test beds for new imaging technologies. They provide an economical method for testing prototype satellite sensors prior to launch and for collecting empirical data to develop and verify new science algorithms. Once a satellite system is in orbit, airborne sensors are frequently used to obtain simultaneous observations for data validation and calibration purposes. A key example is the MODIS (Moderate Resolution Imaging Spectroradiometer) Airborne Simulator (MAS). The MAS system was established to develop and test algorithms for the MODIS instruments (King et al., 1996). When flown at high altitudes airborne data offers a unique capability for validating satellite sensor imagery. For example, at 65,000 ft. the aircraft is above 95% of the Earth's atmosphere. Hence both air- and space-based sensors view the Earth's surface through essentially the same atmospheric column. Therefore, processing algorithms developed using an airborne prototype can conceptually be applied quickly to their space-based analog.

3. Airborne Optical Instruments

There are two basic types of optical multispectral or hyperspectral imaging sensors commonly used in airborne remote sensing: whiskbroom and pushbroom scanners. In a whiskbroom scanner system, a rotating 45 degree scan mirror is placed in front of a telescope. The scan mirror continuously scans the Earth beneath the aircraft perpendicular to the direction of flight. The system collects data one pixel (i.e., defined by the IFOV) at a time sequentially as the mirror scans across the scene to form a scan line of imagery with each mirror rotation. A scan line of pixels is equivalent to the imaged swath. The forward motion of the aircraft used to acquire a scene with sequential scan lines. A collection of scan lines in a single direction is generally defined as a mission flight line. Adjacent parallel flight lines are flown with adequate overlap (typically 20%) to provide complete coverage of the desired area. There are three primary advantages of the whiskbroom design: the spectral bands are spatially co-registered for each pixel, all spectral data are acquired simultaneously, and the spectral response function for each pixel in any band is the same. These features enable very precise spectral and radiometric measurements to be made. The whiskbroom design also allows for a wide total field of view (in some cases up to 100°) that results in large swaths being imaged in one pass. A major disadvantage of a whiskbroom scanner is that the dwell-time (time over which the system integrates the scene radiance) for each pixel is relatively short, resulting in lower theoretical signal-to-noise ratios. Low signal-to-noise ratios are a major problem over water scenes where the total available radiant energy is low, particularly in the blue region of the spectrum in coastal waters where phytoplankton absorb most of the light entering the water column. In addition, short dwell-times produce two inherent spatial distortions in the imagery: the S-bend, and the panoramic distortion. The S-bend distortion results from the forward motion of the platform while the mirror is scanning from one edge of the scene to the other, skewing the resulting imagery. The panoramic distortion, inherent in any wide field-of-view optical system, results in pixels that are larger at the scene edges due to the longer path length relative to the distance at nadir (see Fig. 2). Both the S-bend and panoramic are systematic distortions however, and may therefore be removed using the appropriate post-flight image processing techniques. Lastly, another potential disadvantage of a whiskbroom scanner is that the system is susceptible to platform instabilities due to the amount of time required to acquire each scan line. Variations in aircraft roll, pitch and yaw result in variations in the viewing geometry of the scanner-Earth system that can vary pixel by pixel along a scan line and between scan lines. These variations result in both relative and absolute errors in pixel geographic location.

Many well known airborne instruments are whisk broom scanners such as the Airborne Ocean Color Imager (AOCI), Airborne Visible/Infrared Imaging Spectrometer (AVIRIS, Bagheri and Peters, 2003; Porter and Enmark, 1987; Richardson et al., 1994), and the Calibrated Airborne Multispectral Scanner (CAMS, Miller, 1993; Miller et al., 1994). In addition, most current environmental satellites (e.g., Landsat Thematic Mapper, MODIS, SeaWiFS (Sea-viewing Wide Field-of-view Sensor) employ a whiskbroom scanner.

The pushbroom or along-track scanner is a scanner without a rotating mirror that employs a linear array of solid semiconductive elements to acquire one entire line of spectral data simultaneously. This design also relies on the forward motion of the aircraft to acquire a sequence of imaged lines to map a scene. However, instead of a scanning mirror viewing the surface from side to side across the total swath, the linear array acquires the entire scan line simultaneously as a pushbroom would. Since all pixels for a given scan line are projected onto a detector array at the same time, the S-bend distortion experienced in the whiskbroom design is eliminated. The pushbroom design allows for a longer dwell-time at each pixel hence, increasing the signal to noise performance. The pushbroom design contains no moving parts (e.g. scanning optics), so it is inherently smaller and lighter in weight. Common pushbroom systems include the CASI (Compact Airborne Spectrographic Imager, Anger et al., 1994; Anger et al. 1996; Clark et al., 1997; Hoogenboom et al., 1998), HYDICE (Hyperspectral Digital Imagery Collection Experiment, Rickard, 1993; Rickard et al., 1993) and PHILLS (Hyperspectral Imager for Low Light Spectroscopy, Davis et al., 1999; Fillippi et al., 2002) sensors.

There are, however, two significant disadvantages of a pushbroom sensor. Due to optical constraints, the total field of view of pushbroom systems are relatively narrow, on the order of 20 degrees or less (typical total field of view for whiskbroom scanners is 50 degrees or higher), which results in a narrow swath coverage. Mapping large areas therefore requires from two to four times as many flight lines, and therefore flight hours, as with a whiskbroom system. The most significant challenge of a pushbroom design however, is achieving an accurate calibration. Because there are a large number of detector elements, the radiometric characterization of the imagery is more time consuming. Each detector element has its own response to incoming radiation (i.e., slope and offset) that must be determined and then applied to the resultant imagery. This results in increased post-flight data processing to achieve a science-grade product. More problematic however, is the spectral fidelity of the design. Unlike a whiskbroom system, the spectral response function of a spectral band may vary significantly between pixels along a scan line. This problem is particularly evident in sensors that use a diffraction grating for spectral dispersion. Some highly advanced pushbroom systems, known as wedge spectrometers use a linearly variable filter over the array for spectral differentiation. In this design however, the significant loss of incoming energy offsets the increase in signal-to-noise ratio due to longer dwell-time. Pixel-to-pixel co-registration between bands in these systems may also vary significantly. Although a thorough discussion of these, and other design issues, is beyond the scope of this chapter, potential users should determine the spectral fidelity, band-to-band registration, and radiometric accuracy of any instrument intended for an application over coastal waters.

3.1 INSTRUMENT CALIBRATION

Instrument calibration is critical for any data to be used for scientific studies. Many coastal products are derived from algorithms that are based on radiometric quantities derived from theory (i.e., atmospheric correction) and empirical geo-physical relationships. The effectiveness of these derived products depends largely on accurately calibrated remotely sensed data. The calibration process consists of three basic components: radiometric, spectral, and spatial measurements. Satellite instruments are carefully calibrated prior to launch, and then periodically validated through comparison with ground measurements. Unlike the space environment which is relatively constant, the operating environment of airborne systems can vary significantly during a flight due to changes in weather, humidity, and aircraft flight profiles. In addition, the radiometric response of an airborne imaging system can change due to the deposition of contaminants onto the optical surfaces and small changes in the response of detector circuitry caused by environmental stress. Changes in the relative alignment of optical components can result in both a change in optical throughput (signal strength) and in spectral response. Designs that use diffraction gratings or prisms for spectral dispersion are particularly susceptible to subtle changes in optical alignment. The spatial response of a system can change if the positions of key optical elements move relative to one another that result in a change of focus at the focal plane of the detectors. Changes in optical alignment frequently occur when operating temperatures vary over a wide range due to the differential expansion and contraction of various mechanical structures. Temperature-related changes in the focus or spectral alignment of a system may be difficult to detect as they tend to disappear once the system is returned to room temperature. Therefore, it is important that system operators attempt to isolate instrument components from large environmental changes and perform calibration measurements in a laboratory environmental chamber duplicating flight conditions.

3.1.1 *Radiometric calibration*

Radiometric calibration is the process to numerically relate the output signal from a sensor to radiant energy. Radiometric calibration is often accomplished in an optics laboratory. The general process is have the sensor view a well-characterized external light source coupled to an integrating sphere or view flat panels coated with a highly reflective, spectrally flat, "white" substance (e.g. Barium Sulfate, or any of several commercial products) to produce a lambertian surface. A special tungsten halogen light source is calibrated against a NIST (U.S. National Institute of Standards and Technology) traceable lamp to serve as a secondary standard. Figure 4 shows the radiant output of a typical integrating sphere source, as a function of wavelength. These radiometric sources are in turn typically calibrated against Standard Lamps certified by the NIST providing a measure of scientific traceability. For thermal infrared systems, laboratory calibrations are usually performed by viewing an extended-area blackbody source. There are many types of blackbody sources, but they are all essentially uniform targets of very high emissivity maintained under precise temperature-control. The blackbody sources are usually viewed at several different temperature settings to compare the values observed by the sensor to those measured by the laboratory unit. The conditions and geometry of all radiometric measurements must be tightly controlled to ensure accuracy and repeatability.

Several airborne instruments have integrated calibration systems. For example, the Calibrated Airborne Multispectral Scanner has an onboard integrating sphere and two calibration lamps. Calibration data from the integrating sphere and lamps are recorded for each scan line. This approach enables the user to monitor the stability of the optical and electrical systems and calibrate image data based on real time system performance.

3.1.2 *Spectral characterization*

The second key element in sensor calibration is spectral characterization. Spectral characterization measures the sensitivity of each spectral band across its specified wavelength range (i.e., band-pass). This is usually done by viewing light from a monochromator. A monochromator selectively outputs discrete wavelengths of a reference light source into the aperture of the sensor. The detector output for each band is then recorded as the wavelength of the input energy is gradually changed. The output of this measurement is the Spectral Response Function (SRF) for that band, from which the spectral bandwidth can be defined. Bandwidth is often defined by the wavelengths (both upper and lower) at which 50% of the relative input energy from the monochromator is seen by the sensor. These are usually denoted as the Full Width, Half

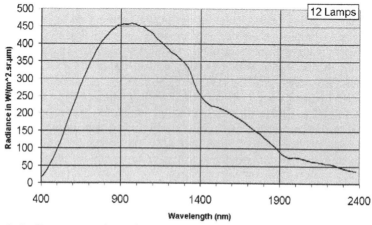

Figure 4. Radiant output from the Ames Research Center 30 in. diameter integrating sphere with twelve 45-Watt Quartz Halogen lamps.

Maximum (FWHM) or Half-Power Points (HPP) for a particular band. Figure 5 shows two SRF measurements of band 25 of the NASA MODIS/ASTER Airborne Simulator (MASTER) system as measured in the laboratory by this method several months apart. The spectral shift between the two curves of approximately eight nm was induced by mechanical stresses on the system during repeated high altitude missions on the WB-57 aircraft in the severe cold of 45,000 ft. altitude. A spectral shift in data collection can have significant implications for the accuracy of parameter retrieval algorithms that rely on relatively narrow absorption or scattering features.

3.1.3 *Spatial characterization*

Beyond spectral calibration, it is critical to accurately characterize or calibrate a sensor's spatial resolution to link individual pixels of an image to the correct ground position. The fundamental spatial response or resolving power of an imaging system is often expressed by the Modulation Transfer Function (MTF). A thorough description of the method to determine the MTF is beyond the scope of this chapter. In general however, the MTF is determined by viewing a set of bar targets or similar sharp black-to-white transition objects. The calculated value of the MTF is fixed by the optical design and quality of the system components, assuming that the instrument optics is at their maximal focus settings. In operational use however, the focus of an airborne imaging system, which usually involves a telescope, may shift from its optimal setting, due to vibration and temperature cycling. Therefore, the focus of the system's optics should be regularly checked by viewing a bar target placed at infinity (or on a collimator) and adjusted as necessary.

4. Deployment Issues

Unlike the general acquisition of satellite data in which a user must order data that was acquired at a predictable time and constant spatial resolution, there is considerable flexibility for a user to define the acquisition parameters associated with an airborne

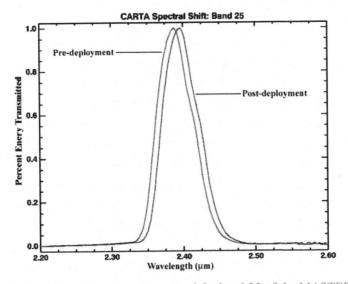

Figure 5. Shift in spectral band-pass measured for band 25 of the MASTER instrument following a deployment over Costa Rica during the CARTA (Costa Rica Airborne Research and Technology Applications) project.

remote sensing mission. Many factors must be considered such as area covered, spatial resolution, solar elevation and position, coordination with field sampling, and cost. The ultimate quality and usability of airborne data often depends on selecting the appropriate parameters for these factors. It is therefore essential that an appropriate level of planning occur to insure that useable data are acquired to support studies of dynamic coastal environments.

4.1 INSTRUMENT OPERATING CONSIDERATIONS

Although the operating environment is beyond the control of the user, knowledge of the operating conditions to which an airborne instrument is exposed will help the user understand the strict requirements for instrument calibration and validation. In general, the operating environment for airborne instruments is harsh and, as stated above, can adversely affect the performance of an airborne system. The necessary process of installing and removing an airborne instrument between deployments can significantly impact the optical alignment and performance of a system. Unfortunately, there is no routine method to fully check or calibrate most instruments after installation.

The ambient operating environment of most instruments places stress on the instrument systems. Outside air temperatures vary radically with altitude, often causing instruments to cool considerably unless mounted inside a pressurized cabin. Instruments with thermal infrared channels, which are constrained to viewing through an open port, usually are subject to severe cooling. When an aircraft subsequently descends into the more humid lower atmosphere, water condensation on the instrument is often a major issue. Extended exposure to salt water condensation in coastal environments is highly corrosive to electrical components and optical coatings. If electrical grounds or shields are allowed to degrade, noise artifacts begin to appear in the image data, or various subsystems may fail. The use of heaters to maintain the temperature of critical electrical and optical components above the local dew point, and hermetic sealing of optics and spectrometers, has proven to increase the science quality of data, as well as the longevity of the systems. A rigorous maintenance program of the instrument optics is required to maintain data quality; deposition of contaminants onto glass surfaces (lenses, windows, etc.) can significantly attenuate visible light, especially in the blue and green portions of the spectrum. A cleaning regimen should be coordinated with the calibration process to avoid un-explained changes in brightness values that could compromise the quantitative quality of the imagery. The user should consider pre- and post-flight calibration of all instrument systems.

The quality of electrical power of many commercial aircraft and associated electrical grounding circuitry may also vary significantly. Unless instruments are designed to thoroughly filter and condition the external aircraft power, the resulting imagery may contain noise and/or fluctuations in brightness levels that can seriously compromise the utility of the data.

Mechanical vibration is another potential problem when operating an optical instrument onboard an aircraft. Each airplane has a vibration frequency spectrum that changes with engine power and control surface settings. Some aircraft have resonant frequencies that occur with certain throttle settings that often can be heard audibly. These flight regimes should be avoided while acquiring data. Excessive vibration can affect the optical alignment of spectrometers, induce microphonic noise in detectors, and cause the failure of electrical connections. Finally, the G-forces associated with landing can also affect an instrument. Some form of vibration and shock isolation between the instrument and airframe is essential and should ideally be tailored to the specific vibration regime of the platform used.

4.2 MISSION PLANNING

Proper mission planning is critical to the success of any airborne remote sensing mission. The user must consider numerous factors when planning a mission and attempting to coordinate an airborne mission with field measurements from ships or small boats. The mission planner must consider local weather patterns, sea state, solar ephemeris data, and operating characteristics of the aircraft. The user must also coordinate all decisions with aircraft and field operations personnel.

4.2.1 *Flight operations and planning*

A major consideration when planning an airborne remote sensing mission is the total cost associated with the deployment, data acquisition and processing. Because the range of a typical remote sensing aircraft is limited to less than 2,000 miles, and often considerably less, the acquisition of airborne data in remote parts of the world often requires multiple transit or "ferry" flights. When transiting to a foreign country, the appropriate flight clearances should be obtained well in advance. This is particularly true when using government aircraft. In this case, overseas military bases are often the best choice, given their extensive support infrastructure. Many systems require the deployment of support personnel. Hence, the costs for missions of this type can quickly become significant. Another cost factor is the number of flight hours required to cover an area of interest. The relatively narrow extent of spatial coverage obtained from most aircraft systems requires multiple flight lines to image most features. This obviously requires more flight time and hence cost. Further, a mission is generally planned so that the user has a flight window or a time period (days) over which to acquire data. That is, a user will frequently delay an aircraft's deployment at the study area within their flight window until clear skies or weather conditions are optimum for their study. These delays generally require funds for the deployment crew only. The total cost of an airborne remote sensing mission varies widely and depends on numerous factors.

Once the geographic area to be imaged has been identified, a flight plan is developed to collect data that meets the scientific or operational requirements of the study in the most cost effective manner. If a set of parallel flight lines is required to cover the study area, adjacent flight lines are constructed to yield a 20 to 25% side overlap of the imagery. The region of overlap insures that data is acquired for the entire study area should the flight path vary by a small amount. It should be noted however that, unlike satellite data, the larger the area to be imaged the longer the elapsed time it will take to accomplish the mission. Missions may require hours to complete. The total time to image an area must be considered carefully especially for highly dynamic coastal environments. The major advantage of remote sensing is that it provides a synoptic view of large areas. In this case however, the processes of interest, such as the dynamics of a river plume boundary, may change during the time required for an airborne instrument to image the total extent of the impacted area. An alternative flight plan may therefore consider a higher altitude mission to increase the swath width of each line. Although the resulting spatial resolution will be smaller, the total time of acquisition, the number of flight lines, and often the flight hour costs will be less.

Airborne sensors are therefore usually most useful when used to observe features within the swath area of a single flight line and the study area can be imaged with a minimum of over-lapping flight lines. Mosaicked images produced by co-registering adjacent flight lines (see section 4.3) frequently contain brightness artifacts in the overlap area (Fig. 6). The artifacts are caused by an increase in atmospheric effects with increasing atmospheric path length at the image edge (Fig. 2) and solar illumination differences experienced with a difference in flight direction between adjacent flight lines. Accurate atmospheric correction algorithms are required to remove the increase in limb brightening observed in airborne images. The variability of these effects however, can be minimized by developing a flight plan that map flight lines so that the aircraft's heading while on a flight line is the solar azimuth (toward the sun) or 180° from (away from the sun) the solar azimuth. Figure 7 shows an example flight plan for a coastal data mission.

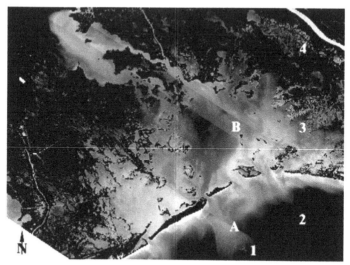

Figure 6. Mosaick image generated from 4 adjacent georeferenced flight lines (1-4) of the Calibrated Airborne Mustispectral Scanner (CAMS) acquired over Barataria Bay, LA. Land is masked to black. Difference in the brightness values of adjacent lines is clearly visible. Fight lines were flown to the southeast (1 and 4, toward the sun) or northwest (2 and 4, away from the sun). The seam between the flight lines is also visible. Two strategies to mosaick adjacent lines are shown: in A, pixels from line 1 are used in the overlapping region; in B, the average brightness values of lines 2 and 3 are used for the overlapping region.

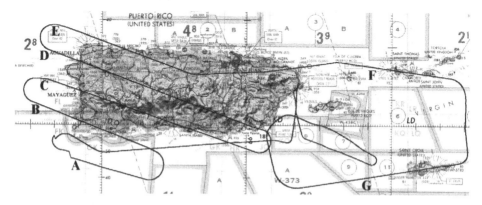

Figure 7. A flight plan developed to study the coastal waters of Puerto Rico using the MODIS Airborne Simulator (MAS) aboard the NASA ER-2. The solid lines indicate the proposed flight path for September 21, 1995. The flight lines (A-E) are oriented relative to the solar azimuth.

4.2.2 *Solar geometry and site conditions*

Perhaps the most important factor to consider when designing a mission is to avoid sun glint. Sun glint is caused by the direct specular reflection of sun light off the sea surface into the viewing aperture of the sensor system. These photons did not enter the water column and hence do not contain any useable information about in-water constituents. Sun glint is manifested as extremely bright (often saturated) areas in the imagery (Fig. 8). Sun glint is best avoided by flying the aircraft directly into, or out of, the solar azimuth during data collections and by acquiring data only in a specific range of solar elevations. The range of acceptable solar elevations will vary somewhat with the viewing geometry of the sensor, but generally data should be collected only within 30 and 50 degrees solar elevation and an aircraft heading within 10 degrees of the solar azimuth or its reciprocal. The heading also avoids a solar, anti-solar edge of the imagery and thus provides a more symmetrical limb brightening due to the increase atmospheric path length (Fig. 8). This symmetrical nature of atmospheric backscatter greatly simplifies the geometric calculations required by most atmospheric correction algorithms. The range of optimum solar elevation equates to two daily data collection periods (morning and an afternoon) centered on local solar noon, when the sun is

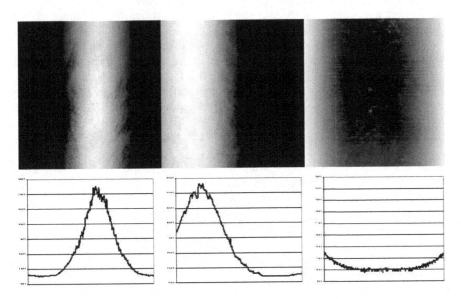

Figure 8. Sun glint patterns obtained in MAS imagery due to high solar elevation angle (left) or viewing angle (middle). No sun glint is observed when solar elevation and flight azimuth are within defined limits (right). Bottom plots show brightness values vs. scan line pixel number. Note the relatively low, symmetrical, limb brightening due to the increase in atmospheric path length (right).

ascending and descending, respectively. Solar elevations above 50 degrees may result in sun glint down the center of the image. Over the course of a long mission, the solar azimuth may change appreciably, so the heading of otherwise parallel data flight lines should be adjusted accordingly. There are several software tools to determine solar geometry, based on scene coordinates, date, and time of day available on the internet. Because the sea surface is rarely flat, there is inevitably some degree of sun glint off of wave fronts as well. For this reason high sea states and white caps should be avoided if possible. Although solar elevations below 30 degrees will avoid sun glint, the upwelling radiance from the water surface is so small that the data may have limited use.

Cloud cover, especially in coastal regions, is often a major consideration. The scene to be imaged must be cloud-free at the time of acquisition. Moreover, clouds close to the area being imaged can reflect light into the scene resulting in abnormally high radiance values over the area of interest. The presence of atmospheric aerosols or haze will also affect the upwelling radiance from the scene, especially in the shorter blue wavelengths, degrading the quality of the data and making atmospheric correction more difficult. An informed decision to fly on any given day should be made based on the best available data, including timely satellite weather imagery and local aviation and marine forecasts, to avoid needless expenditure of resources.

4.3 POST-FLIGHT DATA PROCESSING

Post-flight data processing is an extremely important yet highly variable process among the providers of airborne data. The media type, level of processing, and format of the data received from a supplier should be considered carefully and determined prior to the execution of a mission. Frequently, the data will be available in a proprietary or instrument-specific format. Clearly the user must receive sufficient format information so that the files can be ingested and processed. Many data providers now supply data in a standard image processing format (e.g., ERDAS, ENVI, GEOTIFF) or in a generic scientific data format such as HDF. The HDF (Hierarchal Data Format) is an efficient structure for storing multiple sets of scientific, image and ancillary data, in a single file and is gaining popularity. The level of processing that the data supplier may provide can vary widely. The data may be sent as raw digital image files (no processing), calibrated radiances (radiometric calibrations applied), or calibrated radiances and geoferenced (registered to a map projection). In general, calibrated, georeferenced data are required for subsequent analysis. Processing can be accomplished by the data provider or the data user.

There are several commercially available software packages to process airborne remote sensing data such as ERDAS IMAGINE (Leica Geosystems GIS & Mapping, Atlanta, GA USA) and ENVI (Research Systems International, Boulder, CO USA). The user can convert raw data to radiance (e.g. $W/m^2/\mu m/sr$, or equivalent) by applying predetermined slope and offset factors supplied by the data provider. Other corrections such as instrument temperature response or in-flight calibration data can also be applied. Some sensor-specific calibration algorithms are proprietary however, and are typically performed by the data provider.

Corrections for systematic spatial distortions (e.g. S-bend, panoramic) can be accomplished with most image processing. Geo-correction or georeferencing the imagery is required in most cases, especially when assembling mosaics of multiple flight lines, correlating with other data sets, or integrating *in situ* field measurements. This processing step can be both difficult and time consuming. The standard method to georeference remotely sensed imagery is to "tie" specific pixel locations in the image to their corresponding location on a mapped surface for which the mapped coordinates are well known (ground control points). This approach works well for images of land surfaces in which numerous points such as road intersections, corner of buildings, or parking lots are easily identified in the image and their exact location can be determined using a handheld GPS (Global Positioning System). As there are usually few visual reference points over open water, this process requires external positional data, either from the aircraft navigation system, or imbedded systems on the sensor itself. The primary required parameters for accurately projecting a pixel location onto the surface are platform attitude (pitch, roll, heading,) altitude, latitude, and longitude (usually determined by a Global Positioning System). The positional data are often embedded in the image data for each scanline. This approach is also used by several satellite based systems including the MODIS and AVHRR (Advanced Very High Resolution Radiometer). Whether the data is georeferenced using ground control points or embedded positional data, the image is re-mapped or "warped" using a numerical model of the map projection and imaging geometry. This process can require considerable computational power and data storage. However, common personal desktop computers can now easily accomplish this task using most image processing software. During the process of georeferencing an image the original pixels are often re-sampled when mapped to a new coordinate space. This process may alter a pixel's radiometric and spectral content. In cases where precise surface retrievals are required, especially within areas of high spatial variability, the information should be extracted before the image is re-sampled. Individual fight lines must be georeferenced prior to image mosaicking and atmospheric correction. Georeferencing is an important yet often mysterious aspect of data processing. The reader is directed to Lillesand et al. (2003) as well as numerous tutorials available on the internet to gain an in-depth understanding of the principles of georeferencing remotely sensed imagery.

5. Summary

Remote sensing is a valuable tool for analyzing coastal aquatic environments. Earth observations from airborne and space-based platforms provide a synoptic view that can not be obtained using traditional field sampling techniques. However, the dynamic nature and small-scale features of many coastal regions present a particularly difficult challenge to most satellite systems. In contrast, airborne remote sensing offers unique capabilities that address these issues.

There is considerable diversity in the range of aircraft platforms available and capabilities of remote sensing instruments. The major advantage of airborne remote sensing is the flexibility in the deployment of an instrument. For example, a user can select an airborne system with an operating altitude and optical characteristics that can obtain high spatial resolution data commensurate with most coastal features. More importantly however, the system can be deployed to optimize solar illumination, avoid sun glint, minimize cloud cover, and coordinate with a field sampling program.

There are several issues that potential users of airborne remote sensing should consider. The key to obtaining quality airborne data is proper instrument calibration and establishing appropriate flight parameters. Due to the impact of installation and operation of an instrument during flight, laboratory radiometric, spectral, and spatial calibrations should be performed before and after each series of missions. Proper planning and definition of mission parameters can insure the utility of the data and minimize post-mission data processing and data analysis. Another important issue when considering the use of airborne remote sensing data is the availability of processing and analysis software and algorithms. In general, airborne systems do not have the user community of most satellite instruments. As a consequence, there are fewer algorithms and dedicated processing software compared to major satellite systems.

Hence, airborne remote sensing is often the only practical technology available to adequately study or monitor certain coastal aquatic processes or features. There are clear advantages and disadvantages of airborne compared to satellite systems. With moderate study of the issues presented here, a potential user should be able to assess the application of airborne remote sensing in their work and design a successful mission.

6. References

Anger, C.D., S. Mah, and S.K. Babey. 1994. Technology enhancements to the compact airborne spectrographic imager (CASI). Proceedings of the First International Airborne Remote Sensing Conference and Exhibition, Strasbourg, France, 200-213.

Anger, C.D., S. Achal, T. Ivanco, S. Mah, R. Price, and J. Busler. 1996. Extended operational capabilities of CASI. Proceedings of the Second International Airborne Remote Sensing Conference, San Francisco, California, 124-133.

Bagheri S., and S. Peters. 2003. Retrieval of marine water constituents using atmospherically corrected AVIRIS hyperspectral data. NASA/AVIRIS Workshop, JPL, Pasadena, CA.

Carder, K.L., R.G. Steward, R.F. Chen, S. Hawes, A. Lee, and C.O. Davis. 1993. AVIRIS calibration and application in coastal oceanic environments: tracers of soluble and particulate constituents in the Tampa Bay coastal plume. Photogrammetric Engineering and Remote Sensing, 59(3):339-344.

Clark, C., H. Ripley, E. Green, A. Edwards, and P. Mumby. 1997. Mapping and measurement of tropical coastal environments with hyperspectral and high spatial resolution data. International Journal of Remote Sensing, 18(2):237-242.

Davis, C. O., M. Kappus, J. Bowles, J. Fisher, J. Antoniades, and M. Carney. 1999. Calibration, characterization and first results with the ocean PHILLS hyperspectral imager. Proceedings of the SPIE, 3753:160-168.

Fillippi, A., J.R. Jensen, R.L. Miller, R.A. Leathers, C.O. Davis, T.V. Downes, and K.L. Carder. 2002. Cybernetic statistical learning for hyperspectral remote sensing inverse modelling in the coastal ocean. Proceedings of Ocean Optics XVI, Sante Fe, New Mexico, 9 pp.

Gade, M., W. Alpers, H. Hühnerfuss, V.R. Wismann, and P.A. Lange. 1998. On the reduction of the radar backscatter by oceanic surface films: helicopter measurements and their theoretical interpretation, Remote Sensing of Environment., 66:52-70.

Gordon, H.R., and A.Y. Morel. 1983. Remote assessment of ocean color for interpretation of satellite visible imagery: A Review, Springer-Verlag, New York, 114 pp.

Hamley, M.H. 1994. US LTA 138S airship as an airborne research platform. Proceedings of the First International Airborne Remote Sensing Conference and Exhibition, Strasbourg, France, 341-350.

Harding, L.W.,Jr., E.C. Itsweir, and W.E. Esaias. 1994. Estimates of phytoplankton biomass in the Chesapeake Bay from aircraft remote sensing of chlorophyll concentrations, 1989-92., Remote Sensing of Environment, 49:41-56.

Howland, W. G., 1980. Multispectral Aerial Photography for Wetland Vegetation Mapping, Photogrammetric Engineering & Remote Sensing, 46(1):87-99.

Hoogenboom, H.J., A.G. Dekker, and J.F. De Haa. 1998. Retrieval of chlorophyll and suspended matter in inland waters from CASI data by matrix inversion. Canadian Journal Remote Sensing, 24(2):144-152.

King, M.D., W.P. Menzel, P.S. Grant, J.S. Myers, G.T. Arnold, S.E. Platnick, L.E. Gumley, S.C. Tsay, C.C. Moeller, M. Fitzgerald, K.S. Brown, and F.G. Osterwisch. 1996. Airborne scanning spectrometer for remote sensing of cloud, aerosol, water vapor, and surface properties. Journal of Atmospheric and Oceanic Technology, 13(4):777-794.

Lavender, S.J., and C.R.C. Nagur. 2002. Mapping coastal waters with high resolution imagery: atmospheric correction of multi-height airborne imagery. Journal of.Optics. A: Pure and Applied Optics, 4:S50-S55.

Lillesand, T.M., R.W. Kiefer, and J.W. Chipman. 2003. Remote Sensing and Image Interpretation, Wiley, 784 pp.

Matthews, J. P., V. Wismann, K. Lwiza, R. Romeiser, I. Hennings, and G. P. deLoor. 1997. The observation of the surface roughness characteristics of the Rhine plume frontal boundaries by simultaneous airborne thematic mapper and multifrequency helicopter-borne radar scatterometer. International Journal of Remote Sensing, 18(9):2021-2033.

Miller, R., M.S. Twardowski, C. Moore, and C. Casagrande. 2003. The Dolphin: Technology to Support Remote Sensing Bio-optical Algorithm Development and Applications. Backscatter, 14(2):8-12.

Miller, R.L., G. Carter, J. Sheehy, B. Rock, P. Entcheva, and J. Albrechtova. 1999. Monitoring initial forest recovery in the Krusne hory, Czech Republic, using ground and airborne multispectral digital cameras. Presented at the Fourth International Airborne Remote Sensing Conference and Exhibition/21 Canadian Symposium on Remote Sensing, Ottawa, Ontario, Canada.

Miller, R.L., B. Spiering, A. Peek, J. Hasenbuhler, T. McNamee, R. Lahnemann, and K. Draper. 1999. Analyzing coastal processes using a miniature multispectral imaging system flown on a portable UAV, Proceedings of the Fourth International Airborne Remote Sensing, Ottawa, Canada.

Miller, R.L., G. Carter, M. Seal, and T. Gress. 1997. Analyzing coastal processes using a three-band airborne digital camera system, Proceedings of the Fourth Thematic Conference, Remote Sensing for Marine and Coastal Environments, Orlando, FL.

Miller, R.L., and J.F. Cruise. 1995. Effects of suspended sediments on coral growth: evidence from remote sensing and hydrologic modeling. Remote Sensing Environment, 53:177-187.

Miller, R.L., J.F. Cruise, E. Otero, J.M. Lopez, W.F. Smith Jr., and B.K. Martin. 1994. Monitoring the water quality of Mayaguez Bay Puerto Rico: an integrated program using remote sensing and field measurements. Proceedings of the Second Thematic Conference, Remote Sensing for Marine and Coastal Environments, New Orleans, 204-214.

Miller, R.L. 1993. Mapping coastal suspended sediments using the Calibrated Airborne Multispectral Scanner, Fall Symposium, American Water Resources Association.

Mumby, P.J., E.P. Green, C.D. Clark, and A.J. Edwards. 1998. Digital analysis of multispectral airborne imagery of coral reefs. Coral Reefs, 17:59–69.

Petrie, G. 2001. Robotic aerial platforms for remote sensing robotic aerial platforms for remote sensing: UAVs are now being developed for use as "satellite substitutes". Geoinformatics Magazine, May:12-17.

Porter, W.M., and H.E. Enmark. 1987, System overview of the Airborne Visible/Infrared Imaging Spectrometer (AVIRIS). Proceedings, Society of Photo-Optical Instrumentation Engineers (SPIE), 834:22-31.

Rickard, L.J., 1993. HYDICE: An airborne system for hyperspectral imaging. Proceedings of SPIE Imaging Spectrometry of the Terrestrial Environment, Orlando, 1937:173-179.

Rickard, L.J., R.W. Basedow, E.F. Zalewski, P.R. Silverglate, and M. Landers. 1993. HYDICE: an airborne system for hyperspectral imaging. Proceedings of SPIE, 1937:173-179.

Richardson, L.L, D. Buison, C.J. Lui, and V. Ambrosia. 1994. The detection of algal photosynthetic accessory pigments using Airborne Visible-Infrared imaging Spectrometer (AVIRIS) Spectral Data. Marine Technology Society Journal, 28:10-21.

Richter, R., and D. Schläpfer. 2002. Geo-atmospheric processing of airborne imaging spectrometry data. Part 2: atmospheric/topographic correction. International Journal of Remote Sensing, 23:2631-2649.

Schaepman M., P. Keller, D. Schläpfer, C. Cathomen, and K.I. Itten. 1997: Experimental determination of adjacency effects over an eutrophic lake using a helicopter mounted spectroradiometer for the correction of imaging spectrometer data. Third International Airborne Remote Sensing Conference and Exhibition, Copenhagen, Denmark, 497-504.

Sheppard, C.R.C, Matheson, K., Bythell, J.C., Murphy, P., Blair Myers, C., and Blake, B., 1995, Habitat mapping in the Caribbean for management and conservation: use and assessment of aerial photography. Aquatic Conservation: Marine & Freshwater Ecosystems, 5:277–298.

Siegel, D.A., W. Menghua, S. Maritorena, and W. Robinson. 2000. Atmospheric correction of satellite ocean color imagery: the black pixel assumption. Applied Optics, 39(21):3582- 3591.

Windischbauer, D.I.F., and W. Hans. 1994. The modern airship in the role as an airborne sensor platform. Proceedings of the First International Airborne Remote Sensing Conference and Exhibition, Strasbourg, France. 351-362.

Chapter 4

IN-WATER INSTRUMENTATION AND PLATFORMS FOR OCEAN COLOR REMOTE SENSING APPLICATIONS

[1]MICHAEL S. TWARDOWSKI, [2]MARLON R. LEWIS, [3]ANDREW H. BARNARD AND [3]J. RONALD V. ZANEVELD

[1]*WET Labs, Inc., 165 Dean Knauss Dr., Narragansett, RI, 02882 USA*
[2]*Dalhousie University, Halifax, Nova Scotia B3H 4J1, CANADA*
[3]*WET Labs, Inc., 620 Applegate St., Philomath, OR, 97370 USA*

1. Introduction

Remote sensing of reflected sunlight from the upper ocean is a tremendous tool for studying biological, chemical, geological, and physical processes over a broad range of time and space scales. Global biogeochemical phenomena spanning seasonal (e.g., spring bloom), multi-year (e.g., the El Niño Southern Oscillation), to decadal (e.g., climatic variability) time scales can be resolved by an orbiting satellite imager. Reflected light in the visible domain (wavelengths of ~ 400 to 700 nm) is particularly useful in the study of upper ocean processes, as many important biogeochemical components of seawater absorb and scatter light effectively in this spectral range (the term "ocean color" specifically relates to the spectral character of this water-leaving visible light). These dissolved and particulate seawater components play key roles in the cycling of carbon in the ocean and serve as indicators of ecosystem health. In-water measurements help elucidate the link between these components and the remotely sensed signal.

Down-looking, passive remote sensors in air and space measure sunlight that is reflected upward into the sensor; in addition to the atmospherically scattered photons, a portion of the measured radiance results from photons that have exited the ocean and passed back through the atmosphere to the sensor in orbit. This portion is termed spectral upwelled water-leaving radiance, L_u (W m^{-2} nm^{-1} sr^{-1}) and primarily consists of light scattered in the backward direction off the particles and molecules of seawater (for a complete discussion, refer to Zaneveld et al., Chapter 1). Sunlight incident at the ocean surface is represented as spectral downwelling surface irradiance, E_d (W m^{-2} nm^{-1}), and the so-called remote sensing reflectance, R_{rs}, is derived from L_u/E_d, with L_u strictly defined in the nadir direction (normal to the plane of the ocean surface). Although L_u consists of primarily backscattered light, equally important in terms of its information content is the component of incident sunlight missing in the upwelled light. This is light that has been absorbed (or filtered) by the constituents of seawater in the upper ocean. The dependence of R_{rs} on these optical processes of backscattering and absorption just below the ocean surface (represented as 0$^-$) can be simply written (Morel and Prieur, 1977):

R.L. Miller et al. (eds.), Remote Sensing of Coastal Aquatic Environments, 69–100.
© 2005 *Springer.*

$$R_{rs}(0^-) = \frac{L_u(0^-)}{E_d(0^-)} \cong \Psi\left(\frac{b_b}{a+b_b}\right), \qquad\qquad (1)$$

where the factor Ψ varies within a relatively small range depending on surface illumination conditions and the volume scattering properties of the water body (Morel and Gentili, 1993; 1996). Each of the parameters in Eq. 1 have implicit spectral dependencies. Radiance and irradiance have units of W m^{-2} sr^{-1} and W m^{-2}, respectively, and spectral radiance and spectral irradiance have units of W m^{-2} nm^{-1} sr^{-1} and W m^{-2} nm^{-1}, respectively. For details see Mobley (1994).

Understanding how the different components of seawater alter the path of incident sunlight through backscattering and absorption is essential to using remotely sensed ocean color observations effectively. This is particularly apropos in coastal waters where the different optically significant components (phytoplankton, detrital material, inorganic minerals, etc.) vary widely in concentration, often independently from one another. This understanding is packaged in the form of algorithms that define the relationships between biogeochemical components of seawater and remotely sensed signals. Such algorithms are commonly known as inversions because the forward problem of sunlight being altered by the constituents in the upper ocean to produce a reflected signal is typically inverted in an algorithm to derive in-water constituent(s).

A multitude of algorithms or models have been developed to derive oceanic biogeochemical properties and these continually evolve as technological and theoretical advances clarify optical-biogeochemical relationships. Remote sensing algorithms typically fall into three categories: analytical, semi-analytical, and empirical. Analytical algorithms are based solely on theory; there are, however, very few purely analytical algorithms because they require detailed knowledge of a host of complex and often poorly understood relationships between seawater components and their specific optical properties (Morel, 1980; Morel and Maritorena, 2001). The more popular semi-analytical algorithms are based on theoretical relationships of the underlying physics of ocean color (such as Eq. 1) but include some statistical relationships formulated through data sets of relevant in-water parameters and optical properties. Empirical algorithms are based purely on these statistical regressions and are currently the most common type for oceanic Case 1 waters. For important biogeochemical parameters (e.g., chlorophyll), a multitude of algorithms have been developed.

In-water optical data are required for development, refinement, and validation of these algorithms. As a result, NASA, the U.S. Office of Naval Research, and foreign counterpart agencies maintain large repositories of in-water optical and biogeochemical data for current and future algorithm related needs (e.g., the SeaWiFS Bio-optical Archive and Storage System – SeaBASS; World-wide Ocean Optics Database – WOOD, Smart 2000). For all algorithms, measurements of R_{rs} and the biogeochemical property in question are necessary for validation (note that R_{rs} computed based on below water measurements, and that measured above the water surface have been shown to agree within about 5% when appropriate care is taken with measurement procedures – Hooker et al., 2002, Hooker and Morel, 2003). Measurements of R_{rs} are also necessary to validate and calibrate the signal detected by a remote sensor. These data sets are normally (hopefully) comprehensive, collected in many locations throughout the world's oceans under conditions that cover a large dynamic range in the biogeochemical property. Using such data sets, statistical empirical algorithms have been developed to determine chlorophyll (e.g., O'Reilly et al., 2000), particulate organic carbon (Stramski et al., 1999; Mishonov et al., 2003), calcium carbonate (Gordon and Balch, 2003), macrophyte leaf-area index (Dierssen et al., 2003), oil (De Domenico et al., 1994), nutrients (Goes et al., 2003), and total suspended matter (Kratzer et al., 2000; Clark, 2003).

Semi-analytical algorithms typically use an analytical model (e.g., Morel, 1980; Zaneveld, 1995) to derive the in-water optical properties of backscattering and absorption from R_{rs}, but also require the use of empirical relationships between these optical properties and seawater constituents. As a result, development and validation work require measurements of R_{rs}, the biogeochemical property in question and the in-water optical properties used in the model. Semi-analytical algorithms are typically required in optically complex Case 2 (coastal) waters. Semi-analytical models have been developed to derive chlorophyll (Gordon et al., 1988; Morel, 1988; Roesler and Perry, 1995; Carder et al., 1999; Ciotti et al., 1999; Maritorena et al., 2002; and many others), colored dissolved organic matter (Carder et al., 1999; Siegel et al., 2002), and total suspended matter (Haltrin and Arnone, 2003). Other algorithms such as those for total primary productivity (Platt and Lewis, 1987; Sathyendranath et al., 1989; Morel 1991; Antoine and Morel, 1996; Antoine et al., 1996; Behrenfeld and Falkowski, 1997; Campbell et al., 2002) and new production (e.g., Lewis et al., 1988; Dugdale et al., 1989; Sathyendranath et al., 1991; Siegel et al., 2002) use multiple derived products. So-called Algorithm Theoretical Basis Documents (ATBDs) for the U.S. Moderate Resolution Imaging Spectrometer (MODIS) sensors aboard the Terra and Aqua satellites can be viewed at several NASA web sites.

Some algorithms focus on retrieving only the in-water optical properties from the remote signal (Garver and Siegel, 1997; Gould and Arnone, 1998; Hoge and Lyon, 1999; Loisel and Stramski, 2000; Loisel et al., 2001; Roesler and Boss, 2003). The in-water optical properties themselves can be useful for radiative transfer modeling, general water type classifications, or other applications such as estimating diver visibility (Zaneveld and Pegau, 2003).

In the following sections, the in-water measurements and sensors used in remote sensing applications are discussed in more detail. Because the temporal and spatial characteristics of *in situ* field data relative to data collected by a remote sensor is of fundamental importance for biogeochemical algorithm development and validation, the various types of deployment platforms available to support these measurements are also discussed. Finally, brief comments are offered on strategies for collecting data for various remote sensing applications from a technology perspective.

This work is intended as a review of technology and techniques but the context of the information is intended to help enhance strategies for using in-water instrumentation and platforms for algorithm development and validation in the future. Reviewing current technology as well as current needs should also provide some insight toward pathways for future technology development.

2. In-water Instrumentation

In-water optical properties are classically broken down into two main types, "Inherent" and "Apparent", after Preisendorfer (1976). Both are relevant to remote sensing applications. Inherent Optical Properties (IOPs) are those parameters whose magnitude depend only on the substances in the water and are independent of the ambient light field. Apparent Optical Properties (AOPs) are additionally dependent on the ambient light field and its geometrical structure. Radiative transfer theory describes the relationships between the AOPs and IOPs. Zaneveld et al. (Chapter 1) discuss these relationships in detail and provide definitions and backgrounds for the relevant AOPs and IOPs discussed in this section. The following summarizes sensor technologies for measuring these properties. Specific issues relating to the deployment of optical sensors in the field (e.g., biofouling) are addressed in section 3.

2.1 IN-WATER MEASUREMENT OF INHERENT OPTICAL PROPERTIES

Recent advances in optical instrumentation and methodologies now enable the *in situ* measurement of many of the dissolved and particulate fractional IOP components (Moore, 1994; Pegau et al., 1995; Pegau et al., 1999; Twardowski et al., 1999; Moore et al., 2000; Mueller et al., 2003). The IOPs of pure water (the a_w, c_w, b_w, and b_{bw} terms from Zaneveld et al., Chapter 1) can be considered knowns in the visible range with small error (Morel, 1974; Pope and Fry, 1997). Since no current *in situ* method can physically separate phytoplankton and their pigments from other particles, the absorption and backscattering attributable to phytoplankton, a_ϕ and $b_{b\phi}$, respectively, must be derived from *in situ* bulk particulate measurements (a_p and b_{bp}) using bio-optical spectral decomposition models (e.g., Roesler et al., 1989; Bricaud and Stramski, 1990; Carder et al., 1999).

The conventional methodology for measuring beam attenuation, c, is rooted in the relationship between c and the loss in power of a collimated, unpolarized source due to attenuation ($d\Phi_c$) across an infinitesimally small pathlength (dl) of medium:

$$c\Phi_0 = -\frac{d\Phi_c}{dl},$$

(2)

where Φ_0 is the incident source power. This is a differential equation with solution:

$$\Phi_c = \Phi_0 \exp\left(-\int_0^l c \, dl\right).$$

(3)

If we are interested in the bulk value of c in a medium over the pathlength l, then Eq. 3 reduces to $\Phi_c = \Phi_0 \exp(-\bar{c}l)$. Thus, by measuring the power Φ of a collimated beam at two locations within a medium a distance l apart, we can solve for c.

This is the principle by which conventional c meters (also called transmissometers) function, although there are many details. One of the most important details is that collimating optics are required in front of the detector measuring Φ_c. Collimating optics help to exclude scattered light in the near-forward direction, which is very important since the vast majority of scattered light is deflected into the first few degrees. Remember, the Φ_c we need should ideally be comprised of only the unaltered radiance from the incident collimated source beam. The theoretically ideal c meter would therefore have an acceptance angle for scattered light of ~0°, but practically this is not possible because no light would reach the detector. All c meters consequently suffer from errors due to the acceptance of some near-forward scattered light, with measured attenuation, c_m, related to true c according to:

$$c_m = c - 2\pi \int_0^{\theta_A} \sin(\theta)\beta(\theta)d\theta,$$

(4)

where θ_A is the acceptance angle, and $\beta(\theta)$ the volume scattering function. Note the error term is a function of both meter design and the characteristics of the medium and therefore varies for different natural waters. Acceptance angles should be as small possible while maintaining sufficient signal, although for $\theta_A < \sim0.1°$ turbulence from small refractive index discontinuities in water (Bogucki et al., 1999) can introduce undesirable noise.

There are several possible methodologies for the *in situ* measurement of absorption (e.g., Pegau et al., 1995), but the most popular is based on the reflective tube principle (Zaneveld et al., 1990; Kirk, 1992). In implementing this method, a collimated, unpolarized source is employed, as with the *c* meter. The receiving end in this case, however, is designed to collect as much scattered light as possible. The optical path through the medium is surrounded by a glass reflective tube, which in turn is surrounded by air, effectively reflecting any scattered light out to $\sim42°$ back into the flow cell. The angle $42°$ is the angle of total internal reflection between the glass-air interface. At the collection end, a diffuser and wide-area detector are then employed to gather as much of the light scattered by the medium as possible. Since most of the scattering in the ocean is near-forward, only a small amount of light is lost with a reflective tube device (on the order of 10%) and corrections exist to account for this (Zaneveld et al., 1994).

One of the most significant advancements for the measurement of IOPs has been the development of multi-wavelength combination *a* and *c* meters (Moore et al., 1992; Zaneveld et al., 1992). An example is the WET Labs AC9, measuring *a* and *c* over nine wavelengths in the visible at a sampling rate of 6 Hz using the methodologies described above. The *a* and *c* measurements are blanked to clean water, and thus the absorption and attenuation of the combined dissolved and particulate fractions, a_{pg} and c_{pg}, respectively, are directly measured. Absorption only in the dissolved fraction, a_g, may also be measured by attaching a particle filter to the intake of the absorption tube (Twardowski et al., 1999; 2004). Scattering due to particles, b_p, can be derived by subtracting a_{pg} from c_{pg}. Detailed protocols for using these measurements to obtain the non-water IOPs are described in Mueller et al. (2003). Periodic calibrations with pure water to account for instrument drift (Twardowski et al., 1999), and corrections for the temperature and salinity dependencies of pure water absorption (Pegau et al., 1997) are important aspects of the recommended protocols.

Modern electronics, and a relatively long 25 cm pathlength, allow the AC9 to achieve accuracies sufficient to resolve natural oceanic and coastal levels of *a* and *c* without the need for concentrating samples. Because the magnitudes of many oceanic and coastal IOPs are relatively low, the issue of sufficient accuracy has historically been a challenging problem. The relationship between accuracy and pathlength can be expressed as (Højerslev, 1994):

$$accuracy = \frac{dc}{c} = \frac{e^{cl}}{cl} dT, \tag{5}$$

where the dT term represents the electronic noise of the instrument. The great utility of this relationship is that the function $[e^{cl} / cl]$ is at a minimum when the cl term, known as the optical pathlength, is equal to one. Thus, in clear oceanic regions where attenuations can dip below 0.1 m^{-1}, pathlengths on the order of 10 m would be optimal. For the 25-cm pathlength AC9, accuracy is optimized for *a* and *c* values of ~4 m^{-1}. However, because of stable electronics that make dT very small, the precision in AC9 measurements is ~0.001 m^{-1}, enabling IOP determinations in the clearest waters with accuracies on the order of a few percent (compared to <0.05% in more turbid coastal waters). Calibrations with clean water are very important in achieving these accuracies, as instrument drift introduces bias errors that require correction.

A next-generation, hyperspectral a and c device, the WET Labs ACS, has also been recently developed (Moore et al., 2004; Sullivan et al., 2004). The measurement principles are similar to previous a and c devices, but the spectral resolution is about 4 nm in the visible range, resulting in a total of 84 wavelengths. Increased spectral resolution helps to resolve all the spectral ranges required for algorithms relevant for current and future remote imagers. Sullivan et al. (2004) have been able to document naturally occurring hyperspectral structure in a and c with this device occurring over centimeter scales, not resolved previously.

Another device for measuring hyperspectral absorption that employs a long-pathlength liquid-waveguide capillary cell (Kirkpatrick et al., 2000) was recently adapted from the bench top to an in-water form installed in an autonomous glider vehicle (Schofield et al., 2004). The sensor is designed to determine distributions of the harmful algae *Karenia brevis* through a 4th derivative spectral analysis. Such in-water data combined with remote sensing may prove a powerful tool in assessing the dynamics of harmful algal blooms and may provide the groundwork for the development of remote sensing algorithms based on the *K. brevis* detection model (e.g., Schofield et al. 1999).

For measurements of backscattering, determinations of $\beta(\theta)$ are made in the backward direction (Fig. 1A). Recalling from Zaneveld et al. (Chapter 1)

$$b_b = 2\pi \int_{\pi/2}^{\pi} \sin(\theta)\beta(\theta)d\theta, \tag{6}$$

backscattering coefficients can be derived using measured $\beta(\theta)$ at one (Maffione and Dana 1997; Boss and Pegau 2001) or more (Petzold, 1972; Moore et al., 2000; Zhang et al., 2002; Lee and Lewis, 2003) angles with errors as low as a few percent. With measurements at only one angle, θ_m, a "χ factor" is used in the simplified relationship:

$$b_b = \chi \left[2\pi\sin(\theta_m)\beta(\theta_m) \right]. \tag{7}$$

To minimize errors in estimating b_b from one angle, Maffione and Dana (1997) and Boss and Pegau (2001) recommend measurements of β at θ_m values of 140° and 117°, respectively. Small errors in deriving b_b from $\beta(\theta_m)$ result from the observation that changes in the shape of $\beta(\theta)$ at certain regions of the backward direction are compensated by changes elsewhere, i.e., the $\beta(\theta_m)$ serves as a "pivot point" (Boss and

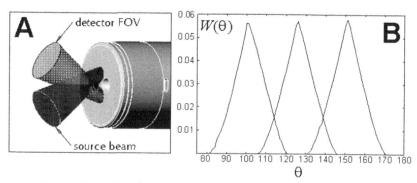

Figure 1. (A) A schematic of a volume scattering measurement with a backscattering sensor and (B) the weighting functions, $W(\theta)$, for a 3 angle (100°, 125°, and 150°) backscattering sensor.

Pegau, 2001). Variability around this "pivot point" is largely a function of the contribution of scattering from particles relative to the water background. As a result, if the water component of $\beta(\theta)$ is removed (see Morel, 1974 for the $\beta(\theta)$ of pure water and seawater), the derivation of b_b is thus simplified and can be more accurate, especially in clearer waters (Boss and Pegau, 2001). When removing the water contribution, the method involves the following basic steps to obtain total b_b from $\beta(\theta_m)$ (see Boss et al., 2004): 1) remove $\beta(\theta)$ due to water; 2) compute b_b for particles only using Eq. 7 and an appropriate χ_p for natural particle populations (see Boss and Pegau, 2001); and, 3) add in the b_b for water.

Recent data suggests χ_p values may vary little for natural particle populations. For example, Sullivan et al. (submitted) observed excellent correlations between measurements of particulate scattering at three angles in the backward direction in a data set that included over a thousand 1-m binned samples from numerous Case 2 water types (linear relationship between $\beta(100°)$ and $\beta(150°)$ exhibited r^2 of 0.97). Consistency in relationships between β values – i.e., a generally representative shape for particulate $\beta(\theta)$ in the backward direction – implies that a consistent set of χ_p values may be used for natural waters and also that the angle θ_m is not so important as long as an appropriate χ_p is applied.

Every measurement of β at one θ in practice resolves a weighted portion of $\beta(\theta)$, where a weighting function, $W(\theta)$, describes the angular dependency for a scattering measurement based on the geometry of the sensor (e.g., source half-angle and detector field-of-view properties):

$$\overline{\beta}(\overline{\theta}, \Delta\theta) = \int_0^\pi \beta(\theta) W(\theta) d\theta . \tag{8}$$

The reported β angle of a scattering sensor (θ_m above) is typically the centroid angle, $\overline{\theta}$, defined by the shape of $W(\theta)$. For example, the weighting functions for the WET Labs ECO-VSF sensor measuring scattering at 100°, 125°, and 150° are provided in Fig. 1B.

Raw backscattering counts can be calibrated to volume scattering coefficients using theoretically defined weighting functions and solutions of particles such as microspherical beads that have known scattering properties (Moore et al., 2000) or by employing a Lambertian-reflecting plaque (Maffione and Dana, 1997). In field studies, sensors calibrated with these methods have agreed within ~10% (Boss et al., 2004). Details of these protocols can be found in Mueller et al. (2003).

For *in situ* measurements of fluorescence, raw fluorescence counts are typically calibrated to a standard such as quinine sulfate, coproporphyrin, or vicariously calibrated to a rigorous bench top spectrofluorometer (Conmy et al., 2004). In the instance that the Raman scattering peak can be resolved in emission spectra, then calibration can be carried out by normalizing emission spectra to the integrated area under the Raman peak (Determann et al., 1998). The Raman-based calibration has the advantage of being independent of excitation and emission spectral bandwidths, and spectral resolution. This technique also accounts for "inner filter effects," or the attenuation of the excitation and emission beams experienced along the optical path within the sample. No *in situ* hyperspectral fluorometers are currently available, but one such device is currently in development with promising preliminary results (R. Miller of NASA Stennis Space Center, and C. Moore of WET Labs, pers. comm., 2004).

A wide variety of single and multiple channel *in situ* fluorometers have been developed for measuring fluorophores such as chlorophyll, phycoerythrin, phycocyanin, and fluorescent DOM (e.g., Desiderio et al., 1993; Moore, 1994). Many single channel sensors now use light-emitting-diode (LED) sources, as the intensity and spectral coverage of commercially available LEDs continue to improve. Notable multi-channel fluorometers include a 6-wavelength excitation, 16-wavelength emission device called the SAFire that employs a xenon flash-lamp to effectively excite fluorescence in the ultraviolet (UV) (Desiderio et al., 1996; Del Castillo et al., 2000; Conmy et al., 2004). Another such device is a UV laser-induced fluorescence (LIF) system with 13-wavelength emission (Sivaprakasam et al., 2003). Flow cytometry technology measuring the fluorescence properties of individual cells has recently been made submersible with promising results as well (Sosik et al., 2002).

A recent focus in new sensor development has been optical sensors compatible with compact, autonomous deployment platforms (discussed in section 3). Such platforms require small sensors that are preferably hydrodynamic with very modest power requirements. For sensors that already have those attributes (e.g., many backscattering devices and single-channel fluorometers), adaptation for deployment on an autonomous platform may not require significant modifications (e.g., Yu et al., 2002). Mechanical installation and data handling are the primary challenges. For sensors that do not have those attributes (e.g., *a* and *c* meters and multi-channel fluorometers), new methods and sensors must be developed.

This gap in technology has led to the development of new methodologies for measuring IOPs and AOPs. One such sensor for determining attenuation uses two measurements of backscattering made at the same angle but over different pathlengths (Twardowski et al., 2002, 2003; Fig. 2). This allows for the rigorous measurement of *c* over relatively long pathlengths (more than 20 cm) with a hydrodynamic sensor only several cm's in length. A simple relationship between *c* and the scattering measurements S_1 and S_2 is theoretically expected and observed:

$$c \propto \ln\left(\frac{S_1}{S_2}\right). \qquad (9)$$

The proportionality can be determined through vicarious calibration with conventional beam attenuation meters. Because the measurement is ratiometric, the device is self-calibrating with respect to fluctuations in source intensity. Developments such as this can extend the capabilities of optical sensors to autonomous platforms capable of sampling time-space scales relevant to remote sensing applications.

2.2 IN-WATER MEASUREMENT OF APPARENT OPTICAL PROPERTIES

Measurements of radiance and irradiance, and the derivation of the diffuse attenuation coefficient and reflectances, have been extensively examined by the oceanographic community and have resulted in a detailed set of protocols and approaches for design, characterization, calibration, at-sea deployment, and data analyses of instruments for the measurement of ocean AOPs. For the majority, these are "passive" instruments that rely on the measurement of radiances resulting from the incident solar beam directly transmitted and scattered by the atmosphere, the sea-surface and the ocean interior. The instruments consist of a set of front-end optics which capture the ambient radiances, spectral filtering or dispersing component to separate the broadband radiances into more or less narrow spectral intervals, detectors which transform the impinging photon energy

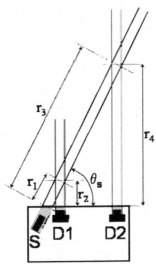

Figure 2. Sensor for determining c from two measurements of backscattering at the same angle (see text). S is the source, and D1 and D2 are photodiode detectors. The effective pathlength is (r3+r4)-(r1+r2).

into electrical signals, and signal processing electronics which condition and digitize the resulting electrical variations into a digital data stream for further analysis.

The front-end optics are fixed depending on the measurement desired. For the measurement of the fundamental radiances, a series of stops or Gershun tubes are generally employed to define the subtended field-of-view (FOV) of the sensor; typical half angles are 10 degrees to 1.5 degrees. Trade-offs between desired narrow FOVs and signal strength/integration time in dark ocean waters are necessary.

For the measurement of irradiances, a variety of collector designs are used. For the measurement of downwelling, $E_d(\lambda)$, and upwelling, $E_u(\lambda)$, irradiances, a diffuser plate is generally used to weight the impinging photons by the cosine of their angle with respect to the surface of the collector. For scalar irradiances, collectors are designed as hemispheres or spheres which weight all incoming photons equally, regardless of the angle of incidence.

Most existing instruments measure a restricted angular distribution, typically downward or upward irradiances, and nadir-viewing radiances. Ideally however, one would like to measure the full radiance distribution, and compute the various irradiances directly from this. Furthermore, and in principle, the change in depth of the radiance distribution should provide sufficient information to not only derive the various irradiances (and associated diffuse attenuation coefficients and reflectances) through integration, but the absorption coefficient and the volume scattering function as well through inversion methods (e.g., Aas and Højerslev, 1999). This type of sensor is particularly of interest in optically shallow regions, where the radiance distribution is strongly modified by sea-bottom interactions (Voss et al., 2003). Although such sensors were used extensively in the 1960's, their complexity and high data rate requirements have meant that very few, apart from the work of Voss, are routinely deployed.

For decomposition of the broadband field into spectral intervals, two fundamental approaches have been taken. For defined wavelength bands, the most effective approach is the use of high-quality Ion Assisted Deposition (IAD) filters which exhibit low levels of fluorescence. These can be manufactured to defined spectral transmittances (albeit at some cost) and can be practically matched to the wavebands of the various satellite sensors for highest accuracy in calibration and validation. Matching diffusers can then be used which are optimised for cosine (or other) response at the center wavelength of these filters. Cut-off filters can be stacked to reduce the out-of-band response to very small levels ($<10^{-6}$). Typically, instruments are manufactured with 1 to 14 defined spectral channels, each carefully chosen for a specific application.

Alternatively, spectral dispersion can be accomplished by prism or grating approaches, and the dispersed beam imaged onto an array of detectors. Finer wavelength resolution and increased spectral channels (~128 to >256) can be achieved with this methodology and these instruments are generally labelled as "hyperspectral" in nature. However, trade-offs arise due to the limited number of photons in the small spectral bands and care must be taken to minimize second-order out-of-band performance. As a general rule, radiometric specifications and performance of hyperspectral instruments are not as rigorous as those for precision filter-based instruments, although the increased spectral resolution confers significant advantages for some remote sensing applications. For example, semi-analytical algorithms that rely on spectral decomposition of IOP components from remote sensing reflectances can generally perform better with more spectral input parameters.

With respect to detectors and associated electronics, a key metric is the signal to noise ratio which must be achieved over the high dynamic range required to cover the range of oceanic conditions. Dynamic ranges of >18 bits can be achieved with individual silicon photodetectors; this results in a capability to profile irradiance reliably to the equivalent depth of the 0.01% light level under cloudy skies, while maintaining a high sampling frequency (6-10 Hz). Hyperspectral instruments, because of their nature and the reduced photon flux into narrow spectral bands, are not as capable, and generally rely on longer integration times (upwards of 8 seconds) to achieve equivalent signal to noise performance.

The accurate characterization of these sensors with respect to FOV (or cosine response), to spectral response, to thermal and pressure variations, to linear response to variations in incident radiance, and to signal to noise is essential. This is in addition to the requirements for instrument calibration. A large body of information on this extensive subject can be found in Mueller et al. (2003) and references found therein. For AOP measurements, a significant advantage is the existence of national standards of irradiance which provide a reference to which instruments anywhere in the world and at any time can be calibrated and intercompared with a high degree of confidence. Measurements of radiance, of attenuation, and of reflectance can thus be regarded as accurate measurements of a physical quantity, which can be rigorously compared over all time/space scales, including those comparisons with similar instruments onboard space-based platforms.

AOP instruments can now be manufactured with a very small form factor, and as with the IOP instruments described above, can be deployed on a range of platforms for the measurement of the apparent optical properties of the ocean, including new autonomous profilers and gliders.

2.3 BIOGEOCHEMICAL PROPERTIES

The ultimate objective of most remote sensing algorithms is usually to derive some biogeochemical property from R_{rs}. For a great many remote sensing algorithms, this derived biogeochemical property is chlorophyll concentration. As discussed earlier, algorithms also exist to derive an extensive and continually growing list of other biogeochemical properties.

So what in-water analytical instrumentation is available to measure these biogeochemical properties directly? With few exceptions, *in situ* determination (using optical or other analytical methods) of biogeochemical properties at accuracies suitable for remote sensing validation work constitutes one of the largest gaps in currently available instrumentation. For example, Table 1 lists several biogeochemical properties that have been derived from optical properties that can be determined *in situ*. Nearly all these derivations have been accomplished via simple empirical relationships. The problem that arises is that many of these relationships are not robust, particularly in coastal Case 2 waters, because of the wide variability in the composition and relative concentrations of the dissolved and particulate components of seawater. For example, while good linear relationships between Total Suspended Matter (TSM) and c_p or b_p are observed at certain times and places in the world's oceans, the relationship is strongly dependent on the size and refractive index distributions of the particles and is therefore variable. Other particle properties such as shape and internal structure can also affect these relationships. Babin et al. (2003) found a more than two-fold variability in the TSM-b_p relationship in coastal waters around Europe. Empirical relationships between IOPs and components of TSM such as POM and POC are even more tenuous because it must also be assumed that the relative proportion of a particular TSM component is constant. This is a poor assumption in coastal waters, where POM : TSM varies from only a few percent to near 100% (Kratzer et al., 2000; Babin et al., 2003).

It is out of the scope of this work to detail the specific problems associated with *in situ* optical characterizations of each property. The important point is that essentially all methodologies for determining biogeochemical parameters with an analytical precision and accuracy suitable for remote sensing applications are laboratory-based at this time. Chlorophyll, for instance, can be estimated from *in situ* fluorescence and/or absorption measurements, but because the concentration-normalized absorption and fluorescence quantum yield of chlorophyll packaged in cells varies, these estimates do not satisfy established standards for algorithm validation work (e.g., Mueller et al., 2003). The lack of suitable, automated, *in situ* instrumentation for biogeochemical parameters can thus be considered a substantial hindrance to algorithm development and validation efforts. For example, consider the potential benefits of a sensor with an *in situ* method for determining chlorophyll concentration with an accuracy comparable to laboratory extraction methods; and this sensor was deployed on a fleet of autonomous platforms collecting data for algorithm development and validation work throughout the world's oceans. As we will see shortly, the autonomous platform technology is nearing maturity; the limitation is the *in situ* biogeochemical sensing technology. If more accurate models are developed in the future to derive these biogeochemical properties from measurements made by optical sensors, the benefit is two-fold. First, accurate, composition-nonspecific determinations of biogeochemical properties may then be made from optical sensors that can indeed be deployed on autonomous platforms in many cases. And second, more rigorous relationships between biogeochemical properties and the optical properties help us move toward potentially more accurate analytical-type remote sensing algorithms are able to account for the changes in the dissolved and particulate composition of seawater that strictly empirical algorithms cannot.

It may be argued that one emerging exception to the lack of *in situ* biogeochemical sensing technology may be *in situ* methodologies for determining nutrients (Johnson and Coletti, 2002; Hanson and Donaghay, 1998; Hanson, 2000), but these techniques are new and have not been rigorously validated. The Johnson and Coletti (2002) technique

Table 1. Some biogeochemical properties derived from optical properties.

Biogeochemical property	Optical Property	Example Reference(s)
Particulate Organic Carbon (POC)	1) c_p or b_p	Peterson 1978; Gardner et al. 1993, 2001; Loisel and Morel 1998; Bishop 1999; Bishop et al. 2002; Claustre et al. 1999, 2000; Mishonov et al. 2003
	2) b_{bp}	Stramski et al. 1999; Balch et al. 1999
Total Suspended Matter (TSM)	1) c_p or b_p	Peterson 1978; Gardner et al. 1993, 2001; Walsh et al. 1995; Prahl et al. 1997
	2) turbidity	Fugate and Friedrichs 2002
Dissolved Organic Matter or Carbon (DOM, DOC)	1) a_g	Pages and Gadel 1990; Vodacek et al. 1997
	2) Fluorescence	Coble et al. 1993; Ferrari et al. 1996; Klinkhammer et al. 2000
DOM composition[a]	1) a_g, spectral shape	Carder et al. 1989; Blough and Green 1995
	2) Fluorescence, multi-spectral shapes	Coble 1996; Del Castillo et al., 1999; McKnight et al. 2001
Chlorophyll	1) a_p	Bricaud et al. 1998; Claustre et al. 2000
	2) Fluorescence	e.g., Yentsch and Menzel 1963; Claustre et al. 1999
Phycobiliproteins	Fluorescence	Cowles et al. 1993; Sosik et al. 2002
Phytoplankon pigment ratios	a_p, spectral shape	Trees et al. 2000; Eisner et al. 2003
Proteins	Fluorescence	Coble et al. 1993; Mayer et al. 1999
Hydrocarbons	Fluorescence	e.g., Holdway et al. 2000
Particle size distribution	1) c_p, spectral shape	Morel 1973; Boss et al. 2001
	2) $\beta(\theta)$	Brown and Gordon 1974; Zaneveld et al. 1974; Agrawal and Pottsmith 2000
Particulate refractive index	1) $\beta(\theta)$	Brown and Gordon 1974; Zaneveld et al. 1974
	2) $c_p(\lambda)$, b_{bp}, and b_p	Twardowski et al. 2001
Sewage	Fluorescence	Petrenko et al. 1997
Nitrate	UV absorption	Johnson and Coletti 2002

[a] For example – ratio of dissolved humic acid to fulvic acid, DOM molecular size distribution, DOM aromaticity, DOM source

determines nitrate via hyperspectral measurements of UV absorption. The Hanson and Donaghay (1998) method determines up to eight different nutrients simultaneously using "wet chemistry," or the *in situ* addition of chromophoric reagents that produce a color (or fluorescence in the case of ammonia) in proportion to the concentration of the nutrient. This latter device is essentially a submersible, digitized autoanalyser.

3. Platforms

In the last several years, an exciting variety of deployment platforms for in-water oceanographic instrumentation have been developed that complement more conventional ship-based measurements. Each platform has a unique niche in terms of the temporal and spatial coverage it provides. Sampling strategies that integrate multiple platforms can therefore be very effective in studying biogeochemical phenomena ranging over large time and space scales (Dickey et al., 2004). This section describes the general types of platforms available and provides some examples of using these platforms for remote sensing related applications.

Obtaining high quality *in situ* optical and biogeochemical data for remote sensing algorithm development and validation can be challenging. In addition to the obvious challenges associated with making accurate measurements on relevant time and space scales, the ocean is inherently a difficult environment to conduct research. Autonomous platforms must contend with occasionally violent weather (e.g., Chang et al., 2001), biological fouling of sensors (e.g., Chavez et al., 2000), various obstacles (e.g., bathymetry and ships), and must rely on wireless communications to send their data and receive instructions or be situated sufficiently close to shore that cabled power and communications can be run to the platform. The ocean environment is also highly corrosive to a wide range of materials. Sensors for these platforms must be compact and have low power requirements. And there is always the challenge of making the needed measurements at a reasonable cost. Nonetheless, it will be apparent in the following that platform technology development efforts have been and continue to be highly successful despite these obstacles.

3.1 STATIONARY VERTICAL PROFILERS

The most common method of *in situ* sampling is vertical profiling from a boat or ship (Fig. 3). Sensors are typically secured in a cage and interfaced with a central data handler/controller that records and time stamps the separate data streams. Power can be provided with underwater batteries and data can be logged on the profiler for downloading later. Alternatively, a cable can be used for data and power to allow real-

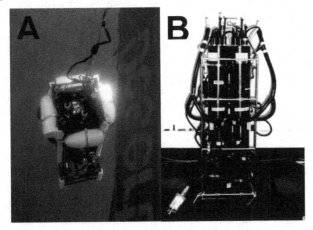

Figure 3. (A) A vertical profiling system during deployment; (B) a vertical profiler in preparation for work in the field. The picture in (B) shows a configuration for simultaneously measuring dissolved and particulate *a* and *c* components of seawater (note capsule filter attached to intakes of meter on the left).

time viewing at the surface. Ballast is often
added to a profiler to bring the net weight
underwater near neutral (see white floats in Fig.
3A). This enables a slow, steady descent rate
when free-falling in order to resolve vertical fine
structure.

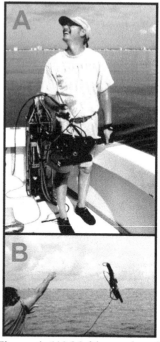

There are several considerations when
measuring AOPs from a ship. Potential
shading/reflection effects from the ship (Waters
et al. 1990) and the package itself (Leathers et al.
2001) must be avoided if possible. AOP
profiling systems consequently have been
developed that allow for profiling tens of meters
from a ship (Fig. 4). These profilers are also
designed so that the radiance (down-looking) and
irradiance (up-looking) sensors are oriented very
close to vertical during profiling, so that L_u in the
nadir direction ($\theta=\pi$) is measured and only the
downwelling photons are included in E_d.

Stationary vertical profiling can provide
excellent resolution of the vertical structure of
optical properties in the water column (e.g.,
Donaghay et al., 1992; Twardowski et al., 1999;
Fig. 5). Data are usually of the highest quality
relative to deployment on autonomous platforms
because cleaning of the optical windows and
calibration protocols can be performed on a

Figure 4. (A) Multi-wavelength
E_d and L_u profiling package and
(B) its deployment.

regular basis. Issues such as power, instrument size, data volume, etc. are also typically
not concerns. Profiling from a ship has the added benefit that discrete samples can be
concurrently collected for laboratory analyses of biogeochemical properties. For these
reasons (as well as the unavailability of other suitable platforms in the past) the vast
majority of data sets for remote sensing algorithm work over the years have been
collected from ships.

While optical and biogeochemical data collected from ships have been enormously
useful in remote sensing applications, there are some important limitations. One
limitation is the relatively high cost. Another is the relatively restricted time-space
domain covered with vertical profiling data from a typical cruise. Nonetheless, in the
foreseeable future it is difficult to envision a sampling strategy for remote sensing
algorithm development and/or validation work that does not heavily rely on vertical
profiling from ships.

Interestingly, optical profiling systems have also been deployed from land-water
planes (A. Petrenko, pers. comm., 1998) and via helicopter during the 1997-1998
European COASTlOOC (coastal surveillance through observation of ocean color)
campaign (Fig. 6). Such aerial platforms are able to sample stations over large spatial
ranges more rapidly than possible with boats.

Autonomous moored profiling technologies have been available since the 1970s (e.g.
Brown et al., 1971; Van Leer et al., 1974). Systems have employed a variety of possible
techniques to profile, but buoyancy manipulations or winches have primarily been used.
Operating power has been supplied by batteries, onshore cable, and even wind-driven
generators. Current incarnations still use these traversal mechanisms (e.g. Provost and

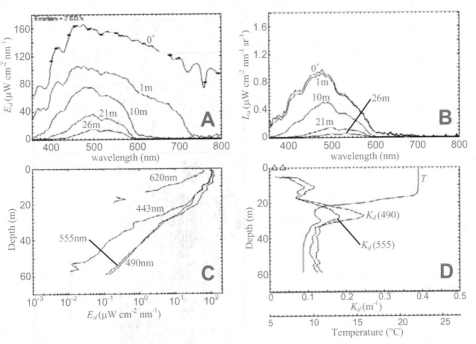

Figure 5. (A) Spectral downwelling irradiance and (B) spectral upwelling radiance at selected depths; vertical profiles of (C) irradiance and (D) the diffuse attenuation coefficients. Collected in the Northwest Atlantic, August 2001. Note phytoplankton layer centered at ~28 m.

du Chaffaut, 1996; Reynolds-Fleming, et al., 2002). In addition, energy from ocean currents and surface waves has also been harnessed to vertically propel the sensor package (Echert et al., 1989 and Rainville and Pinkel, 2001, respectively).

A drawback of the above designs is the fixed presence of a mooring wire and a surface or sub-surface buoy. Surface expression is also a concern in coastal research applications because it can invite vandalism. These problems can be avoided by adopting a bottom-mounted winch design, such as the LEO-15 vertical profiler (Purcell et al., 1997). This system, however, is large (3 x 3 x 1.5 m) and requires a permanent onshore cable for power and data transfer. A cabled underwater winch system with much smaller size will be deployed at the Bonne Bay Cabled Ocean Observatory in Newfoundland (B. de Young, Memorial University, St. Johns, Newfoundland, pers. comm., 2004).

A compact, fully automated profiler termed the Ocean Response Coastal Analysis System (ORCAS) has also recently been developed to resolve finescale vertical structure (Donaghay et al., 2002). This system is designed for shallow water coastal environments and has a sophisticated suite of IOP and AOP sensors. While still not a mature technology, Donaghay et al. (2002) have deployed multiple ORCAS profilers in a network to sample 4-dimensional structure in optical properties with promising results. This work demonstrates the concept of using arrays of platforms separated at the critical scales needed to resolve coastal biogeochemical phenomena.

The Shallow-water Environmental Profiler in Trawl-safe, Real-time configuration (SEPTR) developed by the NATO SACLANT Undersea Research Centre and the University of Rhode Island (Tyce et al., 2000) is another profiler with no surface expression. It consists of a saucer-shaped, trawl-resistant shell (2.0 m diameter at base x 0.5 m height) that encases an Acoustic Doppler Current Profiler (ADCP) and a winch-driven, bottom-up profiling capsule, as well as the associated control electronics and batteries. While the SEPTR profiler has been successfully used in many coastal environments, it has a small payload capacity limiting its utility for remote sensing calibration/validation research. Trawl-resistant structures are critical for coastal autonomous profilers.

Other commercially available moored profiling system designs incorporate a bottom-mounted winch with a slip ring for transfer of power and data to and from the sensor package. This design makes the package vulnerable to several hazards, including rotation which ultimately applies excess torsional stress on the cable. Importantly, bottom-mounted winch profiling systems are highly susceptible to surface waves which alternatively produce conditions of sudden slack and tension in the cable. These

Figure 6. Vertical profiling from a helicopter during the COAST/OOC campaign. Photo courtesy of M. Babin.

systems are also large and heavy, with individual platforms for the winch and data system increasing the complexity of deployment and recovery. Furthermore, the power requirements of these systems including winch are demanding (more than 100 W).

Because of these obstacles, autonomous stationary profiling technology is not mature or operational at this time. The critical current challenges are managing power needs and sustaining reliability for autonomous deployments of 6 months or more. Because of their ability to resolve dynamic vertical structure in coastal waters over a wide range of time scales, these platforms hold particular promise for remote sensing Case 2 algorithm development and validation pursuits.

3.2 FLOW-THROUGH SYSTEMS

Most research ships have built-in flow-through systems that continuously circulate subsurface through on-board laboratories. These systems readily allow the installation of IOP sensors for making continuous measurements while underway. Effective flow-through systems with optical sensors can also be developed for small boats more appropriate for near-shore coastal research (R. Arnone and R. Gould, pers. comm., 2000) or for ferries or ships of opportunity (Schroeder and Petersen, 2000; Balch et al., in press). An important consideration in any flow-through system with optical sensors is the elimination of bubbles. Consequently, holding tanks or in-line debubblers are often employed.

Optical data from flow-through systems effectively resolve small horizontal scales that stationary vertical profiling systems cannot (Pegau et al., 2000). Surface data are the most critical for remote sensing algorithm work, although the underlying vertical structure of optical properties through the euphotic zone along the ship track is required for rigorous comparisons. One of the principle benefits of a flow-through system is the capability to use sensitive bench top instruments that have no submersible analogue.

3.3 TOWED VEHICLES

Since the backscattered signals collected by passive and active remote sensing systems are dependent on the vertical structure of optical properties along the flight path, sufficient resolution in both the vertical and horizontal dimensions are needed to develop effective algorithms for remote sensing applications. While stationary profiling from ships and continuous flow-through systems provide complementary vertical-horizontal coverage, the vertical dimension remains unsampled while underway.

One solution to the problem of synoptically sampling horizontal and vertical dimensions is to use an undulating towed vehicle (Barth and Bogucki, 2000; Hales et al., 2001; Miller et al., 2003). Such a system is able to provide a continuous series of data points for remote sensing applications where transect lines can follow the flight paths of a remote sensor. Historically, the use of towed systems for underway sampling with optical sensors has been pioneered by the Continuous Plankton Recorder (e.g. Hays and Lindley, 1994) and follow-on systems (Aiken and Bellan, 1990).

A towed system specifically developed for remote sensing applications was described by Miller et al. (2003) (Fig. 7). While underway, the towed package can be programmed to automatically undulate through the water column between specified depths. The vehicle is equipped with an *a* and *c* meter, backscatter sensors, fluorometers, a conductivity-temperature-depth (CTD) sensor, and can also be configured with irradiance and radiance sensors. The vehicle and tow cable were designed with a built-in discrete water sampling system, where samples are pumped continuously from the vehicle through a hose embedded in the tow cable to the boat.

Figure 7. (A) A towed vehicle with on-board optical sensors completing a transect, and (B) a close-up of the front of the sensor cage showing backscattering and fluorescence meters, the pump for pumping samples continually to the surface, the end can of an AC9 sensor, and a CTD.

Consequently, the system allows concurrent collection of in-water optical data and discrete sampling for laboratory analyses of biogeochemical properties synoptically in the vertical and horizontal along the ship track (Fig. 8).

While such towed systems are well-suited for remote sensing applications, they still require the use of ships and the associated expense. In coastal regions, however, these systems can be deployed from relatively small boats (less than 10 m) that are inexpensive to operate. The costs in using towed systems for periodic synoptic sampling of coastal regions may therefore be practical in many cases.

3.4 MOORED PLATFORMS

Until recently, moored sensor systems have been the primary means by which long-term, high-frequency optical data have been collected (e.g., Dickey et al., 2004). These fixed position, or Eulerian, platforms provide data streams synoptically with respect to time that match up with satellite-based imagers. Hazardous weather conditions that would normally restrict conventional ship sampling do not affect the performance of properly constructed moorings (Chang et al. 2001). Moored sensors are thus well-suited for calibration and validation of remotely sensed signals (Clark et al., 1997; Pinkerton and Aiken, 1999; Zibordi et al., 2002; Antoine and Guevel, 2000).

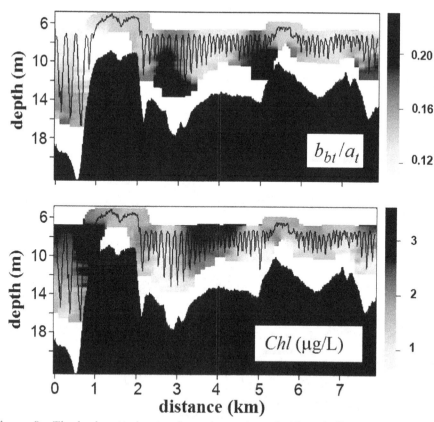

Figure 8. The backscattering to absorption ratio and chlorophyll concentration from fluorescence measurements collected with a towed vehicle platform in Narragansett Bay, RI. The vehicle was programmed to automatically undulate within the range of 5 m from the surface and bottom.

While several moored systems for remote sensing applications have been deployed in the more classical oceanic Case 1 remote sensing environments (Clark et al., 1997; Chavez et al., 1999; Dickey et al., 1998, 2001), few have been deployed in more optically complex Case 2 environments. This has been primarily due to a programmatic emphasis on characterizing the open ocean surface waters using global ocean color imagers. Adapting this technology to the optically diverse and productive nature of most coastal water environments presents additional challenges on the design and use of moored sensor systems. For example, coastal waters are regions of high vertical and horizontal optical variability in comparison to most open ocean environments. In order to fully resolve this variability, coastal mooring systems require an increase in the number of optical sensors in the upper water column (to resolve vertical structure) as well as an increase in temporal sampling (to resolve small scale horizontal variability).

Much progress has been made in using moored optical sensor systems for remote sensing ocean color algorithm development for coastal waters. This has been due to recent advances associated with the miniaturization of optical sensors and the development of anti-biofouling devices and methodologies (Dickey et al., 2001, 2003). Coastal moorings are highly susceptible to biofouling, characterized by the build-up of detrital and living organic matter on optical sensing interfaces (Fig. 9). In order to ensure data integrity, optical sensors on coastal moorings must currently be serviced frequently (order of months) and must include biofouling prevention strategies (McLean et al., 1997; Chavez et al., 2000; Barnard and Roesler, 2003; Manov et al., 2003). The use of copper materials has recently been shown to be effective in mitigating the effects of biofouling, allowing for deployments of up to six months and more for some optical sensors (Barnard and Roesler, 2003; Manov et al., 2003). Various optical sensors can be equipped with copper shutters mounted a few millimeters above the optical face. The slow release of copper ions through dissolution in seawater creates a toxic layer above the sensing face when the sensor is not in use. During sampling, the copper shutter rotates 180 degrees, exposing the optical face. Copper shutters are effective but can be susceptible to mechanical failures due to growth of large marine organisms on or near the shutter that impede the rotation of the shutter. To prevent marine organisms from attaching, various copper materials such as foil tape can be applied to the sensor.

Technological advances such as these have led to the recent proliferation of optical sensing systems in a variety of research and environmental monitoring mooring programs. One such program is the Gulf of Maine Ocean Observing System (GoMOOS; www.gomoos.org). The primary purpose of the *in situ* bio-optical component of the Gulf of Maine Ocean Observing System (GoMOOS) mooring

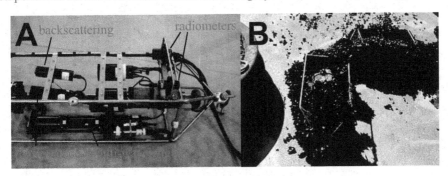

Figure 9. Biofouling of optical sensors at GoMOOS after 5 months at sea. (A) before; (B) after (picture taken in 2002).

program is to provide hourly observations of biogeochemical parameters related to ocean productivity, water clarity, and ecosystem community dynamics. A secondary goal was to provide in situ measurements that could be used to aid interpretations of ocean color remote sensing imagery. GoMOOS operates and maintains four moorings equipped with bio-optical instrumentation in coastal regions of the Gulf of Maine. Two of these moorings contain a robust suite of optical instrumentation near the surface providing radiometric and bio-optical measurements including spectral upwelling radiance, spectral downwelling irradiance, spectral absorption, scattering, beam attenuation and volume backscattering and include anti-biofouling devices such as copper materials and shutters (Fig. 9A). Since 2001, the GoMOOS program has been providing hourly observations of surface optical and radiometric properties which have been used in combination with bio-optical models to develop and validate ocean color inversion algorithms (Barnard and Roesler, 2003; Roesler and Barnard, 2003).

Measurements of normalized water leaving radiance at nadir taken from moored platforms have shown excellent agreement with comparable measurements taken from the SeaWiFS sensor as it passed over the mooring (Dickey et al., 2004) emphasizing the efficacy of such systems for calibration and validation of ocean color satellite sensors. Indeed, the primary means for on-orbit calibration and validation of SeaWiFS and MODIS has been the MOBY moored system (Clark, 2003) and the European MERIS program relies on the BOUSSOLE mooring in the Mediterranean Sea (Antoine and Guevel, 2000), which is now operational.

3.5 PROFILING FLOATS

Over the last few years, the ARGO program has begun to seed the world's oceans with Autonomous Profiling Explorer (APEX) floats (CLIVAR 1999; Wilson 2000). Their goal is to have 3000 floats spaced in a ~300 km grid pattern covering the global ocean. As of October 2003, 947 floats had been deployed (C. Jones, Webb Research, Corp, pers. comm., 2003). These floats are designed to "sleep" at a depth of 1000-2000 m, waking up every 7-10 days to make ascents to the surface while recording CTD measurements. Data is telemetered via satellite when at the surface. A variable buoyancy engine provides the negative and positive buoyancy required to profile. Expected lifetimes for APEX floats are 4-5 years on average. The Scripps Institute of Oceanography Instrument Development Group and the French IFREMER Marine Technology and Information Systems Division make APEX analogue floats called the Sounding Oceanographic Lagrangian Observer (SOLO) and PROVOR, respectively. Recent precursors to the APEX floats were the Autonomous Lagrangian Circulation Explorer (ALACE), and the profiling ALACE (PALACE) (Davis et al. 2001).

Primarily because of ARGO, profiling float platform technology has emerged as one of the most reliable and cost-effective available. To date, the APEX floats used in ARGO have provided a 79% reliability of data return (C. Jones, Webb Research, Corp, pers. comm., 2003). Other cost-effective drifting profiling floats such as the Oceans Sensors, Inc. Autonomous Profiling Vehicle (APV) have also been developed for more coastal applications. Because of their relatively low cost, deploying arrays of floats to address remotely sensed biogeochemical phenomena occurring over large time and space scales can be a practical consideration. Since APEX floats spend the majority of their time out of the photic zone, biofouling is a minor consideration (Bishop et al. 2002). A key obstacle, however, is the availability of compact optical sensors with suitably low power requirements and data volume compatible with satellite communication bandwidths.

There are only a few instances where optical instruments have been deployed on profiling floats. One very effective deployment was carried out by Mitchell et al. (2000) in the Sea of Japan with a SOLO float outfitted with a 3-wavelength irradiance sensor. The float profiled the upper 400 m once every 2 d over about a four month period, capturing the onset of the spring bloom and the accompanying subsurface stratification. Vertical attenuation coefficients, K_d, could be determined for each irradiance profile and showed excellent agreement with the SeaWiFS K490 product. This study demonstrates the potential of expanding the use of such platforms for synoptic remote sensing related applications.

In another recent study, Bishop et al. (2002) equipped an APEX with a customized beam attenuation meter to measure POC concentrations in the North Pacific. POC was derived by a linear empirical relationship and the assumption was made that the composition of the particle population varied by only small amounts during the deployment. These floats, or "Carbon Explorers," were able to resolve vertical distributions of POC over several months. Within the data records, enhanced carbon biomass from a natural iron "fertilization" event associated with an Asian dust storm was documented. It is these kinds of episodic, short-lived phenomena that conventional ship sampling can only document with a great deal of luck. And there is a growing recognition that episodic events not easily sampled discretely from ships – short-lived, intense phytoplankton blooms, dust deposition events, and the passing of storms – may be driving forces behind the global cycling of carbon (Bishop et al., 2002; White et al., 2002; Dickey et al., 2004). Autonomous profiling technology may prove the ideal platform for studying such processes.

3.6 AUTOMATED UNDERWATER VEHICLES

Autonomous Underwater Vehicles (AUVs) provide their own propulsion to allow high-resolution sampling of the ocean's interior in the horizontal spatial dimension as well as the vertical. There are many types of AUVs (Griffiths et al., 2001), but they generally fall into two categories: self-propelled and gliding.

3.6.1 *Self-propelled vehicles*

Autonomous vehicles with propellers have been in development internationally for over 40 years (e.g., Blidberg, 1991). They vary widely in size, depth rating, sensor payload space, rated operation duration, guidance systems, and telemetry modes. Propelled AUVs can rapidly cover relatively large vertical and horizontal regions. For example, the Remote Environmental Monitoring Units (REMUS) made by Hydroid, Inc. can cover 100 km in about 20 hours. While most propelled vehicles are still rather expensive to realize widespread use for oceanographic research (typically cost several US$100,000), cost has come down substantially over the last several years. Some small AUVs that are produced in high volume, such as the "disposable" AUVs made by Sippican that are used as targets in military exercises, can be purchased for as little as a few thousand US dollars.

Battery power with AUVs is currently one of the key limiting factors in their long-term use. Most propelled AUVs are designed for deployments of not more than a day or two. Efficient, energy dense storage technologies such as solid oxide fuel cells are currently a key development area (e.g., Singhal, 2000). As an illustration of the current problem, one of the largest AUVs, the U.S. Navy's Seahorse, uses 9,216 common alkaline "C" batteries for power.

Optical sensors routinely used on propelled AUVs are backscatter sensors and chlorophyll fluorometers (e.g., Yu et al., 2002). Sophisticated spectral upwelling and downwelling radiometers have been integrated in a REMUS AUV and successfully deployed in a number of operating scenarios in coastal regions off New Jersey (Brown et al. 2004). *In situ* "wet chemistry" nutrient analysers have also been adapted for AUV use (A. Hanson, SubChem, Inc., pers. comm., 2003).

Technology is currently being developed that will allow AUV networking and adaptive sampling so features of interest such as biological layers or river plumes can be intensively sampled. More advanced technologies such as node docking and equipment deployment capabilities are being pursued in industries such as oil exploration and cable laying that may be transferable to oceanographic studies in the future.

3.6.2 *Gliders*

Gliders are buoyancy regulated like APEX floats but use wings to convert vertical velocity into forward velocity. They are suitable for long duration sampling (weeks to months), and typically travel in a "sawtooth" pattern. Although traveling velocity in the horizontal is relatively slow (< 0.5 m s^{-1}), total distances traveled during a mission can be thousands of kilometers. A variety of two-way wireless communication methods are supported, including satellite-based Iridium. Because gliders can be programmed to surface on a frequent basis, their sampling mission can be altered at any time. Like propelled AUVs, technology enabling adaptive sampling is also in development.

Four primary glider technologies have been developed: the Slocum Littoral (Webb Research Corp), the Slocum Thermal (Webb Research Corp), the Spray (Scripps Institute of Oceanography), and the Seaglider (University of Washington). All use power from lithium or alkaline batteries except for the Slocum Thermal. The Thermal harnesses energy from a chemical change-of-state reaction that occurs from ambient temperature changes as it glides through the ocean's thermocline.

Like other autonomous platforms, power and size constraints are of paramount importance for sensors deployed on gliders. A non-technical problem with deployment of the Slocum Littoral glider has also been fishermen, who have picked up the platforms while they are at the surface transmitting data. Figure 10A shows a deployment by the Rutgers University glider team of a Slocum Littoral glider equipped with optical sensors (Fig. 10B) measuring spectral backscattering, fluorescence, and beam attenuation. The beam attenuation meter employs the recently developed dual-backscattering method described in section 2.1. Data from this sensor is shown in Fig. 11.

3.7 DIVERS AND NEKTON

Divers have been used to collect optical data for remote sensing applications when the exact proximity of the sensing element and/or its intake is critical (Zaneveld et al., 2001; Dierssen et al., 2003; Fig. 12). Divers are often necessary when studying the optical properties of the bottom, such as in the validation of the leaf-area index for seagrasses (Dierssen et al., 2003). Optical phenomena that occur over very small spatial fields (e.g., the scattering properties of the sea-bottom, the particle attenuation around corals, or the absorption by dissolved materials in close proximity to seagrass beds or in sediment pore waters) can only be effectively sampled by divers (Zaneveld et al., 2001; Boss and Zaneveld, 2003, Voss et al., 2003).

Finally, large fish and whales should be considered as possible platforms. The fisheries research community is already using light sensors on fish as a method of estimating geolocation (Sibert and Nielsen, 2001). This work demonstrates that these platforms could perhaps be suitable for remote sensing related applications if the relevant *in situ* sensing technology can be appropriately miniaturized.

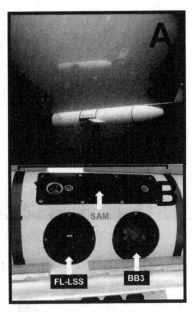

Figure 10. (A) Slocum Littoral gliders, and (B) optical sensors integrated in the payload compartment of the glider; SAM = Scattering-Attenuation Meter (see Fig. 2), BB3 = 3-wavelength backscattering sensor, and FL-LSS = DOM fluorescence and Light Scattering Sensor (broadly weighted side scatter). Photos courtesy of E. Creed, Rutgers U. glider team.

4. Considering Sampling Strategy

The usefulness of in-water measurements with respect to remote sensing can be broken down into two broad categories: 1) helping interpret ocean color measurements from remote sensors; and, 2) filling the gaps in data along time-space axes not resolved by remote sensors. While a detailed discussion of sampling strategies for these sets of applications is out of the scope of this review, some comments from a technology perspective may be useful.

Ocean color interpretations via algorithms require comprehensive data sets covering broad dynamic ranges of biogeochemical and optical properties. Often important endmember data points for these ranges can only be collected during short-lived, episodic events such as intense blooms (that may perhaps be harmful), dust deposition, and vigorous mixing from passing storms. These events are also usually the most interesting from a science perspective, and may be critical in understanding the more long term dynamics of ocean ecosystems. Conventional ship sampling is not well-suited to resolving such events. And on the serendipitous occasion such an event is observed, conditions before and after the event are rarely well documented. These observations lead to the conclusion that automated platforms should play a more significant role in ocean color work. Preliminary investigations strongly support this notion (Mitchell et al., 2000; Bishop et al., 2002; Barnard and Roesler, 2003; Roesler and Barnard, 2003).

The need for comprehensive data sets at reasonable cost also suggests that the implementation of *in situ* instrumentation and techniques in ocean color sampling strategies should continue to be pressed. *In situ* measurements typically have high sampling rates, are less labor-intensive than lab-based methods, and many can now be

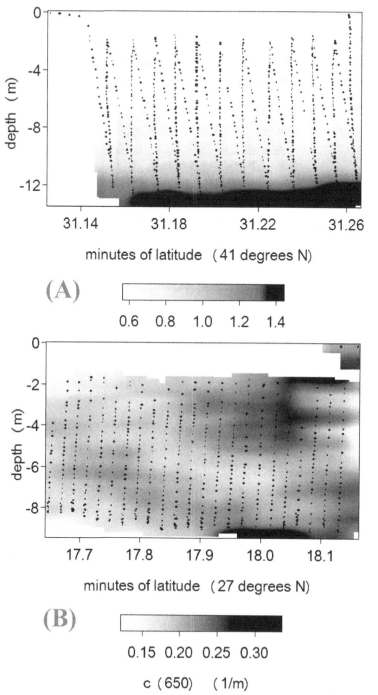

Figure 11. Attenuation data from a glider deployed by the Rutgers glider team in (A) Buzzards Bay, MA, 8/19/2003, and (B) the west Florida coast, 11/06/2003.

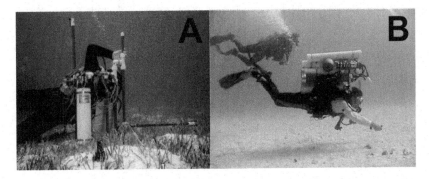

Figure 12. Diver-controlled in-situ measurement of (A) AOPs with a Diving Operated Benthic Bio-optical Spectrometer (DOBBS) and (B) IOPs with an AC9. Sensor intakes in (B) held in diver's right hand. Photos courtesy of R. Zimmerman (A) and E. Boss (B).

made autonomously. There is, however, a fundamental lack of *in situ* sensors for making measurements of the needed biogeochemical properties with the accuracy required for algorithm development and validation. Most biogeochemical determinations require laboratory analysis on discrete samples. Overcoming the necessity to collect discrete samples could substantially accelerate future algorithm development/validation efforts. Developing accurate *in situ* biogeochemical sensing technology compatible with automated remote operation should thus be a continued area of focus by the oceanographic community.

At the present, the problem of sampling strategy is (whether we accept it or not) at least constrained by several factors, including the availability of sensing technology, the availability of platform technology, and cost. While the latter was not discussed in detail, by reviewing the first two factors we hope a better understanding of feasible sampling strategies for remote sensing applications may be realized. Areas where the programmatic top-down approaches to addressing global biogeochemical phenomena do not overlap with this more bottom-up approach should be the focus of development efforts in the future.

5. Acknowledgments

We are indebted to many who helped provide material directly or though collaborations, including M. Babin, E. Boss, E. Creed, J. Cullen, T. Dickey, H. Dierssen, P. Donaghay, S. Freeman, S. Glenn, C. Jones, J. Kerfoot, S. McLean, R. Miller, C. Moore, O. Schofield, J. Sullivan, and I. Walsh. Support for this work is gratefully acknowledged from the Office of Naval Research Optics and Biology and Small Business Initiative for Research programs, the National Aeronautics and Space Administration Small Business Initiative for Research programs at Stennis Space Center and Goddard Space Flight Center, the Natural Sciences and Engineering Research Council of Canada, Satlantic Inc., and WET Labs, Inc.

6. References

Aas, E., and N. K. Højerslev. 1999. Analysis of underwater radiance observations: Apparent optical properties and analytical functions describing the angular radiance distribution. Journal of Geophysical Research, 104: 8015–8024.
Agrawal, Y.C., and H.C. Pottsmith. 2000. Instruments for Particle Size and Settling Velocity Observations in Sediment Transport. Marine Geology, 168:89-114.

Aiken, J. and I. Bellan. 1990. Optical oceanography: an assessment of towed measurement. In P. J. Herring, A. Campbell, M. Whitfield, and L. Maddock, editors, Light and Life in the Sea, Cambridge University Press. Pg. 39-57.

Antoine, D., and P. Guevel. 2000. Calibration and validation of satellite ocean color observations: the BOUSSOLE project. Proceedings of Ocean Optics XV, 16-20 October, Monaco, Office Naval Research, USA, CD-ROM.

Antoine, D. and A. Morel. 1996. Oceanic primary production: I. Adaptation of a spectral light-photosynthesis model in view of application to satellite chlorophyll observations. Global Biogeochemical Cycles, 10:43-55.

Antoine, D., J.M. Andre, and A. Morel. 1996. Oceanic primary production: II. Estimation at global scale from satellite (Coastal Zone Color Scanner) chlorophyll. Global Biogeochemical Cycles, 10:57-69.

Babin, M., A. Morel, V. Fournier-Sicre, F. Fell, and D. Stramski. 2003. Light scattering properties of marine particles in coastal and open ocean waters as related to the particle mass concentration. Limnology and Oceanography, 48(2):843-859.

Balch, W.M., D. Drapeau, B. Bowler, E. Booth, J. Goes, A. Ashe, and J. Fry. 2004. A multi-year record of optical properties in the Gulf of Maine: I. Spatial and temporal variability. Progress in Oceanography, In Press.

Balch, W.M., D.T. Drapeau, T.L. Cucci, R.D. Vaillancourt, K.A. Kilpatrick, and J.J. Fritz. 1999. Optical backscattering by calcifying algae: separating the contribution by particulate inorganic and organic carbon fractions. Journal of Geophysical Research, 104:1541-1558.

Barnard, A.H. and C. S. Roesler. 2003. Temporal variability in the remotely sensed reflectance in the Eastern Maine Coastal Current as observed by the Gulf of Maine Ocean Observing System (GOMOOS). ASLO Aquatic Science Meeting. Salt Lake City, UT.

Barth, J.A., and D.J. Bogucki. 2000. Spectral light absorption and attenuation measurements from a towed undulating vehicle. Deep-Sea Research I, 47:323-342.

Bartz, R., J.R.V. Zaneveld, and H. Pak. 1978. Transmission for profiling and moored observation in water. Ocean Optics V, Interntational Society For Optical Engineering. pg. 102-108.

Behrenfeld, M., and P. Falkowski. 1997. A consumer's guide to phytoplankton primary productivity models. Limnology and Oceanography, 42(7):1479-1491.

Bishop, J., S.E. Calvert, and M.Y.S. Soon. 1999. Spatial and temporal variability of POC in the northeast Subarctic Pacific. Deep-Sea Research II, 46:2699-2733.

Bishop, J., R.E. Davis, and J.T. Sherman. 2002. Robotic observations of dust storm enhancement of carbon biomass in the North Pacific. Science, 298:817-821.

Blidberg, D.R. 1991. Autonomous underwater vehicles: a tool for the ocean. Unmanned systems, 9:10-15.

Blough, N.V., and Green, S.A., 1995. Spectroscopic characterization and remote sensing of non-living organic matter. In: R.G. Zepp and Ch. Sonntag (Editors), The Dahlem workshop on the role of nonliving organic matter in the Earth's carbon cycle, Berlin, 1993. John Wiley and Sons, New York, pg. 23-45.

Boss, E., and W.S. Pegau. 2001. Relationship of light scattered at an angle in the backward direction to the backscattering coefficient. Applied Optics, 40:5503-5507.

Boss, E. and J.R.V. Zaneveld. 2003. The effect of bottom substrate on inherent optical properties: evidence of biogeochemical processes. Limnology and Oceanography, 48(1, part 2):346-354.

Boss, E., M.S. Twardowski, and S. Herring. 2001. Shape of the particulate beam attenuation spectrum and its inversion to obtain the shape of the particulate size distribution. Applied Optics, 40:4885-4893.

Boss, E., W. S. Pegau. M. Lee, M. S. Twardowski, E. Shybanov, G. Korotaev, and F. Baratange. 2004. The particulate backscattering ratio at LEO-15 and its use to study particles composition and distribution. Journal of Geophysical Research – Oceans, In Press.

Bricaud, A., and D. Stramski. 1990. Spectral absorption coefficients of living phytoplankton and nonalgal biogenous matter: A comparison between Peru upwelling area and the Sargasso Sea. Limnology and Oceanography, 35:562–582.

Brown, O.B., and H.R. Gordon. 1974. Size-refractive index distribution of clear coastal water particulates from light scattering, Applied Optics, 13:2874-2881.

Brown, D., J. Isaacs, and M. Sessions. 1971. Continuous temperature-depth profiling deep-moored buoy system. Deep-Sea Research, 18:845-849.

Brown, C.A., Y. Huot, M.J. Purcell, J.J. Cullen, and M.R. Lewis. 2004. Mapping coastal bio-optical properties with an autonomous underwater vehicle (AUV) and a spectral inversion model. Limnology and Oceanography Methods, Submitted.

Campbell, J., and others. 2002. Comparisons of algorithms for estimating ocean primary production from surface chlorophyll, temperature, and irradiance. Global Biogeochemical Cycles, 16(3): 10.1029/2001GB001444

Carder, K. L., Chen, F. R., Lee, Z. P., Hawes, S., & Kamykowski, D. 1999. Semi-analytic MODIS algorithms for chlorophyll *a* and absorption with bio-optical domains based on nitrate-depletion temperatures. Journal of Geophysical Research, 104(C3):5403-5421.

Carder, K.L., R.G. Steward, G.R. Harvey, and P.B. Ortner. 1989. Marine humic and fulvic acids: their effects on remote sensing of ocean chlorophyll, Limnology and Oceanography, 34:68-81.

Chang, G.C., T.D. Dickey, and A.J. Williams, III. 2001. Sediment resuspension over a continental shelf during Hurricanes Edouard and Hortense. Journal of Geophysical Research, 106:9517-9531.

Chavez, F.P., P.G. Strutton, G. Friederich, A. Feely, G.C. Feldman, D.C. Foley, and M.J. McPhaden. 1999. Biological and chemical response of the equatorial Pacific Ocean to the 1997-98 El Nino. Science, 286:2126-2131.

Chavez, F.P., D. Wright, R. Herlien, M. Kelley, F. Shane, and P.G. Strutton. 2000. A device for protecting moored spectroradiometers from fouling. Journal of Atmospheric and Oceanic Technology, 17:215-219.

Ciotti, A.M., Cullen, J.J., and Lewis, M.R. 1999. A semi-analytical model of the influence of phytoplankton community structure on the relationship between light attenuation and ocean color. Journal of Geophysical Research, 104:1559-1578.

Clark, D. 2003. MODIS Ocean Data Product MOD23, ATBD 18. Can be viewed at http://modis-ocean.gsfc.nasa.gov/qa/dataproductmap.html.

Clark, D., H.R. Gordon, K.J. Voss, Y. Ge, W. Broenkow, and C. Trees. 1997. Validation of atmospheric correction over the oceans. Journal of Geophysical Research, 102:17,209-17,217.

Claustre, H., F. Fell, K. Oubelkheir, L. Prieur, A. Sciandra, B. Gentili, and M. Babin. 2000. Continuous monitoring of surface optical properties across a geostrophic front: biogeochemical inferences. Limnology and Oceanography, 45(2):309-321.

Claustre, H., A. Morel, M. Babin, C. Cailliau, D. Marie, J-C. Marty, D. Tailliez, and D. Vaulot. 1999. Variability in particle attenuation and chlorophyll fluorescence in the tropical Pacific: scales, patterns, and biogeochemical implications. Journal of Geophysical Research, 104(C2):3401-3422.

CLIVAR. 1999. The design and implementation of Argo – A global array of profiling floats. Report 21, International CLIVAR Project Office, Southampton, UK, pg. 1-35.

Coble, P.G. 1996. Characterization of marine and terrestrial DOM in seawater using excitation-emission matrix spectroscopy. Marine Chemistry, 51:325-346.

Coble, P.G, C.A. Schultz, and K. Mopper. 1993. Fluorescence contouring analysis of DOC Intercalibration Experiment samples: a comparison of techniques. Marine Chemistry, 41:173-178.

Conmy, R.N., Coble, P.G., and Del Castillo, C.E. 2004. Calibration and performance of a new in-situ multi-channel fluorometer for measurement of colored dissolved organic matter in the ocean. Continental Shelf Research, In Press.

Davis, R. E., J. T. Sherman and J. Dufour, 2001: Profiling ALACEs and other advances in autonomous subsurface floats. Journal of Atmospheric and Oceanic Technology, 18:982-993.

De Domenico, L., E. Crisafi, G. Magazzu, A. Puglisi, and A. La Rosa. 1994. Monitoring of petroleum hydrocarbon pollution in surface waters by a direct comparison of fluorescence spectroscopy and remote sensing techniques. Marine Pollution Bulletin, 28:587.

Del Castillo, C.E., P.G. Coble, R.N. Conmy, F.E. Muller-Karger, L. Vanderbloemen, and G.A. Vargo. 2000. Multispectral in situ measurements of organic matter and chlorophyll fluorescence in seawater: Documenting the intrusion of the Mississippi River plume in the West Florida Shelf. Limnology and Oceanography, 46(7):1836-1843.

Del Castillo, C.E., P.G. Coble, J.M. Morell, J.M. Lopez, and J.E. Corredor. 1999. Analysis of the optical properties of the Orinoco River plume by absorption and fluorescence spectroscopy. Marine Chemistry, 66:35-51.

Desiderio, R.A., T.J. Cowles, J.N. Moum, and M.L. Myrick. 1993. Microstructure profiles of laser-induced chlorophyll fluorescence spectra: evaluation of backscatter and forward-scatter fiber-optic sensors. Journal of Atmospheric and Oceanic Technology, 10:209-224.

Desiderio, R.A., T.J. Cowles, J.R.V. Zaneveld, and C. M. Moore. 1996. Multi-excitation spectral absorption and fluorescence: a new in situ device for characterizing dissolved and particulate material. AGU meeting abstract OS21C-10, p. OS52, San Diego, CA, Feb. pg. 12-16.

Determann, S., J.M. Lobbes, R. Reuter, and J. Rullkotten. 1998. UV fluorescence excitation and emission spectroscopy of marine algae and bacteria. Marine Chemistry, 62:137-156.

Dickey, T. and G. Chang. 2001 New technologies and their roles in advancing recent biogeochemical studies. Oceanography, 14(4):108-120.

Dickey, T., and D.A. Siegel [Eds.]. 1993. Bio-optics in U.S. JGOFS. Report of the bio-optics workshop, June 17-19, 1991, Boulder, CO.

Dickey, T., M. Lewis, and G. Chang. 2004. Optical oceanography: recent advances and future directions using global remote sensing and in situ observations. In J.F. Gower (Ed.), Manual for Remote Sensing of the Oceans. In Press

Dickey, T., C. Moore and O-SCOPE Group. 2003. New sensors monitor bio-optical / biogeochemical ocean changes. Sea Technology, 44(10):17-24

Dickey, T., D. Frye, H. Jannasch, E. Boyle, D. Manov, D. Sigurdson, H. McNeil, M. Stramska, A. Michaels, N. Nelson, D. Siegel, G. Chang, J. Wu, and A. Knap. 1998. Initial results from the Bermuda Test bed Mooring Program. Deep-Sea Research I, 45:771-794.

Dierssen, H.M., R.C. Zimmerman, R.A. Leathers, T.V. Downes, and C.O. Davis. 2003. Ocean color remote sensing of seagrass and bathymetry in the Bahamas Banks by high-resolution airborne imagery. Limnology and Oceanography, 48(1, part 2):444-455.

Donaghay, P.L., H.M. Rines, and J.McN. Sieburth. 1992. Simultaneous sampling of fine scale biological, chemical, and physical structure in stratified waters. Arch. Hydrobiol. Beih. Ergebn. Limnol, 36:97-108.

Donaghay, P.L., J.M. Sullivan, C. Moore, and B. Rhoades. 2002. 4-D measurement of the finescale structure of inherent optical properties in the coastal ocean using the Ocean Response Coastal Analysis System (ORCAS). Proceedings from Ocean Optics XVI, 18-22 November, Santa Fe, NM, Office Naval Research, USA, CD-ROM.

Echert, D., J. Morison, G. White, and E. Geller. 1989. The autonomous ocean profiler: A current-driven oceanographic sensor platform. IEEE Journal Oceanic Engineering, 14:195-202.

Eisner, L., M.S. Twardowski, T.J. Cowles, and M.J. Perry. 2003. Resolving phytoplankton photoprotective:photosynthetic carotenoid ratios on fine scales using in situ spectral absorption measurements. Limnology and Oceanography, 48:632-646.

Ferrari, G.M., Hoepffner, and M. Mingazzini. 1996. Optical properties of the water in a deltaic environment: prospective tool to analyze satellite data in turbid waters. Remote Sensing of the Environment, 56:69-80.

Fugate, D.C., and C.T. Friedrichs. 2002. Determining concentration and fall velocity of estuarine particle populations using ADV, OBS, and LISST. Continental Shelf Research, 22:1867-1886.

Gardner, W.D., I.D. Walsh, and M.J. Richardson. 1993. Biophysical forcing of particle production and distribution during a spring bloom in the North Atlantic. Deep-Sea Research II, 40:171-195.

Gardner, W.D., J.C. Blakey, I.D. Walsh, M.J. Richarson, S. Pegau, J.R.V. Zaneveld, C. Roesler, M.C. Gregg, J. A. MacKinnon, H.M. Sosik, and A.J. Williams III. 2001. Optics, particles, stratification, and storms on the New England continental shelf. Journal of Geophysical Research, 106(C5):9473-9497.

Garver, S.A. and D.A. Siegel. 1997. Inherent optical property inversion of ocean color spectra and its biogeochemical interpretation: 1. Time series from the Sargasso Sea. Journal of Geophysical Research, 102:18,607-18,625.

Goes, J.I., T. Saino, H. Oaku, J. Ishizaka, C.S. Wong, and Y. Nojiri. 2000. Basin scale estimates of sea surface nitrate and new production from remotely sensed sea surface temperature and chlorophyll. Geophysical Research Letters, 27:1263-1266.

Gordon, H. and W. Balch. 2003. MODIS Ocean Data Product MOD25, ATBD 23. Can be viewed at http://modis-ocean.gsfc.nasa.gov/qa/dataproductmap.html.

Gordon, H.R., O.B. Brown, R.H. Evans, J.W. Brown, R.C. Smith, K.S. Baker, and D.K. Clark. 1988. A semianalytical radiance model of ocean color. Journal of Geophysical Research, 93:10,909-10,924.

Gould, R.W., and R.A. Arnone. 1998. Three-dimensional modeling of inherent bio-optical properties in a coastal environment: coupling ocean colour imagery and in situ measurements. International Journal of Remote Sensing, 19:2141-2159.

Griffiths, G., R. Davis, C. Erikson, D. Frye, P. Marchand, and T. Dickey. 2001. Towards new platform technology for sustained observations, In: Observing the Ocean for Climate in the 21st Century, C.J. Koblinsky and N.R. Smith [Eds.]. GODAE, Bureau of Meteorology, Australia, Melbourne, pg. 324-338.

Hales, B. 2001. Small-scale variability in the Ross Sea. Oceanography, 14:90.

Haltrin, V.I. and R. A. Arnone. 2003. An algorithm to estimate concentrations of suspended particles in seawater from satellite optical images. Proceedings of the II International Conference "Current Problems in Optics of Natural Waters," ONW 2003, I. Levin and G. Gilbert [Eds.], St. Petersburg, Russia.

Hanson, A.K. 2000. A new in situ chemical analyzer for mapping coastal nutrient distributions in real time. Proceedings from Oceans 2000 MTS/IEEE, 3:1975-1982.

Hanson, A.K. and P.L. Donaghay. 1998. Micro- to fine-scale Chemical Gradients and Thin Plankton Layers in Stratified Coastal Waters. Oceanography, 11(7):10-17.

Hays, G.C. and Lindley, J.A. 1994. Estimating chlorophyll a abundance from the 'phytoplankton colour' recorded by the Continuous Plankton Recorder survey: Validation with simultaneous fluorometry. Journal of Plankton Research, 168: 23-34.

Hoge, F.E. and P.E. Lyon. 1999. Spectral parameters of inherent optical property models: Methods for satellite retrieval by matrix inversion of an oceanic radiance model. Applied Optics, 38:1657-1662.

Holdway, D., A. Radlinski, N. Exon, J-M. Auzende, and S. Van de Beuque. 2000. Continuous multi-spectral fluorescence and absorption spectroscopy for petroleum hydrocarbon detection in near-surface ocean waters: ZoNeCo5 Survey, Fairway Basin area, Lord Howe Rise. Australian Geological Survey Organisation, Record 2000/35, Canberra, Australia, ACT 2601.

Hooker, S.B. and A. Morel. 2003. Platform and environmental effects on above-water determinations of water-leaving radiances. Journal of Atmospheric and Oceanic Technology, 20:187-205.

Hooker, S.B., G. Lazin, G. Zibordi, and S. McLean. 2002. An Evaluation of Above- and In-Water Methods for Determining Water-Leaving Radiances. Journal of Atmospheric and Oceanic Technology, 19:486-515.

IOCCG Report 3. 2000. Remote sensing of ocean colour in coastal, and other optically-complex waters. Report 3 of the International Ocean Colour Coordinating Group, S. Sathyendranath [Ed.]. Dartmouth, Nova Scotia, Canada, 140 pp.

Johnson, K.S., and L.J. Coletti. 2002. In situ ultraviolet spectrophotometry for high resolution and long-term monitoring of nitrate, bromide and bisulfide in the ocean. Deep-Sea Research I, 49:1291-1305.

Kirk. J.T.O. 1992. Monte Carlo modeling of the performance of a reflective tube absorption meter. Applied Optics, 31(30):6463-6468.

Kirk, J.T.O. 1994. Light and photosynthesis in aquatic ecosystems., 2nd ed. Cambridge, 509 pp.

Kirkpatrick, G.J., D.F. Millie, M. Moline, and O. Schofield. 2000. Optical discrimination of a phytoplankton species in natural mixed populations. Limnology and Oceanography, 45(2):467-471.

Klinkhammer, G.P., J. McManus, D. Colbert, and M.D. Rudnick. 2000. Behavior of terrestrial dissolved organic matter at the continent-ocean boundary from high-resolution distributions. Geochimica Cosmochimica Acta, 64:2765-2774.

Kratzer, S., D. Bowers, and P.B. Tett. 2000. Seasonal changes in colour ratios and optically active constituents iin the optical case-2 waters of the Menai Strait, North Wales. International Journal of Remote Sensing, 21(11):2225-2246.

Leathers, R.A., T.V. Downes, and C.D. Mobley. 2001. Self-shading correction for upwelling sea-surface radiance measurements made with buoyed instruments. Optics Express, 8:561-570.

Lee, M.E., and M.R. Lewis. 2003. A new method for the measurement of the optical volume scattering function in the upper ocean. Journal of Atmospheric and Oceanic Technology, 20(4):563-571.

Lewis, M.R., N. Kuring, and C.S. Yentsch. 1988. Global patterns of ocean transparency: Implications for the new production of the open ocean. Journal of Geophysical Research, 93:6847-6856.

Lewis, M.R., M.E. Carr, G. Feldman, W. Esaias, and C. McClain. 1990. The influence of penetrating irradiance on the heat budget of the equatorial Pacific Ocean. Nature, 347:543-545.

Loisel, H., and A. Morel. 1998. Light scattering and chlorophyll concentration in case 1 waters: A reexamination. Limnology and Oceanography, 43:847-858.

Loisel, H. and D. Stramski. 2000. Estimation of the inherent optical properties of natural waters from irradiance attenuation coefficient and reflectance in the presence of Raman scattering. Applied Optics, 39:3001-3011.

Loisel H., D. Stramski, B.G. Mitchell, F. Fell, V. Fournier-Sicre, B. Lemasle and M. Babin. 2001. Comparison of the ocean inherent optical properties obtained from measurements and inverse modeling. Applied Optics, 40:2384-2397.

Maffione, R.A., and D.R. Dana. 1997. Instruments and methods for measuring the backward-scattering coefficient of ocean waters. Applied Optics, 36:6057.

Manov, D.V., G.C. Chang, and T.D. Dickey. 2004. Methods for reducing biofouling of moored optical sensors. Journal of Atmospheric and Oceanic Technology, In Press.

Maritorena, S., D.A. Siegel, and A.R. Peterson. 2002. Optimization of a semianalytical ocean color model for global-scale applications. Applied Optics, 41(15):2705-2714.

Mayer, L.M., L.L. Schick, and T. Loder, 1999. Dissolved protein fluorescence in two Maine estuaries, Marine Chemistry, 64:171-179.

McKnight, D.M., E.W. Boyer, P.K. Westerhoff, P.T. Doran, T. Kulbe, and D.T. Andersen. 2001. Spectrofluorometric characterization of dissolved organic matter for indication of precursor organic material and aromaticity. Limnology and Oceanography, 46(1):38-48.

McLean, S., B. Scofield, G. Zibordi, M. Lewis, S. Hooker, and A. Weidemann. 1997. Field evaluation of anti-biofouling compounds on optical instrumentation. SPIE, 2963:708-713.

Miller, R., M.S. Twardowski, C. Moore, and C. Casagrande. 2003. The Dolphin: Technology to Support Remote Sensing Bio-optical Algorithm Development and Applications. Backscatter, 14(2):8-12.

Mishonov, A.V., W.D. Gardner, and M.J. Richardson. 2003. Remote sensing and surface POC concentration in the South Atlantic. Deep-Sea Research II, 50(22-26):2997-3015.

Mitchell, B.G., M. Kahru, and J. Sherman. 2000. Autonomous temperature-irradiance profiler resolves the spring bloom in the Sea of Japan. Proceedings of Ocean Optics XV, 16-20 October, Monaco, Office Naval Research, USA, CD-ROM.

Mobley, C.D. 1994. Light and water: radiative transfer in natural waters. Academic, San Diego, CA, 592 pp.

Moore, C.M.. 1994. In situ biochemical, oceanic optical meters: spectral absorption, attenuation, fluorescence meters – a new window of opportunity for ocean scientists. Sea Technology, 35:10-16.

Moore, C., B. Rhodes, A. Derr, and R. Zaneveld. 2004. An instrument for hyperspectral characterization of inherent optical properties in natural waters. ALSO/TOS Ocean Research Conference, 15-20 February, Honolulu, HA.

Moore, C., M.S. Twardowski, and J.R.V. Zaneveld. 2000. The ECO VSF - A multi-angle scattering sensor for determination of the volume scattering function in the backward direction. Proceedings of Ocean Optics XV, 16-20 October, Monaco, Office Naval Research, USA, CD-ROM.

Moore, C., J.R.V. Zaneveld, and J.C. Kitchen. 1992. Preliminary results from an in situ spectral absorption meter. Ocean Optics XI, Proceedings of SPIE, 1750:330-337.

Morel, A. 1974. Optical properties of pure water and pure seawater, In: Optical aspects of Oceanography, N.G. Jerlov and E. Steemann Nielson [Eds.], pp. 1-24, Academic, New York.

Morel, A. 1980. In-water and remote measurements of ocean color. Boundary-Layer Meteorology, 18:177-201.

Morel, A. 1988. Optical modeling of the upper ocean in relation to its biogenous matter content (case I waters). Journal of Geophysical Research, 93:10,749-10,768.

Morel, A. 1991. Light and marine photosynthesis: A spectral model with geochemical and climatological implications. Progress in Oceanography, 26:263-306.

Morel, A. and B. Gentili. 1996. Diffuse reflectance of oceanic waters, III: implications of bidirectionality for the remote sensing problem. Applied Optics, 35:4850-4862.

Morel, A. and S. Maritorena. 2001. Bio-optical properties of oceanic waters: a reappraisal. Journal of Geophysical Research, 106(C4):7163-7180.

Morel, A. and J.L. Mueller. 2003. Normalized water-leaving radiance and remote sensing reflectance: Bidirectional reflectance and other factors. In: Mueller, J.L., G.S. Fargion, and C.R. McClain [Eds]. Ocean Optics Protocols For Satellite Ocean Color Sensor Validation, Revision 4, Volume III: Radiometric Measurements and Data Analysis Protocols. NASA, Goddard Space Flight Center, Greenbelt, MD, pg. 32-59.

Morel, A. and L. Prieur. 1977. Analysis of variations in ocean color. Limnology and Oceanography, 22:709-722.

Morel, A., K. J. Voss, and B. Gentilli. 1995. Bi-directional reflectance of oceanic waters: A comparison of model and experimental results. Journal of Geophysical Research, 100:13,143-13,150.

Mueller, J.L., G.S. Fargion, and C.R. McClain [Eds]. 2003. Ocean Optics Protocols For Satellite Ocean Color Sensor Validation, Revision 4, Volume IV. NASA, Goddard Space Flight Center, Greenbelt, MD.

O'Reilly, J.E. and others. 2000. Ocean color chlorophyll a algorithms for SeaWiFS, OC2, and OC4: Version 4, In: SeaWiFS Postlaunch Calibration and Validation Analyses, Part 3. NASA Tech. Memo. 2000-206892, Vol. 11, S.B. Hooker and E.R. Firestone [Eds.], NASA Goddard Space Flight Center, Greenbelt, MD.

Pages, J., and F. Gadel. 1990. Dissolved organic matter and UV absorption in a tropical hyperhaline estuary. The Science of the Total Environment, 99:173-204.

Pegau, W.S., and others. 1995. A comparison of methods for the measurement of the absorption coefficient in natural waters. Journal of Geophysical Research, 100(C7):13,201-13,220.

Pegau, W.S., J. Barth, and M. Kosro. 2000. Optical variability off the Oregon coast. Proceedings of Ocean Optics XV, 16-20 October, Monaco, Office Naval Research, USA, CD-ROM.

Pegau, W.S., A. Barnard, E. Boss, and M. Twardowski. 1999. Ac9 protocols. At: http://photon.oce.orst.edu/ocean/instruments/ac9/ac9.html.

Peterson, R.E.. 1978. A study of suspended particulate matter: Arctic Ocean and northern Oregon continental shelf. Ph.D. thesis, Oregon State University. 134 pp.

Petrenko, A.A., B.H. Jones, T.D. Dickey, M. LeHaitre, and C. Moore. 1997. Effects of a sewage plume on the biology, optical characteristics, and particle size distributions of coastal waters. Journal of Geophysical Research, 102(C11):25,061-25,071.

Petzold, T.J. 1972. Volume scattering functions for selected ocean waters. Rep. 72-78, Scripps Institution of Oceanography, La Jolla, CA.

Pinkerton, M.H. and J. Aiken. 1999. Calibration and validation of remotely sensed observations of ocean color from a moored data buoy. Journal of Oceanic and Atmospheric Technology, 37:8710-8723.

Platt, T. and M.R. Lewis. 1987. Estimation of phytoplankton production by remote sensing. Advanced Space Research, 7:131-134.

Pope, R.M., and E.S. Fry. 1997. Absorption spectrum (380-700 nm) of pure water. II. Integrating cavity measurements. Applied Optics, 36:8710-8723.

Prahl, F.G., L.F. Small, and B. Eversmeyer. 1997. Biogeochemical characterization of suspended particulate matter in the Columbia River estuary. Marine Ecology Progress Series, 160:173-184.

Preisendorfer, R.W. 1976. Hydrologic Optics, Volumes 1-6. Department of Commerce, National Oceanic and Atmospheric Administration, Washington, D.C.

Provost, C. and M. du Chaffaut. 1996. 'Yoyo Profiler': an autonomous multisensor. Sea Technology X:39-45.

Purcell, M., T. Austin, R. Stokey, C. von Alt, and K. Prada. 1997. Measurements in coastal waters. Proceedings of Oceans '97 MTS/IEEE, pg. 219-24.

Rainville, L. and R. Pinkel. 2001. Wirewalker: An autonomous wave-powered vertical profiler. Journal of Atmospheric and Oceanic Technology, 18:1048-1051.

Reynolds-Fleming, J., J. Fleming, and R. Luettich, Jr. 2002. Portable autonomous vertical profiler for estuarine applications. Estuaries, 25:142-147.

Roesler, C.S. and A.H. Barnard. 2003. Temporal variability in ecosystem structure in the Eastern Maine Coastal Current as observed by the Gulf of Maine Ocean Observing System (GOMOOS). ASLO Aquatic Science Meeting. Salt Lake City, UT.

Roesler, C.S., and E. Boss. 2003. Ocean color inversion yields estimates of the spectral beam attenuation coefficient while removing constraints on particle backscattering spectra. Geophysical Research Letters, 30(9):1468.

Roesler, C.S., and M.J. Perry. 1995. In situ phytoplankton absorption, fluorescence emission, and particulate backscattering spectra determined from reflectance. Journal of Geophysical Research, 100:13,279-13,294.

Roesler, C.S., M.J. Perry, and K.L. Carder. 1989. Modeling in situ phytoplankton absorption from total absorption spectra. Limnology and Oceanography, 34:1512-1525.

Sathyendranath, S., T. Platt, C.M. Caverhill, R.E. Warnock, and M.R. Lewis. 1989. Remote sensing of oceanic primary productivity: Computations using a spectral model. Deep-Sea Research I, 36:431-453.

Sathyendranath, S., T.Platt, E. P. W. Horne, W. G. Harrison, O. Ulloa, R. Outerbridge, and N. Hoepffner. 1991. Estimation of new production in the ocean by compound remote sensing. Nature, 353:129-133.

Schofield, O., S. Glenn, G. Kirkpatrick, M. Moline, C. Jones. 2004. Mapping red tide using autonomous underwater Webb gliders. ALSO/TOS Ocean Research Conference, 15-20 February, Honolulu, HA.

Schofield, O., J. Gryzmski, P. Bissett, G. Kirkpatrick, D.F. Millie, M.A. Moline, and C. Roesler. 1999. Optical monitoring and forecasting systems for harmful algal blooms: Possibility or pipedream? Journal of Phycology, 35:125-145.

Schroeder, F. and W. Petersen. 2000. The Ferry-Box as a monitoring tool for marine waters: Concepts and technical solutions for European systems. http://coast.gkss.de/projects/ferrybox/ferrypages/pubs/Techno Ocean-manuser.pdf.

Shifrin, K. 1988. Physical optics of ocean water. American Institute of Physics, New York, 285 pp.

Sibert, J.R., and J.L. Nielsen [Eds.]. 2001. Electronic Tagging and Tracking in Marine Fisheries, Proceedings of the Symposium on Tagging and Tracking Marine Fish With Electronic Devices, February 7-11, 2000, East-West Center, University of Hawaii, Honolulu, Hawaii. Kluwer Academic Publishers. 484 pp.

Siegel, D.A., S.C. Doney, and J.A. Yoder. 2002. The North Atlantic spring phytoplankton bloom and Sverdrup's critical depth hypothesis. Science, 296:730-733.

Siegel, D.A., S. Maritorena, N.B. Nelson, D.A. Hansell, and M. Lorenzi-Kayser. 2002. Global distribution and dynamics of colored dissolved and detrital organic materials. Journal of Geophysical Research, 107:3228-3242.

Singhal, S.C. 2000. Science and Technology of Solid-Oxide Fuel Cells. MRS Bulletin, 25(3):16-21.

Sivaprakasam, V., R.F. Shannon, Jr., C. Luo, P.G. Coble, J.R. Boehme, and D.K. Killinger. 2003. Development and initial calibration of a portable laser-induced fluorescence system used for in situ measurements of trace plastics and dissolved organic compounds in seawater and the Gulf of Mexico. Applied Optics, 42:6747-6757.

Smart, J. 2000. World-wide Ocean Optics Database (WOOD). Oceanography, 13(3):70-74.

Sosik, H.M., R.J. Olson, M.G. Neubert, A. Shalapyonok, and A.R. Solow. 2002. Time series observations of a phytoplankton community monitored with a new submersible flow cytometer. Proceedings of Ocean Optics XVI, 18-22 November, Santa Fe, NM, Office Naval Research, USA, CD-ROM.

Stramski, D., E. Boss, D. Bogucki, and K. Voss. The role of seawater constituents in light backscattering in the ocean. Progress in Oceanography, Submitted.

Stramski, D., R. A. Reynolds, M. Kahru, and B. G. Mitchell. 1999. Estimation of particulate organic carbon in the ocean from satellite remote sensing. Science, 285:239-242.

Sullivan, J.M., Twardowski, M.S., Donaghay, P.L., and Freeman, S.A. 2004. Using scattering characteristics to discriminate particle types in U.S. coastal waters. Applied Optics, In Review.

Sullivan, J.M., M.S. Twardowski, C. Moore, B. Rhodes, R. Zaneveld, R. Miller, and S. Freeman. 2004. Field and laboratory characterization of a new hyperspectral ac meter (the acs). ALSO/TOS Ocean Research Conference, 15-20 February, Honolulu, HA.

Trees, C.C., D.K. Clark, R.R. Bidigare, M.E. Ondrusek, and J.L. Mueller. 2000. Accessory pigments versus chlorophyll *a* concentrations within the euphotic zone: A ubiquitous relationship. Limnology and Oceanography, 45(5):1130-1143.

Twardowski, M.S., E. Boss, J.B. Macdonald, W.S. Pegau, A.H. Barnard, and J.R.V. Zaneveld. 2001. A model for estimating bulk refractive index from the optical backscattering ratio and the implications for understanding particle composition in Case I and Case II waters. Journal of Geophysical Research, 106(C7):14,129-14,142.

Twardowski, M.S., E. Boss, J.M. Sullivan, and P.L. Donaghay. 2004. Modeling spectral absorption by chromophoric dissolved organic matter (CDOM). Marine Chemistry, In Press.

Twardowski, M.S., J.M. Sullivan, P.L. Donaghay, and J.R.V. Zaneveld. 1999. Microscale quantification of the absorption by dissolved and particulate material in coastal waters with an ac-9. Journal of Atmospheric and Oceanic Technology, 16(12):691-707.

Twardowski, M.S., J.R.V. Zaneveld, and C. Moore. 2002. A novel technique for determining beam attenuation compatible with a small sensor form factor and compact deployment platforms. Proceedings from Ocean Optics XVI, 18-22 November, Santa Fe, NM, Office Naval Research, USA, CD-ROM.

Twardowski, M.S., J.R.V. Zaneveld, and C. Moore. 2003. Light attenuation meter using multiple path transmission and scattering. Provisional U.S. Patent App No. 453,739.

Tyce, R., F. de Strobel, V. Grandi, and L. Gualdesi. 2000. Trawl-safe profiler development at SACLANT Centre for shallow water environmental assessment and read time modeling. Proceedings of Oceans '00 MTS/IEEE, pg. 99-104.

Van Leer, J., W. Duing, R. Erath, E. Kennelly, and A. Speidel. 1974. The Cyclesonde: an unattended vertical profiler for scalar and vector quantities in the upper ocean. Deep-Sea Research, 21:385-400.

Vodacek, A., N.V. Blough, M.D. DeGrandpre, E.T. Peltzer, and R.K. Nelson. 1997. Seasonal variation of CDOM and DOC in the Middle Atlantic Bight: terrestrial inputs and photooxidation. Limnology and Oceanography, 42:674-686.

Voss, K.J., C.D. Mobley, L.K. Sundman, J.E. Ivey, and C.H. Mazel. 2003. The spectral upwelling radiance distribution in optically shallow waters. Limnology and Oceanography, 48:364-373.

Walsh, I.D., S.P. Chung, M.J. Richardson, and W.D. Gardner. 1995. The diel cycle in the integrated particle load in the equatorial Pacific: a comparison with primary production. Deep-Sea Research II, 42:465-477.

Walsh, I.D., W.D. Gardner, and M.J. Richarson. 1992. Transmissometer profiles in the equatorial Pacific during the JGOFS EQPAC program: Estimates of primary production. Eos, 73:295.

Waters, K., R.C. Smith, and M.R. Lewis. 1990. Avoiding ship-induced light-field perturbation in the determination of oceanic optical properties. Oceanography, 3:18-21.

White, P., S. Honjo, T. Dickey, and H. Weller. 2002. Episodic primary production and export carbon fluxes in the Arabian Sea. Eos, 83(4), Ocean Sciences Meeting Supplement, Abstract OS12D-170.

Wilson, S. 2000. Launching the ARGO armada. Oceanus, 42:17-19.

Yentsch, C.S., and D.W. Menzel. 1963. A method for the determination of phytoplankton chlorophyll and phaeophytin by fluorescence. Deep-Sea Research, 10:221-231.

Yu, X., T. Dickey, J. Bellingham, D. Manov, and K. Streitlien. 2002. The application of autonomous underwater vehicles for interdisciplinary measurements in Massachusetts and Cape Cod Bays. Continental Shelf Research, 22:2225-2245.

Zaneveld, J.R.V. 1995. A theoretical derivation of the dependence of the remotely sensed reflectance of the ocean on the inherent optical properties. Journal of Geophysical Research, 100:13,135.

Zaneveld, J.R.V., and W.S. Pegau. 2003. Robust underwater visibility parameter. Optics Express, 11(23):2997-3009.

Zaneveld, J.R.V., R. Bartz, and J.C. Kitchen. 1990. Reflective-tube absorption meter. Ocean Optics X, Proceedings of SPIE, 1302:124-136.

Zaneveld, J.R.V., E. Boss, and C.M. Moore. 2001. A diver-operated optical and physical profiling system. Journal of Atmospheric and Oceanic Technology, 18:1421-1427.

Zaneveld, J.R.V., J.C. Kitchen, and C.M. Moore. 1994. The scattering error correction of reflecting-tube absorption meters. Ocean Optics XII, Proceedings of SPIE, 2258:44-55.

Zaneveld, J.R.V., J.C. Kitchen, A. Bricaud, and C.M. Moore. 1992. Analysis of in situ spectral absorption meter data. Ocean Optics XI, Proceedings of SPIE, 1750:187-200.

Zaneveld, J.R.V., D.R. Roach, and H. Pak. 1974. The determination of the index of refraction distribution of oceanic particulates. Journal of Geophysical Research, 79:4091.

Zibordi, G., S.B. Hooker, J.F. Berthon, and D. D'Alimonte. 2002. Autonomous above water radiance measurements from stable platforms. Journal of Atmospheric and Oceanic Technology, 19:808-819.

Zhang, X., M.R. Lewis, M. Li, B. Johnson, and G. Korotaev. 2002. The volume scattering function of natural bubble populations. Limnology and Oceanography, 47:1273-1282

Chapter 5

THE COLOR OF THE COASTAL OCEAN AND APPLICATIONS IN THE SOLUTION OF RESEARCH AND MANAGEMENT PROBLEMS

[1]FRANK E. MULLER-KARGER, [1]CHUANMIN HU, [2]SERGE ANDRÉFOUËT, [3]RAMÓN VARELA, AND [4]ROBERT THUNELL

[1]*Institute for Marine Remote Sensing, College of Marine Science, University of South Florida 140 7th Ave. S., St. Petersburg FL, 33701 USA*
[2]*Institut de Recherche pour le Développement (IRD), Nouvelle Calédonie*
[3]*Fundación La Salle de Ciencias Naturales, Venezuela*
[4]*University of South Carolina, Columbia, SC, 29208 USA*

1. Introduction

The coastal ocean, which we define as extending from the coast seaward to the edge of the continental margin (approximately 500 m depth), is the most productive area of the global ocean (Chen et al., 2001; Koblenz-Mishke et al., 1970). It experiences significant carbon fixation by photosynthesis, which in turn supports abundant and diverse pelagic and benthic fauna relative to the deep ocean. For humans, the coastal ocean has important recreation and transportation functions, and it is the locus of significant living and non-living resource extraction activities. Because of these factors and because of the beauty of the coastal ocean, approximately 25% of the global human population resides in coastal watersheds in densities that exceed those of inland communities by a factor of three (Small and Nichols, 2003). In the United States, over 50 percent of the population resides in coastal watershed counties which occupy less than 25 percent of the country's land area. Around the globe, human population density continues to increase faster along the coast than in inland areas. However, the coastal ocean is affected by the close proximity of growing urban areas, commercial and industrial interests, and sediment, nutrient, or other pollutant discharges generated from our inland and marine activities. The result is stress of coastal ocean resources, with undesired consequences of significant, perceptible, and quick degradation in water quality, biodiversity, and fish abundance on a global scale.

Climate variability further complicates this situation, but the impacts of climate change on ocean resources, their utilization, and on human development are still unclear. We know that the ocean plays an important role in defining geochemical imbalances at the Earth's surface over geological time scales through carbon sequestration by marine organisms. However, whether the ocean helps mitigate the increase of anthropogenic greenhouse gases over shorter time scales (years to decades) or not remains one of the most pressing questions in ocean biogeochemistry.

These issues present difficult challenges for stewards of the marine environment, who currently have limited ways to assess conditions over these sensitive environments. Similarly, scientists studying climate change, associated feedbacks, and other natural ocean processes have found that traditional techniques are inadequate to cover the range

R.L. Miller et al. (eds.), Remote Sensing of Coastal Aquatic Environments, 101–127.
© 2005 *Springer.*

of space and time scales involved. The public also remains largely unaware of the scale of the ocean, of natural processes and their impact on our lives and vice-versa, and of opportunities for conservation of resources in coastal waters. In this chapter, we illustrate how ocean color satellite sensors are used to address some of these concerns.

2. Coastal Ocean Color

Since 1970, there have been significant advances in the development of ocean color satellites that afford rapid, repeated, synoptic, and concurrent assessment of environmental parameters in oceanic areas during daytime (Figure 1). The objectives are to gain insight on processes such as the distribution and dispersal of dissolved organic carbon, oceanic primary productivity, targeting and identifying coastal phytoplankton blooms, global biogeochemical assessments, understanding changes in benthic communities in clear, shallow tropical coral reefs, and monitoring coastal water quality.

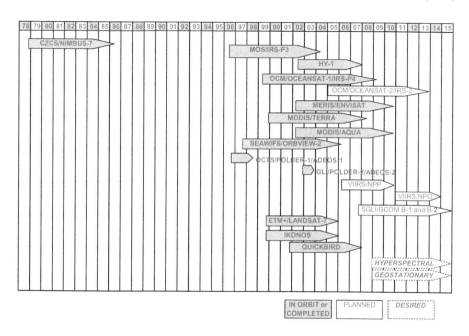

Figure 1: Ocean color satellite sensors and satellites. Figure based on information from W. Patzert and M. vanWoert (pers. comm.) and IGOS, 2001.

Attempts have been made to map coastal and oceanic water quality parameters using Landsat sensors (Amos and Alfoldi, 1979; Braga et al., 1993; Dwivedi and Narain, 1987; Khorram, 1981; Maul and Gordon, 1975; Tassan, 1987 and 1993; and others). Landsat has been most useful to map broad categories of bottom types at medium spatial resolution in shallow clear coastal waters (Lyzenga, 1981; see Andréfouët et al., Chapter 13) and to detect change over the past two decades in coastal benthos or lands. However, Landsat data does not permit systematic studies of water-borne constituents such as phytoplankton concentration, either in coastal zones or elsewhere, because of the limited radiometric, geometric, and revisit characteristics of the sensors, and data costs.

Several attempts have been made to use the reflected visible radiation channels on the Advanced Very High Resolution Radiometer (AVHRR) flown on the NOAA Polar Orbiters to estimate coastal and oceanic water quality parameters (Ackleson and Holligan, 1989; Froidefond et al., 1993; Gagliardini et al., 1984; Gallegos et al., 1989; Stumpf and Pennock, 1991; Stumpf and Tyler, 1988). These attempts have not led to accurate assessments and have not developed into widely-used operational tools. The best promise for remote sensing of biogeochemical processes in the coastal ocean remains with high sensitivity, high dynamic range, narrow spectral band, medium spatial and spectral resolution, well-calibrated visible radiation sensors, i.e. ocean color instruments.

Ocean color observations are of particular interest because of their utility in estimating the spatial distribution of stocks and rate of change of various organic carbon pools in the ocean, including those associated with living and detrital particulate materials and with colored dissolved organic matter. In addition, these data are useful for coastal applications including monitoring of suspended particulate and dissolved materials associated with river discharge and resuspension from the bottom, and in general for assessing coastal water quality. This is accomplished primarily by retrieving absorption and backscattering properties of the water, as well as chlorophyll concentration, from spectral measurements of ocean surface reflectance.

In 1978 the Coastal Zone Color Scanner (CZCS, Hovis et al., 1980) was launched and began an important set of global observations of visible radiation reflected by the ocean. The CZCS mission stimulated significant research in aquatic optics and bio-optics, the study of factors and constituents that affect the color of water. An important issue that also received significant attention is the estimation of the atmospheric radiance seen by the satellite. Over 80-90% of the total visible radiance received by a satellite sensor looking at the ocean is due to atmospheric scattering of solar radiation or specular reflection of light by the ocean's surface. These sources of radiance need to be estimated with high precision and accuracy exceeding 1%, since an error of this order will lead errors that can exceed 10% in the water-leaving radiance. The water-leaving radiance is light backscattered by the water column, not by the surface, and which is detected by a downward-looking sensor just above the surface. Thus, the CZCS studies established the foundation for such "atmospheric corrections" and for algorithms to assess water-leaving radiance, phytoplankton chlorophyll concentration, and the diffuse attenuation coefficient (a measure of how fast light attenuates during propagation) in oceanic waters. Many related studies examined spatial and temporal variability of constituents, and explored ways to estimate rates, such as primary productivity, from space (Bricaud et al., 1981; Campbell et al., 2002; Gordon et al., 1982; Gordon et al., 1983a and 1983b; Kirk, 1983; Morel, 1980; Morel and Prieur, 1977; Platt and Herman, 1983; Smith and Baker, 1978, 1982; Smith et al., 1982; Tassan and Sturm, 1986; Yoder et al., 1987; and others).

The CZCS, nevertheless, provided limited global coverage, spectral resolution and radiometric quality. It lacked sufficient sensitivity and dynamic range for useful coastal observations, but most importantly it was impossible to effect an atmospheric correction in near-shore, turbid waters that reflect significant light in the red CZCS bands that were used to derive the atmospheric parameters. It was quickly determined that CZCS data were most applicable to oceanic waters, and less so to study turbid coastal waters.

As we enter the 21st century, oceanic remote sensing has moved toward complex and multidisciplinary studies that require cross-referencing and merging observations from various satellites and understanding coastal and turbid waters. Ocean color observations are at the center of these efforts (IOCCG, 2000).

2.1 OCEAN COLOR SATELLITE MISSIONS

Figure 1 shows major ocean color missions designed to obtain global coverage of coastal and oceanic areas. Landsat was designed for land studies, but the data are useful for benthic and suspended sediment mapping (see Andrefouet et al., Chapter 13). Landsat satellites have flown since 1972 and Thematic Mapper sensors have operated since 1982, but it was not until 1999 and Landsat-7, with its Enhanced Thematic Mapper (ETM+), that systematic coverage of global coastal areas was initiated.

There was a 10-year gap between the end of the CZCS mission in 1986 and the resumption of ocean color observations with the Japanese OCTS on ADEOS-I. The second-generation Sea-viewing Wide Field-of-view Sensor (SeaWiFS or Orbview-II) was launched in August 1997 shortly after the failure of the ADEOS-I satellite.

The differences between sensors and missions are considerable, ranging from orbit and ground revisit time to swath width, spectral and spatial resolution, sensitivity, and accuracy. Examining each is beyond the scope of this chapter. To illustrate applications, we focus on NASA's Moderate Resolution Imaging Spectroradiometer (MODIS, Esaias et al., 1998) and Orbimage's SeaWiFS (McClain et al., 1998) because of their multi-year time series, global coverage, and their high quality observations. SeaWiFS is a commercial satellite (Orbimage), but its high-quality data are available free for research purposes, and data older than five years are public. All MODIS data from both the Terra and Aqua satellites are public. The MODIS model is of particular interest because it is the prototype for the Visible/Infrared Imager/Radiometer Suite (VIIRS), which will fly on the National Polar-Orbiting Operational Environmental Satellite System (NPOESS) and the NPOESS Preparatory Mission (NPP).

2.2 CASE 1 AND 2 WATERS: OPTICAL CLASSIFICATION

The oceanographic community has developed an extensive theoretical basis for assessment of ocean color and parameters that affect it. Water itself absorbs and scatters light, but otherwise the color of marine waters depends on the concentration of dissolved and particulate materials (constituents) that absorb and scatter light. The constituents of interest are phytoplankton, other organic and inorganic particulate materials (detritus and inorganic sediment), and colored dissolved organic materials (CDOM, yellow substance, or Gelbstoff). Frequently, CDOM and detritus are lumped together because they have similar light absorption properties. In shallow coastal regions, light reflected off the bottom may affect the color of water (optically shallow waters). The contribution of bacteria and zooplankton are often ignored (IOCCG, 2000).

Classification schemes have been proposed to guide when to apply specific optical, bio-optical, and atmospheric correction algorithms. The dependence of color on water quality led Jerlov (1971) to develop a scheme based on comparisons of water samples to a color scale. Morel and Prieur (1977) defined two broad categories of optical water types. The simplest was "Case 1", when watercolor is primarily a function of phytoplankton concentration. Detritus and CDOM are assumed to co-vary with chlorophyll concentration. Case 1 water can be oligotrophic or eutrophic. Remote sensing prior to the mid-1990's focused on Case 1 waters and areas away from the coast.

The condition where water color is also strongly affected by terrigenous particulate and dissolved materials, resuspended sediment, CDOM, or highly concentrated phytoplankton blooms such as coccolithophores or red tides, is generally referred to as "Case 2". In Case 2 waters, constituents are independent of each other and don't co-vary with chlorophyll. Each constituent or the bottom represents possible end members that contribute to the color in varying proportion. These may change with time or geographic location, due to a variety of factors such as river input and wind.

More recently, work has focused on understanding the optics of coastal and other turbid Case 2 waters. Significant effort has been placed on algorithms that automatically detect whether a specific region is Case 1 or Case 2. The objective is to help select the atmospheric and in-water algorithms and assess the accuracy of ocean color products.

2.3 ATMOSPHERIC CORRECTION

Two general steps are required to obtain quantitative estimates of water constituents from calibrated radiances measured by a satellite sensor. The first is removal of the color of the atmosphere (atmospheric correction). The second is the derivation of water properties based on water-leaving radiance or surface reflectance estimates. We outline these only briefly below since they are discussed in detail in other chapters of this book.

Because the deep ocean is dark, to estimate the color of the ocean the accuracy of the atmospheric correction needs to be of the order of ±0.001 in the retrieved surface reflectance. This is one order of magnitude smaller than required for assessments of land reflectance. Such accuracies are achieved by two means. One is a sophisticated atmospheric correction algorithm designed especially for the ocean, which takes into account multiple scattering in the atmosphere, molecular-aerosol interaction, polarization, surface roughness, whitecaps, and sun glint effects (Ding and Gordon, 1995; Gordon and Wang, 1994; Wang and Bailey, 2001). A vicarious calibration is also required and is done using well-characterized ocean sites and consistent atmospheric correction. The Gordon and Wang (1994) approach, for example, assumes zero water-leaving radiance in the near-IR (less than one digital count). Atmospheric properties at these wavelengths are derived and then extrapolated to the visible part of the spectrum using pre-computed lookup tables. For clear Case 1 waters, retrieved water-leaving radiance is generally within 5-10% of the in situ "truth" (McClain et al., 1998).

The situation is more complicated in coastal and Case 2 waters. Turbid water or the bottom can contribute to observed reflectance in the red and near-IR. This requires an independent assessment of the color of the atmosphere, along one of the following lines:

1) An iterative approach (Arnone et al., 1998; Stumpf et al., 2003) based on an assumed relationship between surface reflectance at 670, 765, and 865 nm. Processing seeks convergence toward this relationship (used by NASA in SeaDAS4 SeaWiFS processing);
2) A fixed relationship between water-leaving reflectances in 765 and 865 nm, based on statistical analysis from SeaWiFS (Ruddick et al., 2000);
3) A chlorophyll-driven iterative approach (Siegel et al., 2000) where the optical properties in the near-IR are functions of chlorophyll only. This approach has been applied in the operational processing of MODIS data by NASA; and,
4) A nearest-neighbor approach (Hu et al., 2000), which uses aerosol type but not aerosol thickness from an adjacent (nearest) clear-water areas.

Each approach has its advantages and disadvantages. Improvements have been made in the retrieved coastal water-leaving radiance, as evidenced by a reduced number of negative water-leaving radiance pixels relative to the "standard" (Case 1) approaches.

2.4 COASTAL OCEAN COLOR AND BIO-OPTICAL ALGORITHMS

Water constituents are estimated based on the water-leaving radiance. Algorithms related to biological oceanographic properties are referred to as "bio-optical" algorithms.

For Case 1 waters, where one variable (phytoplankton) dominates color, a band-ratio (blue/green) bio-optical algorithm is often effective because phytoplankton absorbs more blue light than green. Band-ratio algorithms have been used for CZCS (Gordon et al., 1983), SeaWiFS (O'Reilly et al., 2000), and MODIS (Carder et al., 1999).

In Case 2 waters, band-ratio algorithms often fail (e.g., Hu et al., 2003a). Semi-analytical algorithms may help in these cases (Carder et al., 1999; Hu et al., 2002; Lee et al., 1999; Lee et al., 2002; Maritorena et al., 2002). The objective is to differentiate between various optically active constituents, namely chlorophyll, colored dissolved organic matter (CDOM), and total suspended solids (TSS). While some progress has been made (e.g., Hu et al., 2003a), assumptions may need to be fine-tuned regionally, as Case 2 waters vary in space and time. Examples of parameters that are difficult to estimate but that are required in these optical models are the spectral slope of CDOM absorption, particle backscattering, bottom reflectance, and depth.

Chlorophyll is a parameter desired by many coastal managers, but it is very difficult to estimate with sufficient accuracy in coastal waters. A small error in CDOM assessment significantly affects blue absorption estimates, and causes large errors in chlorophyll retrievals. For example, the CDOM absorption coefficient at 443 nm in an estuarine area may be 0.3-0.4 m^{-1} or larger, while a chlorophyll concentration ~1 mg m^{-3} has an absorption coefficient of 0.03-0.04 m^{-1} at 443 nm. Thus, a 10% error in CDOM estimates leads to an error of the order of 100% in chlorophyll estimates.

Chomko and Gordon (2001) and Chomko et al. (2003) attempted to deconvolve the atmosphere and water contribution to the at-sensor radiance simultaneously by using a database of spectral signatures to determine the best set of constituents to yield the measured spectrum. The approach focused on Case 1 waters and was limited by computational requirements. Another optimization approach, designed for interpreting hyperspectral water-leaving radiance (Lee et al., 1999) has been adapted and tested for SeaWiFS processing (Hu et al., 2002). The computational speed is satisfactory for operational use (~10 minutes for a 1200×999 SeaWiFS scene covering the Gulf of Mexico using computational facilities in 2004). Accuracy was generally satisfactory, based on a limited validation dataset in CDOM-rich waters, but the technique is insensitive to changes in chlorophyll and led to unacceptable speckling in the chlorophyll image. The approach of Lee et al. (1994) uses water-leaving radiance as input, and thus results are also affected by atmospheric correction skill.

The overall accuracy of retrieved water leaving radiance and constituents thus depends on the performance of the atmospheric correction and in-water algorithms. However, of primary concern are the sensor calibration and characterization information. If these contain large uncertainties or are incomplete, errors in water-leaving radiance estimates are exacerbated in coastal and estuarine waters, because the blue water-leaving radiance is often much smaller than that in clear Case 1 waters. A very large effort in characterizing and calibrating the SeaWiFS and MODIS sensors has much improved the quality of data derived in coastal waters. Such efforts are essential.

Significant research has focused on improving the traditional multi-band chlorophyll retrievals. However, MODIS is also equipped with spectral bands specifically designed for measuring chlorophyll fluorescence relative to a baseline (chlorophyll fluorescence line height algorithm; Letelier and Abbott, 1996). The chlorophyll fluorescence efficiency varies in time and space, which is an area of research. Aggregating such data with those from other bio-optical algorithms would help address coastal waters.

Clearly, there are many potential applications of ocean color data even while the in-water and bio-optical algorithms continue to be improved. The study of spatial patterns and of temporal anomalies is particularly useful to assess changes in ecosystems (e.g. Hu et al., 2003b; Hu and Muller-Karger, 2003; SWFDOG, 2002).

3. Applications

The word "application" in the context of oceanographic remote sensing generally refers to one of two broad topics. One broad category includes studies that require satellite data, such as coral reef mapping (see Andrefouet et al., Chapter 13), assessments of spatial and temporal variability of phytoplankton blooms, red tide identification or monitoring (see Stumpf and Tomlinson, Chapter 12), river plume or upwelling assessments, global productivity analyses, and oil spill detection and tracking.

The other category of applications includes the software tools used to process or derive information from remote sensing data. Over the past two decades, access to sophisticated computational and graphics desktop capabilities has increased. Similarly, internet access and bandwidth has grown. The research community has successfully demonstrated the applicability of remote sensing data to oceanographic studies. However, a number of factors including the variety of algorithms, data archival and delivery mechanisms, diversity in data formats, lack of information about product quality, and a distinct separation between the research, management, and public education communities have slowed down interdisciplinary scientific research and particularly the migration of key space-based technologies into resource management and education communities. Traditionally, software applications that deal with this diversity are developed by small groups and have a limited user base of scientists.

3.1 COASTAL OCEAN MONITORING AND OCEANOGRAPHIC APPLICATIONS

A few examples of applications of ocean color data in coastal and continental shelf environments illustrate the range of studies that is possible with such data. Many new applications are continuously being developed by researchers around the globe.

3.1.1 *River plumes and their utility as coastal ocean circulation tracers*

River discharge plays an important role in the productivity, hydrological cycle and thermodynamic stability of the coastal ocean. Knowledge of river plumes, of variations in their extent and dispersal patterns, and of their mixing rates with oceanic water is critical in all aspects of continental shelf and regional oceanography (e.g., Lohrenz et al., 1990). Rivers also carry sediments to the coastal ocean, and in areas of active land use increasing and excessive sediment load has an impact on coastal ecosystems such as coral reefs (McCulloch et al., 2003).

Several studies have addressed the dispersal of some of the largest rivers in the world using ocean color. For example, Muller-Karger et al. (1995) and Hu et al. (2004) examined the complex patterns of dispersal of the Amazon plume in the Atlantic Ocean. Muller-Karger et al. (1989) and Muller-Karger and Varela (1990) examined the seasonal dispersal patterns of the Orinoco River discharge in the Caribbean Sea. Muller-Karger et al. (1991) and Del Castillo et al. (2000) used CZCS and SeaWiFS images to demonstrate that the Mississippi plume does not always disperse to the west of the Mississippi Delta, but that it also disperses to the east and south in the Gulf of Mexico (Figure 2), reaching the Florida Keys, and then along the eastern seaboard of the U.S.

Impacts of small rivers on coral reefs off Honduras and Belize. Understanding the biological connection between coastal and reef areas in terms of transport of larvae or organisms is a fundamental requirement for managing fisheries and marine biodiversity (Ogden, 1997; Roberts, 1997). This connectivity over scales of tens to hundreds of kilometers by physical processes of water transport is also important with respect to the

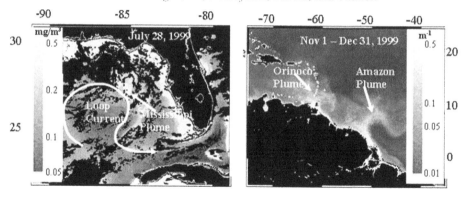

Figure 2. Examples of large river plumes. Left: SeaWiFS chlorophyll image showing the Mississippi plume. Right: SeaWiFS CDOM absorption image (Carder et al., 1999 algorithm) showing the Amazon and Orinoco river plumes.

dispersal of pollutants and pathogens. Satellite observations of ocean color hold great potential to study connectivity by tracing the dispersal of small river plumes. Here we illustrate this potential with some examples of SeaWiFS images.

The Mesoamerican Barrier Reef System extends over 600 km and includes coral reefs in waters of Mexico, Belize, and Honduras. Until 1998, the three atolls off Belize were considered to be oceanic and not affected by terrestrial run-off. Glovers Reef, for example, is located over 40 km from the Belize coast and 100 km from the Honduras-Guatemala coasts. In late October 1998, hurricane Mitch helped to change this perception.

Mitch reached the Caribbean coast of Honduras as a category 5 storm and caused severe loss of property and human life (Hellin et al., 1999). After the hurricane, residents and scientists at Glovers Reef noticed a marked decline in coastal water quality. While this was initially attributed to storm-induced local upwelling and sediment resuspension, satellite data revealed that river water from Honduras reached Glovers Reef (Fig. 3). A turbid plume due to discharge from the Sico, Paulaya, Platano, and Patuca rivers in Honduras extended over 300 km, reaching Chinchorro Bank atoll (Mexico) in three days. Turbid river waters from Honduras/Guatemala rivers reached the Bay Islands in less than one day and Belize's offshore atolls in about five days. All islands experienced high sediment and chlorophyll concentrations for two weeks in November 1998 (Andréfouët et al., 2002).

Presumably the general circulation patterns traced by the colored river plumes also exist when river plumes are not present. This has implications for larval transport and colonization of reefs. The effects of high sediment load also include possible mortality in adult corals and destruction of settlement space for larvae. Fertilizers and pesticides used in the coastal zone of Central America may be advected far offshore via coastal plumes. Satellite sensors help understand these patterns.

"Black water" off Florida. Coastal waters off west Florida are optically complex. Phytoplankton, river-derived CDOM, suspended particles, and shallow bathymetry all play important roles in affecting optical properties such as transparency. These waters regularly experience red tides and significant river discharge.

In early 2002, a massive patch of dark water (Figure 4) caused significant anxiety among local coastal residents, divers, and fishermen (SWFDOG, 2002). The media called this a "black water" event. Daily SeaWiFS data showed that this event lasted over four months. The patch evolved from a senescent bloom stimulated into activity by local river input (SWFDOG, 2002). The color was due to high concentrations of CDOM (CDOM absorption at 440 nm was about 0.1-0.3 m^{-1}, versus ~0.02 m^{-1} in clear waters) and decreased backscattering. A survey showed 5-10 mg chlorophyll m^{-3} in a non-toxic diatom (*Rhizosolenia*) bloom, and red tide (*Karenia brevis*) counts were medium to low ($<1 \times 10^5$ cells l^{-1}). Surveys at two sites in the Florida Keys found that corals and sponges were decimated when the patch flowed between the Keys (Hu et al., 2003b).

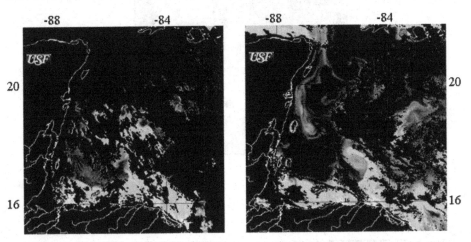

Figure 3. SeaWiFS images of Mesoamerica shortly after hurricane Mitch. Left: river plume (light grey) extending north from the Honduras coast bathed Honduras coral reefs (3 November 1998). Right: river plume extended from Honduras to Belize and Guatemala (14 November 1998).

The optical complexity of this region leads to great uncertainty in SeaWiFS chlorophyll estimates. In such cases, the normalized water-leaving radiance (L_{wn}) can be used as an ocean color index to detect change (Hu et al., 2003b). Figure 5 shows a dark water event recorded by MODIS in 2003. The dark plume originates at the coast south of Charlotte Harbor. The MODIS fluorescence line height (FLH) image (Fig. 5) shows that the plume contained phytoplankton. A field survey conducted near the Dry Tortugas also detected high CDOM levels in the plume, indicating riverine influence.

3.1.2 *Red tide detection*

Some phytoplankton blooms can be toxic, and in many coastal areas harmful algal blooms (HABs) occur repeatedly. Some of these blooms contain very high concentrations of dinoflagellates or diatoms and they are commonly referred to as "red tide" or "brown tide" because they change the color of marine waters. In the west Florida shelf (WFS), toxic red tides of the dinoflagellate *Karenia brevis* (previously *Gymnodinium breve*) occur almost annually. Reports dating back to the late 1800's show that *Karenia* blooms are a natural phenomenon, yet the frequency, duration, and distribution appear to have increased in recent years (Anderson, 1995; Morton and Burklew, 1969; Steidinger and Haddad, 1981; Steidinger and Ingle, 1972; Steidinger and Joyce, 1973; Williams and Ingle, 1972; Walsh and Steidinger, 2001). Red tides lead to fish and marine mammal deaths, and illness and discomfort in humans, representing a hazard to local economies (Anderson, 1995).

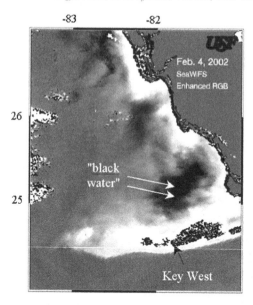

Figure 4. SeaWiFS enhanced RGB image (R: 555 nm, G: 490 nm, B: 443 nm) of the West Florida Shelf (4 February 2002).

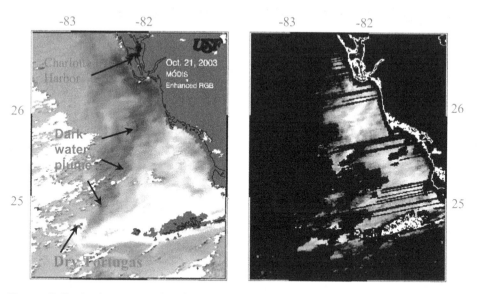

Figure 5. Dark plume extended from Charlotte Harbor to Dry Tortugas (MODIS, 21 October 2003). Left: Enhanced RGB composite (R: 551 nm, G: 488 nm, B: 443 nm); Right: fluorescence line height (FLH) data reveals bloom.

Satellite ocean color sensors have shown great advantage over traditional means to monitor the spatial extent and duration of HABs and the environmental conditions maintaining them. Figure 6 shows an enhanced RGB composite of SeaWiFS bands (R: 555 nm, G: 490 nm, B: 443 nm) that shows a broad band of dark water patches along the

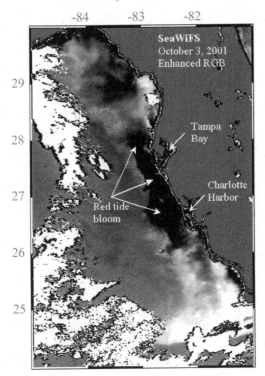

Figure 6. Enhanced SeaWiFS RGB image (R: 555 nm, G: 490 nm, B: 443 nm) shows extensive red tide bloom along the coast from Tampa Bay to Charlotte Harbor, Florida.

WFS. Ship surveys confirmed the presence of a red tide. The dark color is consequence of the strong absorption of blue light by chlorophyll.

At present there is no reliable algorithm to discriminate between red tides and other blooms, such as one dominated by diatoms. For example, during *Alexandrium* blooms that occur in the Gulf of Maine off the northeastern United States, these organisms co-occur with significant quantities of other phytoplankton and therefore waters don't seem to have a particular spectral signature that would allow identification of this HAB.

Chlorophyll concentration has been used as a proxy to build a preliminary operational monitoring tool for *Karenia* blooms off Florida (e.g., Stumpf, 2001). However, reliable estimates of chlorophyll concentration in these coastal waters have not been achieved and bio-optical algorithms are still the focus of intense research. A fundamental difficulty is the presence of CDOM, since a small error in CDOM estimates generates large errors in computed chlorophyll values. A positive chlorophyll concentration anomaly also does not represent proof positive of a red tide bloom. Therefore, false positives are inevitable, particularly during periods of higher river discharge or when other blooms occur. At this stage, the capability to detect an anomalous chlorophyll concentration provides an important advantage in planning field surveys and responses.

Great expectations revolve around identification of phytoplankton species by spectral analyses combined with other environmental observations (Huang and Lou, 2003; Zhang, 2002). A recent study found that the backscattering to chlorophyll ratio is generally lower in *K. brevis* blooms relative to other blooms (Cannizzaro et al., 2004). The proposed explanation is that this species experiences less grazing. How to take this scientific discovery to an algorithm and into an operational system is being investigated.

3.1.3 *Oil spill detection and estuarine water quality monitoring*

Traditional remote sensing techniques to address oil spills in aquatic environments include optical (passive visible and infrared, laser fluorosensors), and passive and active microwave (e.g., Synthetic Aperture Radar, SAR) using aircraft or satellites (see Fingas and Brown, 1997 and 2000). Most of these have been deemed expensive and don't provide the required high-frequency coverage. The MODIS sensors now include bands that generate images at 250 and 500 m spatial resolution. These have great unexploited potential for coastal monitoring because they can be used to study small-scale features. One application is in assessments of oil spills and water quality where 1-km satellite data are often inadequate. Exploring the promise of this technology is important to understand the potential of the VIIRS imaging bands on the NPOESS platforms.

Starting December 2002, the oil industry operating in and around Lake Maracaibo in Venezuela suffered a series of accidents that resulted in oil being spilled from floating oil storage and transfer stations. The spills were recorded in December 2002 by official photography and video of leaking infrastructure, and unofficial recordings continued in January and February 2003. The medium-resolution (250- and 500-m) MODIS Level-1 total radiance imagery from 1 December 2002 to 9 March 2003 showed patterns within Lake Maracaibo related to these spills (Hu et al., 2003c). Figure 7 shows an example MODIS image illustrating the oil slicks. We estimated the oil-contaminated area at ~133 km^2. MODIS provided near-daily coverage at low cost, and allowed an estimate of spill size that would have been difficult otherwise.

3.1.4 *Total suspended sediment estimates in estuaries*

Initial examination of the MODIS medium-resolution bands revealed that they have sufficient sensitivity for coastal water studies. Their sensitivity is comparable to or higher than CZCS, and 2-4 times higher than Landsat-7/ETM+. This high sensitivity suggested that these MODIS data could be used to assess constituents in smaller bodies of water such as estuaries. Figure 8 shows the distribution of total suspended sediment (TSS) in Tampa Bay, Florida, based on MODIS/Aqua data from 22 October 2003. The image was based on a simple empirical algorithm relating reflectance to *in situ* TSS observations. In very shallow areas the bottom probably affected the reflectance, yet the TSS distribution pattern appeared reasonable.

The main problem in automating this process is uncertainties in the MODIS sensor calibration/characterization, since this changes with each image. Maximum benefits will be achieved when remote sensing data are paired with concurrent field data (Hu et al., submitted). A limited number of observations from buoys or other platforms can provide the ground calibration and validation information needed to extrapolate spatially over an estuarine or coastal environment using synoptic MODIS-class imagery. This provides a reasonable means to derive sequences of water quality (chlorophyll, CDOM, TSS) maps of coastal and estuarine waters. With such ancillary data, sophisticated atmospheric correction or bio-optical inversions are not required immediately, even for sensors not designed for ocean color such as Landsat-7/ETM+ (Ouillon et al., 2004).

3.1.5 *Ocean dumping of wastewaters*

During the 2003 rainy season, the southern shores of Tampa Bay, Florida, and particularly Bishop's Harbor were threatened with contamination by nutrient-rich wastewater accumulating within the evaporation ponds of an abandoned phosphate mine. The threat of spillage over or rupture of the containment dykes of the Piney Point

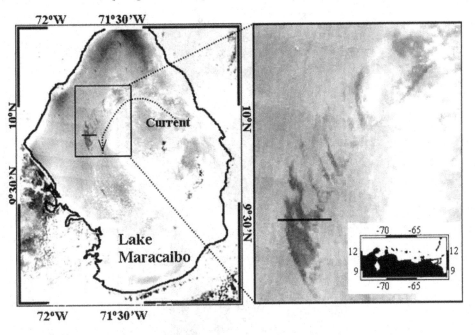

Figure 7. Oil spill slicks in Lake Maracaibo, Venezuela. Left: MODIS 250-m (620-670 nm) image taken on 18 January 2003 (white represents clouds, land, or extreme turbidity; arrow indicates current direction according to observations and numerical simulations). Right: enlarged image of the slicks.

mine led the Florida Department of Environmental Protection to begin an emergency operation on July 20, 2003 to remove water from the abandoned plant, treat it, and disperse it over a wide area offshore in the Gulf of Mexico (Figure 9) under a special permit granted by the Environmental Protection Agency. Preliminary work had determined that the treated waste water would be effectively dispersed in a rapid oceanic current, as opposed to remaining stagnant or possibly threatening the coast if dumped in slow-moving waters nearshore or over the shelf. Water was therefore transported offshore to an area in which MODIS ocean color satellite images detected the Loop Current or its influence.

MODIS direct broadcast (DB) data were used within 1-2 hours of a satellite overpass to monitor the dispersal of the dumped wastewater, and in particular to establish that the circulation patterns did not return the material to shallower waters. Over the period of several days after a discharge event, streamers of slightly enhanced chlorophyll concentration were detected downstream of the discharge area (Fig. 9; Hu and Muller-Karger, 2003). The highly sensitive MODIS allowed discrimination of these slightly higher concentrations (0.01-0.02 mg m^{-3} over surrounding oceanic waters). These small differences would likely not have been detected with ship-board instruments.

3.1.6 *Chlorophyll and primary productivity in an upwelling regime*

In general, little is known about carbon fluxes along continental margins and in the tropics. Studies that focus on small areas are helping shed light on these processes. Here we review some of the findings of a study in the Cariaco Basin, on the continental margin off Venezuela. Cariaco provides a convenient setting in which to study the

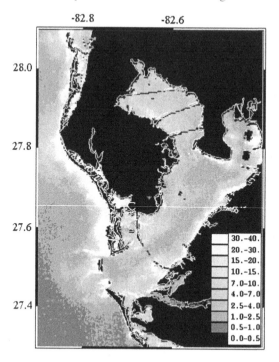

Figure 8. Total suspended solids (TSS, mg L^{-1}) in Tampa Bay, Florida, obtained from the 250-m resolution MODIS/Aqua data (645- and 859-nm; 22 October 2003).

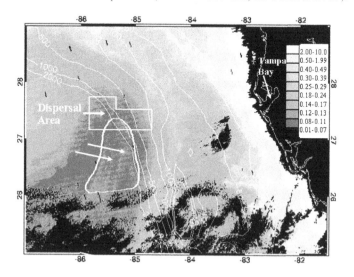

Figure 9. MODIS chlorophyll-a concentration for 5 October 2003. Overlaid on the images are bathymetry contours and location of the dispersal (white box). Faint chlorophyll "streamers" might be due to wastewater discharged.

connection between surface ocean processes and the settling of particulate organic carbon to depth. An oceanographic time series was established within 20 nautical miles of the coast at 10.5°N, 64.66°W to examine such connections. This is the Carbon Retention In A Colored Ocean (CARIACO) time series (Muller-Karger et al., 2004).

Astor et al. (2003) described the variability in wind and hydrographic conditions at the CARIACO station. Muller-Karger et al. (2004) described the temporal variation in phytoplankton concentration. Briefly, the wind is predominantly westward and varies seasonally in intensity. Scalar and westward wind speeds <6 m s^{-1} occurred between approximately July and November. Seasonal maxima >6 m s^{-1} occurred between November and May. The change in the Trade Wind and transient eddies moving through the southern Caribbean Sea lead to variability in the intensity and timing of upwelling and associated blooms. Figure 10 shows the spatial extent of a typical phytoplankton bloom during upwelling and the close correspondence with cold waters in the Cariaco Basin (but not in the Orinoco plume around Gulf of Paria).

We used satellite data to estimate temporal variability in sea-surface temperature (SST) and chlorophyll-a in the region and the size of the upwelling plume forming along the coast and extending over the Cariaco Basin. The SST images were derived from the Advanced Very High Resolution Radiometer (AVHRR) sensors. Each of the SeaWiFS and SST images was mapped to a congruent cylindrical equidistant projection at a spatial resolution of 9x9 km^2 per pixel. Time series of these parameters were derived by sampling a single pixel at the CARIACO station. The upwelling plume area was estimated by counting all pixels where SST was <26°C in AVHRR images, and >0.4 mg Chl-a m^{-3} in SeaWiFS images, and multiplying the area of one pixel by the number of pixels counted. Figure 11 shows the variation of the extent of the upwelling plume, and Figure 12 shows the cycle in phytoplankton growth.

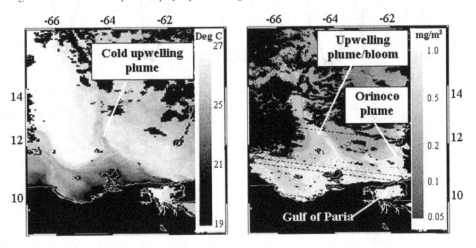

Figure 10. Satellite images of the southeastern Caribbean Sea. Left: AVHRR Sea Surface Temperature (March 1998); Right: SeaWiFS chlorophyll-a (March 1998). Clouds and turbid waters are shown in black.

On average, SeaWiFS overestimated *in situ* chlorophyll, particularly at low concentrations. The reason for this was not clear. The large river plumes from the Orinoco and Amazon Rivers have minimal influence within the Cariaco Basin (Muller-Karger et al., 1989; Muller-Karger and Varela, 1990). During upwelling, it is possible that the upwelling source water is colored by high background CDOM, as suggested for other upwelling zones (Coble et al., 1998; Siegel et al., 2002). Alternatively, CDOM from small local rivers may have affected the basin during September and October (the rainy season).

To assess regional primary productivity, we used the empirical Vertical Generalized Production Model (VGPM; Behrenfeld and Falkowski, 1997a and 1997b; Falkowski et al., 1998). This is considered to be one of the most robust, yet simple to apply, remote sensing primary productivity models (Campbell et al., 2002). We first tested the VGPM using *in situ* observations (optics and Photosynthetically Active Radiation or PAR data, chlorophyll-a, and temperature) and found a need to reformulate the Assimilation Number (PB^{opt}) algorithm included in the model.

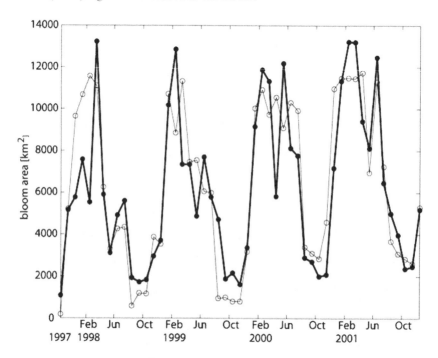

Figure 11. Surface area [km^2] covered by the upwelling plume at the ocean's surface in waters contained by the 100 m isobath within the Cariaco Basin, based on AVHRR (open circles) and SeaWiFS (close circles).

SSTs in the Cariaco Basin, as in most of the tropics, are typically above 21°C. The Behrenfeld and Falkowski (1997a) model generated depressed PB^{opt} at SST>21°C relative to our field data, which showed a positive relationship between PB^{opt} and SST (Muller-Karger et al., 2001). We computed a new 7th order polynomial to model PB^{opt} based CARIACO data for 1996-2001, and merged this with the model of Behrenfeld and Falkowski (1997a) for temperatures <21°C (we arbitrarily weighted each of the points shown in Figure 7 of Behrenfeld and Falkowski (1997a) by a factor of 5 and binned these points with the CARIACO data). This new relationship spans a broader SST range, reaching 29.5°C. Waters outside Cariaco may reach temperatures >30°C, but we had no data to assess the applicability of the polynomial in those conditions.

The new polynomial is:

$$PB^{opt} = A*SST^7 + B*SST^6 + C*SST^5 + D*SST^4 + E*SST^3 + F*SST^2 + G*SST + H,$$

where:

A= (-0.00000019075),   B= 0.00001847528,   C= (-0.00068620340),
D= 0.01212491779,   E= (-0.10165597657), F= 0.33672550197,
G= 0.10336875858,   H= 1.00022697060.

Significant digits are retained. SST range is between 21°C and 29.5°C. Outside this range we revert to the original Behrenfeld and Falkowski (1997a) model.

The VGPM was then run using remotely sensed PAR, SST, and chlorophyll-a, and our PB^{opt} algorithm. We used NASA's SeaWiFS version 4 products (Summer 2002 processing), which included spatial filtering to reduce digitization noise (e.g., Hu et al., 2001), as well as calibration corrections of the blue bands, sophisticated atmospheric correction and a bio-optical algorithm to estimate chlorophyll-a concentration.

The results were tested with independent *in situ* and satellite data from 2002. Figure 12 shows the modifications led to better estimates. Since the VGPM results depend on accurate input biomass, the effects of a biased SeaWiFS chlorophyll-a estimate is confounded here with the new relationship between SST and PB^{opt}.

Both *in situ* and satellite data showed one large 1-3 month long production event per year (Figure 12). Average 6-year integrated annual production was 525 gC m^{-2}.

3.1.7 *The coastal ocean and global carbon cycles*

Satellite ocean color data also provide an important means to examine large-scale phenomena, such as changes in phytoplankton biomass and productivity in the global coastal ocean. Attempts to estimate global production invariably conclude that continental margins (shelves, slopes, rise) play a key role in the global carbon budget (Chen et al., 2001; Jahnke, 1996; Liu et al., 2000). Global new production and particle export from continental margins has been estimated to range anywhere between 3 and 64% of global ocean new production (Liu et al., 2000; Walsh, 1991). Because of scant field data, a systematic assessment of the role of margins is required.

How can we approach this problem? Advances in deriving global estimates of production from satellite data have been substantial, as indicated above. However, other questions like the magnitude of the flux of settling particulate organic carbon remain to be addressed. Ultimately the question of the role of different parts of the ocean as sources or sinks of carbon dioxide requires understanding of this vertical process.

We computed global ocean net primary production (NPP) for 1998 through 2001 at a 9x9 km^2 spatial resolution using the VGPM as described above. Sinking POC flux was estimated using an exponential decay model (Pace et al., 1987), with flux projected to the bottom. Margins were defined as between 50 and 500 m depth.

Figure 13 shows an example of various input parameters and results for August 2001 and global bathymetry. Over the global ocean (>50 m deep), annual NPP averaged 48.2 Pg C (1 Pg = 10^{15} g) in 1998-2001, varying less than ±2% from year to year (the average for waters >20 m was 50.8 Pg C y^{-1}; Table 1). NPP over continental margins averaged 5.2 Pg C y^{-1} (or 7.8 Pg C y^{-1} for 20-500 m), with interannual variability < ±1%.

Previous global ocean estimates derived with the VGPM using the CZCS, SeaWiFS, and AVHRR satellite data are within about 20% those obtained here (see Gregg et al., 2003; Morel and Antoine, 2002). The SeaWiFS data we used were released by NASA in 2003 (version 4 reprocessing) and included corrections to the original calibration, atmospheric correction, and bio-optical algorithms. Behrenfeld et al. (2001) obtained

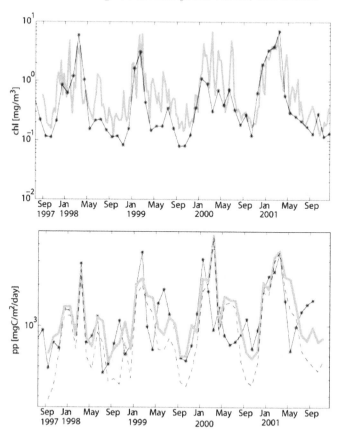

Figure 12. TOP: In situ (thin solid line, asterisks) and SeaWiFS-derived (version 4) chlorophyll-a (thick grey line) at the surface at the CARIACO station. BOTTOM: *In situ* (thin solid line, asterisks) and modeled depth-integrated (100 m) daily NPP, derived with the original (broken line) and the modified (thick grey line) VGPM.

Table 1. Net primary production (Net PP) and bottom particulate organic carbon flux over the global ocean. (1 Pg=10^{15} g.)

	Year	Global [50 m-max]	Global [20 m-max]	Margins [50-500 m]	Margins [20-500 m]	% Margins NPP and Flux [>50 m]	% Margins NPP and Flux [>20 m]
Net PP [Pg]	1998	47.26	49.84	5.19	7.77	11%	16%
	1999	48.89	51.49	5.27	7.87	11%	15%
	2000	48.57	51.11	5.20	7.74	11%	15%
	2001	48.08	50.75	5.24	7.91	11%	16%
Flux [Pg]	1998	0.93	1.62	0.55	1.24	59%	76%
	1999	0.95	1.65	0.56	1.25	59%	76%
	2000	0.94	1.62	0.55	1.23	58%	76%
	2001	0.94	1.66	0.55	1.27	59%	77%

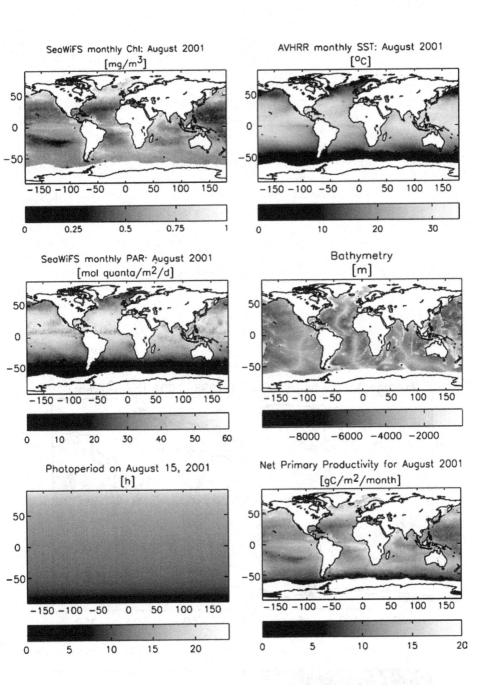

Figure 13. Chlorophyll, sea surface temperature (SST), photosynthetically active radiation (PAR), and photoperiod input for the VGPM and global euphotic-zone NPP for August 2001 (units imply monthly integration).

global NPP values of 54 to 59 Pg C y^{-1} using SeaWiFS data released in 2000. Gregg et al. (2003) estimated NPP for areas >200 m to be around 42.5 Pg C y^{-1} based on SeaWiFS (version 3 reprocessing) for September 1997-June 2002, and 45.3 Pg C y^{-1} based on CZCS for January 1979-June 1986. They suggested that the differences represented a change in marine ecosystems over the intervening period. Conversely, we consider the differences to be associated with disparity in methods to process the data.

Estimates of annual POC flux to the ocean bottom for 1998-2001 were obtained using the Pace et al. (1987) model (Table 1; Figure 14). The four-year average estimate of global POC flux was 0.94 Pg C y^{-1} (or 1.6 Pg C y^{-1} for >20 m), with ±1% variation between 1998 and 2001. Over this period, continental margins exported 0.55 Pg C y^{-1} (or 1.25 Pg C y^{-1} for 20-500 m), with ±1% interannual variation. The Pace et al. (1987) relationship may not be applicable globally (Lampitt and Antia, 1997; Lutz et al, 2002) and factors such as ballasting influence the efficiency of organic carbon flux to the sea floor (Armstrong et al., 2002), but these results provide a first-order approximation of the role of margins.

We estimate that the total amount of particulate organic carbon deposited on the sea floor at depths greater than 500 m is about 0.34 Pg C y^{-1}, with very little interannual variation (Table 1). Most of the deep sea (> 3,000 m) received very small amounts of POC (i.e. $\ll$ 1 g C m^{-2} y^{-1}). There were only small differences in the spatial patterns of POC flux to the ocean bottom from year to year. Clearly, these variations may be significant locally for benthic biota but they don't affect the total global flux. The largest values of POC flux (i.e. > 2 g C m^{-2} y^{-1}) were observed under major divergences (equatorial upwelling zones and upwelling areas along eastern ocean margins), under the South Atlantic and Southern Indian Ocean Currents, beneath the north Pacific

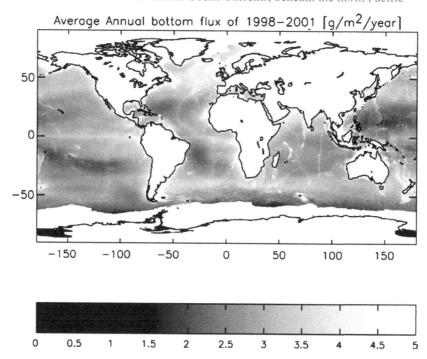

Figure 14. Average annual POC flux to the bottom (1998-2001).

convergence, and on the continental slopes and shelves. Seamounts and substantial portions of the mid-ocean ridges were apparent in the bottom POC flux images; these topographic highs received over twice the amount of POC derived from surface production as the surrounding sea floor, because they intercepted sinking POC. Particularly high flux values were calculated for the mid-ocean ridges, especially south of Iceland (> 3 gC m^{-2} y^{-1}). The POC flux to the bottom of continental margins (values > 4 gC m^{-2} y^{-1}) exceeded anything estimated for the deep ocean. Margins showed both high production and high deposition of POC.

We find that while net primary production on continental margins (50-500 m) contributes ~11% to total global ocean net production (~15% for 20-500 m), POC flux to the ocean bottom on margins accounts for nearly 60% of total global POC flux reaching the sea floor (76% if margins are defined as 20-500 m). Only ~0.8% of the overlying net primary production reaches the sea floor in the open ocean (>500 m depth), in agreement with previous studies (Lutz et al., 2002). Even with the 2.8 Pg C decade^{-1} decrease in open ocean NPP estimated by Gregg et al. (2003) between the late 1980's and the early 2000's, this would lead to a decrease of no more than 0.002 Pg C sequestered at the bottom every year - or a total decrease of 0.02 Pg C over 10 years. By comparison, nearly 10% of the NPP over continental margins, or over 0.5 Pg C y^{-1}, reaches the sea floor.

In addition to phytoplankton production, coastal waters are a source of dissolved organic carbon to the ocean. Walsh et al. (1992) estimated that ~1.7x10^{15} gC y^{-1} DOC may be delivered via rivers and phytoplankton excretion to shelves and slopes, which contributes to the large color signal seen through satellites along margins (Behrenfeld and Falkowski, 1997b; Hochman et al., 1993, 1995; Hojerslev et al., 1996; Muller-Karger et al., 1989).

The satellite data suggest that significant carbon inputs occur to the ocean and ocean bottom along continental margins. Margins probably are not simply a "dirty ring around the oceanic bath tub" of little importance in global carbon budgets. This implies that a systematic review of the importance of margins in global biogeochemical and climate models is required.

4. Conclusions

Coastal ecosystems experience high spatial and temporal variability of physical, chemical, biological and geological parameters. These extensive systems are linked with land ecosystems and also the open ocean. Therefore, coastal zone observation tools need to span synoptic scales and provide data rapidly and repeatedly. Sensors that measure visible radiation reflected by the Earth provide such data and help assess a number of environmental indicators in coastal zones. Many sensors have helped expand our understanding of what can be done with ocean color data. The enhanced sensitivity and coverage of the MODIS sensors on the Terra and Aqua satellites provides a number of significant improvements and an excellent pathway to prepare for future sensors.

In particular, SeaWiFS, MODIS, and a several other international sensors provide near daily, global coverage at 1-km nominal resolution. The MODIS 250 and 500-m sharpening bands provide unprecedented capability for studying small-scale features. This capability opens new possibilities to examine environments such as moderate-sized estuaries. The MODIS model serves as a prototype for the Visible/Infrared Imager/Radiometer Suite (VIIRS), which will fly on the National Polar-Orbiting Operational Environmental Satellite System (NPOESS) and the NPOESS Preparatory Mission (NPP), to be started approximately in 2006. The continuity of these missions ensures an uninterrupted series of observations for the scientific community and the general public. They represent tools to monitor the marine weather in real time, similar to the weather services.

We expect that the emerging integrated ocean observing system, that includes the coastal ocean and, which encompasses global scales, will include mechanisms to assess bio-optical and carbon flux parameters. An important lesson is that the most productive approach to solving scientific or monitoring problems is to combine data from various space- and ground-based sensors. Used with robust models that incorporate processes on the globe's coastal oceans, this emerging observing system will help gain more than the rudimentary understanding that is possible today, and coastal resource management will undergo a revolution.

Of great interest is finding the proper role for remote sensing experts in addressing local problems. This is not easy. The variety of algorithms, data archival and delivery mechanisms, diversity in data formats, lack of information about product quality, and a distinct separation between the research, management, commercial, and public education communities have slowed down migration of key technologies within the scientific community, and especially across to the resource management or educational communities. Continued attempts to provide limited products generated by a centralized process and the lack of tools tailored for use of the data by the general public have led to inconsistent quality and availability of products, difficulty in understanding and using the information, and inadequate feedback regarding the information needed. Typically, satellite images are used as pretty pictures, and their full value as data is not realized.

Our understanding of the earth as a system (ocean, land, and atmosphere), as well as our ability to better monitor and understand the coastal ocean, will continuously improve as advances are made with sensor calibration/characterization, algorithms, data products, and infrastructure to operationally process, archive, and distribute the various data products in a user-friendly environment and format. However, this process requires particular attention to education and to the development of application software. Planning for the new observing systems needs to include the formal and informal education and training programs required to enable design and production of the sensors, the products, and the synthesis of information required by managers, educators, industry, and scientists. The software that facilitates access and use of the data needs to be more widely available and built in a way that users can tailor their analysis and save application settings, examine the environment in real time and from a historical perspective, and construct early warning systems that issue alerts (such as e-mail) when the environment crosses prescribed thresholds. These are necessary steps for better understanding the environment that sustains our life.

5. Acknowledgements

The work described here has been possible with the support of our families and the U.S. National Science Foundation (NSF), the National Aeronautics and Space Administration (NASA), the National Oceanic and Atmospheric Administration (NOAA), the U.S. Geological Survey (USGS) and various state agencies including the Florida Department of Environmental Protection and the University of South Florida. Significant contributions to the analyses presented here were made by Remy Luerssen, Haiying Zhang, Yrene Astor, Glenda Arias, and a long list of colleagues to whom we apologize for not including them here explicitly.

6. References

Ackleson, S.G. and P.M. Holligan. 1989: AVHRR observations of a Gulf of Maine coccolithophorid bloom. Photogrammetric Engineering and Remote Sensing, 55, 473-474.

Anderson, D. M. 1995. ECOHAB: The Ecology and Oceanography of Harmful Algal Blooms, a national research agenda. Woods Hole Oceanog. Institute. Woods Hole, MA. 67 pp.

Andréfouët S., P.J. Mumby M. McField, C. Hu, F.E. Muller-Karger. 2002. Revisiting coral reef connectivity. Coral Reefs, 21:43-48.

Amos, C.L. and T.T. Alfoldi. 1979. The determination of suspended sediment concentration in a macrotidal system using Landsat data, Journal Sedimentary Petrology, 49:159-174.

Arnone, R.A., P. Martinolich, R.W. Gould Jr., R. Stumpf, and S. Ladner. 1998. Coastal optical properties using SeaWiFS. Proceedings Ocean Optics XIV, Kailua Kona, Hawaii, USA, November 10-13.

Armstrong R.A., C. Lee, J.I. Hedges, S. Honjo, and S. G. Wakeham. 2002. A new, mechanistic model for organic carbon fluxes in the ocean based on the quantitative association of POC with ballast minerals. Deep-Sea Research II, 49(1-3):219-236.

Behrenfeld, M.J. and P.G. Falkowski. 1997a. Photosynthetic rates derived from satellite-based chlorophyll concentration. Limnology and Oceanography, 42(1):1-20.

Behrenfeld, M.J. and P.G. Falkowski. 1997b. A Consumer's Guide to Phytoplankton Primary Productivity Models. Limnology and Oceanography, 42(7):1479-1491.

Behrenfeld, M. J, Randerson, J. T., McClain, C. R., Feldman,G. C., Los, S. O., Tucker, C. J., Falkowski, P. J., Field, C. B., Frouin, R., Esaias, W. E., Kolber, D. D., PollackN., H. 2001. Biospheric Primary Production During an ENSO Transition. Science, 291:2594-2597.

Bricaud, A., A. Morel, and L. Prieur. 1981. Absorption by dissolved organic matter in the sea (yellow substance) in the UV and visible domains. Limnology and Oceanography, 26:43-53.

Braga, C. Z. F., Setzer, A. W., and Drude de Lacerda, L., 1993. Water Quality Assessment with Simultaneous Landsat-5 TM Data at Guanabara Bay, Rio de Janeiro, Brazil, Remote Sensing of Environment, 45:95-106.

Campbell, J; Antoine, D; Armstrong, R; Arrigo, K; Balch, W; Barber, R; Behrenfeld, M; Bidigare, R; Bishop, J; Carr, ME; Esaias, W; Falkowski, P; Hoepffner, N; Iverson, R; Kiefer, D; Lohrenz, S; Marra, J; Morel, A; Ryan, J; Vedernikov, V; Waters, K; Yentsch, C; Yoder, J. 2002. Comparison of algorithms for estimating ocean primary production from surface chlorophyll, temperature, and irradiance. Global Biogeochemical Cycles, 16 (3):1035-1035.

Cannizzaro, J.P., K.L. Carder, F.R. Chen, C.A. Heil, and G.A. Vargo. 2004. Bio-optical signatures of red tides on the west Florida shelf. Continental Shelf Research, In Press.

Carder, K. L., Chen, F. R., Lee, Z. P., Hawes, S., and Kamykowski, D. 1999. Semi-analytic MODIS algorithms for chlorophyll a and absorption with bio-optical domains based on nitrate-depletion temperatures. Journal of Geophysical Research, 104(C3), 5403-5421.

Chen, C.-T. A., K. K. Liu, and R. MacDonald. 2001. Continental margins and seas as carbon sink. Scientific Committee on Problems of the Environment - SCOPE. XI General Assembly & Scientific Symposia. September 24-28, 2001. Germany.

Chomko, R. M., and H. R. Gordon. 2001. Atmospheric correction of ocean color imagery: Test of the spectral optimization algorithm with the Sea-viewing Wide Field-of-View Sensor. Applied Optics, 40:2973-2984.

Chomko, R. M., H. R. Gordon, S. Maritorena, and D. A. Siegel. 2003. Simultaneous retrieval of oceanic and atmospheric parameters for ocean color imagery by spectral optimization: a validation. Remote Sensing Environment, 84:208-220.

Coble, P. G., Del Castillo, C. E., and Avril, B., 1998. Distribution and optical properties of CDOM in the Arabian Sea during the 1995 Southwest Monsoon. Deep-Sea Research Part II, 45:2195-2223.

Del Castillo, C., F. Gilbes, P. Coble, and F. E. Muller-Karger. 2000. On the dispersal of riverine colored dissolved organic matter over the West Florida Shelf. Limnology and Oceanography, 45(6):1425-1432.

Ding, K., and Gordon, H. R., 1995. Analysis of the influence of O2 A-band absorption on atmospheric correction of ocean-color imagery. Applied Optics, 34:2068-2080.

Dwivedi, R. M., and Narain, A., 1987. Remote sensing of phytoplankton: An attempt from the Landsat Thematic Mapper, International Journal Remote Sensing, 8(10):1563-1569.

Esaias, W. E., M. R. Abbott, I. Barton, O. B. Brown, J. W. Campbell, K. L. Carder, D. K. Clark, R. H. Evans, F. E. Hoge, H. R. Gordon, W. M. Balch, R. Letelier, and P. J. Minnett. 1998 An overview of MODIS capabilities for ocean science observations, IEEE Transactions Geoscience Remote Sensing, 36:1250-1265.

Falkowski, P.G., R.T. Barber, V. Smetacek. 1998. Biogeochemical controls and feedbaks on ocean primary production. Science, 281:200-206.

Fingas, M., and C. Brown. 1997. Remote sensing of oil spills. Sea Technology, 38:37-46.

Fingas, M., and C. Brown. 2000. Oil-spill remote sensing – An update. Sea Technology, 41:21-26..

Froidefond, J. M., Castaing, P., Jouanneau, J. M., Prud'Homme, R., and Dinet, A., 1993. Method for the quantification of suspended sediments from AVHRR NOAA-11 satellite data, International Journal Remote Sensing, 14(5):885-894.

Gagliardini, D.A., H. Karszenbaum, H., R. Legeckis, and V. Klemas, 1984: Application of LANDSAT MSS, NOAA/TIROS AVHRR, and Nimbus CZCS to study the La Plata River and its interaction with the ocean. Remote Sensing Environment, 15, 21-36.

Gallegos, S.C., T.I. Gray, and M.M. Crawford, 1989: A study into the responses of the NOAA-n AVHRR reflective channels over water targets. Quantitative remote sensing: An economic tool for the Nineties: Proc. IGARSS '89 and 12th Canadian Symp. on Remote Sens., Vol. 2, IEEE Inc., New York, New York, 712-715.

Gordon, H. R., D. K. Clark, J. W. Brown, O. B. Brown, and R. H. Evans. 1982. Satellite measurement of the phytoplankton pigment concentration in the surface waters of a warm core Gulf Stream ring. Journal Marine Research, (40):491-502.

Gordon, H. R., D. K. Clark, J. W. Brown, O. B. Brown, R. H. Evans and W. W. Broenkow. 1983a. Phytoplankton pigment concentrations in the Middle Atlantic Bight: Comparison of ship determinations and CZCS estimates. Applied Optics, 22:20-35.

Gordon, H. R., J. W. Brown, O. B. Brown, R. H. Evans and D. K. Clark. 1983b. Nimbus 7 CZCS: Reduction of its radiometric sensitivity with time. Applied Optics. (22:24):3929 3931.

Gordon, H. R., and Wang, M. 1994. Retrieval of water-leaving radiance and aerosol optical thickness over the oceans with SeaWiFS: a preliminary algorithm. Applied Optics, 33:443-452.

Gregg, W.W., M. E. Conkright, P. Ginoux, J. E. O'Reilly, and N. W. Casey. 2003. Ocean Primary Production and Climate: Global Decadal Changes. Geophysical Research Letters, 30(15):1809, doi:10.1029/2003GL016889.

Hellin, J., M. Haigh, and F. Marks. 1999. Rainfall characteristics of Hurricane Mitch. Nature, 399:316.

Hochman, H.T., F.E. Muller-Karger, and J.J. Walsh. 1993. Interpretation of the Coastal Zone Color Scanner (CZCS) signature of the Orinoco River. Journal Geophysical Research, (99:C4):7,443-7,455.

Hochman, H. T., John J. Walsh, Kendall L. Carder, A. Sournia, Frank E. Muller-Karger. 1995. Analysis of ocean color components within stratified and well-mixed water of the western English Channel. Journal of Geophysical Research, 100(C6):10,777-10,787.

Hojerslev, N. K., N. Holt, and T. Aarup. 1996. Continental Shelf Research, 16:1329-1342.

Hovis, W.A., D. K. Clark, F. Anderson, R. W. Austin, W. H. Wilson, E. T. Baker, D. Ball, H. R. Gordon, J. L. Mueller, S. El-Sayed, B. Sturm, R. C. Wrigley, and C. S. Yentsch. 1980. Nimbus-7 coastal zone color scanner: System description and initial imagery. Science, 210:60-63.

Hu, C., Carder, K. L., and Muller-Karger, F. E. 2000. Atmospheric correction of SeaWiFS imagery over turbid coastal waters: a practical method. Remote Sensing Environment, 74:195-206.

Hu, C., Carder, K. L., and Muller-Karger, F. E., 2001. How precise are SeaWiFS ocean color estimates? Implications of digitization-noise errors. Remote Sensing of Environment, 76(2):239-249.

Hu, C., Z.P. Lee, F. E. Muller-Karger, and K. L. Carder. 2002. Application of an optimization algorithm to satellite ocean color imagery: A case study in Southwest Florida coastal waters. SPIE proceedings 4892. (Ocean Remote Sensing and Applications, edited by R. J. Frouin, Y. Yuan, and H. Kawamura), p 70-79.

Hu, C., Muller-Karger, F. E., Biggs, D. C., Carder, K. L., Nababan, B., Nadeau, D., and Vanderbloemen, J. 2003a. Comparison of ship and satellite bio-optical measurements on the continental margin of the NE Gulf of Mexico. International Journal Remote Sensing, 24:2597-2612.

Hu, C., Hackett, K. E., Callahan, M. K., Andréfouët, S., Wheaton, J. L., Porter, J. W., and Muller-Karger, F. E. 2003b. The 2002 ocean color anomaly in the Florida Bight: A cause of local coral reef decline? Geophys. Res. Lett. 30(3), 1151, doi:10.1029/2002GL016479.

Hu, C., F. E. Muller-Karger, C. Taylor, D. Myhre, B. Murch., A. L. Odriozola, and G. Godoy. 2003c. MODIS detects oil spills in Lake Maracaibo, Venezuela. Eos. Transactions, American Geophysical Union. 84(33), pg. 313 and 319.

Hu, C., E.T. Montgomery, R.W. Schmitt, and F.E. Muller-Karger. 2004. The Amazon and Orinoco River plumes in the tropical Atlantic and Caribbean Sea: Observation from space and S-PALACE floats. Deep-Sea Research Part II, In Press.

Hu, C., and F. E. Muller-Karger. 2003. MODIS monitors the Florida's ocean dispersal of the Piney Point Phosphate treated wastewater. The Earth Observer. Vol. 15. No. 6. p.21-23.

Hu, C., Z. Chen, T. Clayton, P. Swarzenski, and F. E. Muller-Karger. 2004. On the operational monitoring of estuary water quality with MODIS medium-resolution bands: Initial results from Tampa Bay, Florida. Remote Sensing Environment, Accepted.

Huang, W. G., and X. L. Lou, 2003. AVHRR detection of red tides with neural networks. International Journal Remote Sensing, 24: 1991-1996.

IGOS. An Ocean Theme for the IGOS Partnership. Final Report from the Ocean Theme Team. Integrated Global Observing Strategy Partnership. January 2001. 37 pp.

IOCCG. 2000. Remote Sensing of Ocean Colour in Coastal,and Other Optically-Complex Waters. International Ocean-Colour Coordinating Group,No.3: General Introduction. (S. Sathyendranath, Ed.). International Ocean-Colour Coordinating Group (IOCCG) Report. pg. 5-22.

Jahnke, R. A. 1996. The global ocean flux of particulate organic carbon: Areal distribution and magnitude. Global Biogeochemical Cycles, (10:1):71-88.

Jerlov, N.G., 1971. Optical Studies of Ocean Waters. Reports of the Swedish Deep-Sea Expedition, Vol. III. Physics and Chemistry, No. 1.

Khorram, S., 1981. Water quality mapping from Landsat digital data, International Journal Remote Sensing, 2(2):145-153

Kirk, J. T. O. 1983. Light and Photosynthesis in Aquatic Environments. Cambridge University Press. 401 pp.

Koblenz-Mishke, O. J., V. V. Volkovinsky, and J. G. Kabanova. 1970. Plankton primary production of the world ocean. In: W. S. Wooster (ed.). Scientific Exploration of the Southern Pacific. National Academy of Science, Washington, DC. 183-193.

Lampitt R. S., and A. N. Antia. 1997. Particle flux in deep seas: regional characteristics and temporal variability. Deep-Sea Research I, 44(8):1377-1403.

Lee, Z., Carder, K. L., Mobley, C. D., Steward, R. G., and Patch, J. S.. 1999. Hyperspectral remote sensing for shallow waters: 2. Deriving bottom depths and water properties by optimization. Applied Optics, 38:3831-3843.

Lee, Z., K. L. Carder, and R. A. Arnone. 2002. Deriving inherent optical properties from water color: a multiband quasi-analytical algorithm for optically deep waters. Applied Optics, 41:5755-5772.

Letelier, R. M., and M. R. Abbott. 1996. An analysis of chlorophyll fluorescence algorithms for the Moderate Resolution Imaging Spectroradiometer (MODIS). Remote Sensing of Environment, 58:215-223.

Liu, K. -K., K. Iseki, and S.-Y. Chao. Continental margin carbon fluxes. 2000. In: The Changing Ocean Carbon Cycle. International Geosphere-Biosphere Programme Book Series. R. B. Hanson, H. Ducklow, and J. G. Field (Eds.), University Press, Cambridge. Vol. 5. Ch. 7. 187-239.

Lohrenz, S. E., Dagg, M. J., and Whiteledge, T. E., 1990. Enhanced primary production at the plume/oceanic interface of the Mississippi River. Continental Shelf Research, 10:639-664.

Lutz, M., R.L. Dunbar, and K. Caldeira. 2002. Regional variability in the vertical flux of particulate organic carbon in the ocean interior. Global Biogeochemical Cycles, 16:91-110.

Lyzenga, D. R., 1981. Remote sensing of bottom reflectance and water attenuation parameters in shallow water using aircraft and Landsat data, International Journal Remote Sensing,, 2(1):71-82.

Maritorena, S., D. A. Siegel, and A. R. Peterson. 2002. Optimization of a semianalytical ocean color model for global-scale applications. Applied Optics, 41:2705-2714.

Maul, G. A., and H.R. Gordon. 1975. On the use of the Earth Resources Technology Satellite (LANDSAT-1) in optical oceanography, Remote Sensing of Environment, 4:95-128.

McClain, C. R., Cleave, M. L., Feldman, G. C., Gregg, W. W., Hooker, S. B., and Kuring, N., 1998. Science quality SeaWiFS data for global biosphere research. Sea Technology, 39:10-16.

McCulloch M., S. Fallon, T. Wyndham, E. Hendy, J. Lough, and D. Barnes. 2003. Coral record of increased sediment flux to the inner Great Barrier Reef since European settlement. Nature, 421:727-730.

Morel, A. 1980. In water and remote measurements of ocean color. Boundary Layer Meteorology, (18):177 201.

Morel, A., and D. Antoine. 2002. Small Critters-Big Effects. Science, 296:1980-1982.

Morel A., and L. Prieur. 1977. Analysis of variations in ocean color. Limnology and Oceanography, 22:709-722.

Morton, R.A. and M.A. Burklew, 1969: Florida shellfish toxicity following blooms of dinoflagellate *Gymnodium breve*. Florida Department of Natural Resources Marine Laboratory, Techical Series No. 60, 26 pp.

Muller-Karger, F. E., R. Varela, R. Thunell, Y. Astor, H. Zhang, and C. Hu. 2004. Processes of Coastal Upwelling and Carbon Flux in the Cariaco Basin. Deep-Sea Research II, In Press.

Muller-Karger, F. E., R. Varela, R. Thunell, M. Scranton, R. Bohrer, G. Taylor, J. Capelo, Y. Astor, E. Tappa, T. Y. Ho, and J. J. Walsh. 2001. Annual Cycle of Primary Production in the Cariaco Basin: Response to upwelling and implications for vertical export. Journal of Geophysical Research, 106(C3):4527-4542.

Muller-Karger, F. E., P. L. Richardson, and D. McGillicuddy. 1995. On the offshore dispersal of the Amazon's Plume in the North Atlantic. Deep-Sea Research I, 42(11/12):2127-2137.

Muller-Karger, F. E., and R. Varela. 1990. Influjo del Rio Orinoco en el Mar Caribe: observaciones con el CZCS desde el espacio. Memoria. Sociedad de Ciencias Naturales La Salle. Caracas, Venezuela. Tomo IL, numero 131-132; Tomo L, numero 133-134. 361-390.

Muller-Karger, F.E., J.J. Walsh, R.H. Evans, and M.B. Meyers. 1991. On the Seasonal Phytoplankton Concentration and Sea Surface Temperature Cycles of the Gulf of Mexico as Determined by Satellites. Journal of Geophysical Research, 96(C7):12645-12665.

Muller-Karger, F. E., C. R. McClain, T. R. Fisher, W. E. Esaias, and R. Varela. 1989. Pigment distribution in the Caribbean Sea: observations from space. Progress in Oceanography, 23:23-69.

Ogden, J. C. 1997. Marine managers look upstream for connections. Science, 278:1414–1415.

O'Reilly, J.E., S. Maritorena, D.A. Siegel, M.C. O'Brien, D.Toole, F.P. Chavez, P. Strutton, G.F. Cota, S.B. Hooker, C.R. McClain, K.L. Carder, F. Muller-Karger, L. Harding, A. Magnuson, D. Phinney, G.F. Moore, J. Aiken, K.R. Arrigo, R.Letelier, and M. Culver. 2000. Ocean Chlorophyll-a Algorithms for SeaWiFS, OC2 and OC4: version 4. SeaWiFS Postlaunch Calibration and Validation Analyses, Part 3. NASA Tech. Memo. 2000-206892, Vol. 11, S.B. Hooker and E.R. Firestone, Eds., NASA Goddard Space Flight Center, Greenbelt, Maryland, 9-23.

Ouillon, S., P. Douillet, and S. Andréfouët. 2004. Coupling satellite data with in situ measurements and numerical modeling to study fine suspended-sediment transport: A study for the lagoon of New Caledonia. Coral Reefs, DOI: 10.1007/s00338-003-0352-z.

Pace, M., G. Knauer, D. Karl and J. Martin. 1987. Primary production, new production and vertical flux in the eastern Pacific Ocean. Nature, 325:803-804.

Platt, T., and A. W. Herman. 1983. Remote sensing of phytoplankton in the sea: surface-layer chlorophyll as an estimate of water-column chlorophyll and primary production. International Journal Remote Sensing, 4(2):343-351.

Roberts, C. M. 1997. Connectivity and management of Caribbean coral reefs.Science, 278:1454-1457.

Ruddick, K.G., F. Ovidio, and M. Rijkeboer. 2000. Atmospheric correction of SeaWiFS imagery for turbid coastal and inland waters. Applied Optics, 39:897-912.

Siegel, D.A., S. Maritorena, N.B. Nelson, D.A. Hansell, and M. Lorenzi-Kayser, 2002. Global distribution and dynamics of colored dissolved and detrital organic materials. Journal of Geophysical Research, 107(C12):3228-3242.

Siegel, D.A., M. Wang, S. Maritorena, and W. Robinson. 2000. Atmospheric correction of satellite ocean color imagery: The black pixel assumption. Applied Optics, 39:3582-3591.

Small, C., and R. Nicholls. 2003. Global analysis of human settlements on coastal zones. Journal of Coastal Research., 19(3):584-599.

Smith, R. C., and K. S. Baker. 1978. The bio-optical state of ocean waters and remote sensing. Limnology and Oceanography, 23(2):247-259.

Smith, R. C., and K. S. Baker. 1982. Oceanic chlorophyll concentrations as determined by satellite (Nimbus-7 Coastal Zone Color Scanner). Marine Biology, 66:269-279.

Smith, R. C., R. W. Eppley, and K. S. Baker. 1982. Correlation of primary production as measured aboard ship in southern California coastal waters and as estimated from satellite chlorophyll images. Marine Biology, 66:281-288.

Steidinger, K.A. and K. Haddad. 1981: Biologic and hydrographic aspects of red tides. Bioscience, 31(11):814-819.

Steidinger, K.A. and E.A. Joyce. 1973: Florida red tide. Florida Department of Natural Rescources Report No. 17, Marine Research Department, St. Petersburg, Florida, 26 pp.

Steidinger, K.A. and R.M. Ingle, 1972: Observations on the 1971 summer red tide in Tampa Bay. Florida Environmental Letters, 3(4):271-277.

Stumpf, R. P. and M. A. Tyler. 1988. Satellite detection of bloom and pigment distributions in estuaries, Remote Sensing of Environment, 24:385-404.

Stumpf, R.P. and J.R. Pennock. 1991. Remote estimation of the diffuse attenuation coefficient in a moderately turbid estuary, Remote Sensing of Environment, 38:183-191.

Stumpf, R. P. 2001. Applications of satellite ocean color sensors for monitoring and predicting Harmful Algal Blooms, Human and Ecological Risk Assessment, 7:1363-1368.

Stumpf, R. P., R. A. Arnone, R. W. Gould, Jr., P. M. Martinolich, and V. Ransibrahmanakul. 2003. A partially coupled ocean-atmosphere model for retrieval of water-leaving radiance from SeaWiFS in coastal waters. SeaWiFS postlaunch technical report series. Vol. 22: Algorithm updates for the fourth SeaWiFS data processing. (S. B. Hooker and E. R. Firestone, Eds). pg. 51-59.

SWFDOG. 2002. Satellite images track 'black water' event off Florida coast. EOS Trans. AGU 83:281-285.

Takahashi, et al., 1997. Global air-sea flux of CO2: An estimate based on measurements of sea-air pCO2 difference. Volume 94, Proceedings of the National Academy of Sciences, USA, pg. 8929-8299, August 1997.

Tassan, S. 1987. Evaluation of the potential of the Thematic Mapper for marine application, International Journal Remote Sensing, 8(10):1455-1478.

Tassan, S. 1993. An improved in-water algorithm for the determination of chlorophyll and suspended sediment concentration from Thematic Mapper data in coastal waters, International Journal Remote Sensing, 14(6):1221-1229.

Tassan, S. and B. Sturm. 1986. An algorithm for the retrieval of sediment content in turbid coastal waters from CZCS data, International Journal Remote Sensing, 7(5):643-655.

Walsh, J. J. 1991. Importance of continental margins in the marine biogeochemical cycling of carbon and nitrogen. Nature, 350:53-55.

Walsh, J.J., K.L. Carder, and F.M. Muller-Karger. 1992. Meridional fluxes of dissolved organic matter in the North Atlantic Ocean. Journal Geophysical Research, 97:15625-15637.

Walsh, J. J., and K. A. Steidinger, 2001, Saharan dust and Florida red tides: The cyanophyte connection, Journal Geophysical. Research, 106:11,597-11,612.

Wang, M., and S.W. Bailey. 2001. Correction of sun glint contamination on SeaWiFS ocean and atmosphere products. Applied Optics, 40:4790-4798.

Williams, J., and R.M. Ingle. 1972: Ecological note on *Gonyaulax monilata* (Dinophyceae): Blooms along the West Coast of Florida. Florida Department of Natural Resources Marine Laboratory Leaflet Series, 1(5), 12 pg.

Yoder, J. A., C. R. McClain, J. O. Blanton, and L.-Y. Oey. 1987. Spatial scales in CZCS-chlorophyll imagery of the southeastern U.S. continental shelf. Limnology and Oceanography, 32(4):929-941.

Zhang, H. 2002. Detecting Red Tides on the West Florida Shelf by Classification of SeaWiFS Satellite Imagery. Master's thesis, Department of Computer Science and Engineering, University of South Florida.

Chapter 6

BIO-OPTICAL PROPERTIES OF COASTAL WATERS

[1]EURICO J. D'SA AND [2]RICHARD L. MILLER
[1]*Lockheed Martin Space Operations, Remote Sensing Directorate, Stennis Space Center, MS, 39529 USA*
[2]*National Aeronautics and Space Administration, Earth Science Applications Directorate, Stennis Space Center, MS, 39529 USA*

1. Introduction

   Coastal waters are generally characterized by large variations in biological, physical and chemical properties and are an important source of dissolved and particulate matter for the open ocean. In highly productive coastal waters, a significant fraction of organic matter produced in surface waters reach the bottom sediments of coastal margins and become buried (Hedges and Keil, 1995). River dominated coastal margins present additional complexities due to large inputs of freshwater that vary seasonally and carry constituents such as suspended sediments, organic matter, and nutrients. As such, uncertainties exist in the contribution of coastal carbon flux rates to the biogeochemical carbon cycle (Wollast, 1991). Repeated, synoptic coverage with satellite remote sensing provides the capability for understanding and monitoring of many coastal processes. The development of numerous bio-optical algorithms provides a bridge between ocean color remote sensing and concentrations of seawater constituents and has been useful, for example, in studies of oceanic carbon cycling through its ability to obtain estimates of phytoplankton biomass and its derived primary productivity (Platt et al., 1988; Muller-Karger et al., 1991). Monitoring water quality parameters such as clarity, suspended sediment concentrations and presence of harmful algal blooms are also some of the other applications of ocean color remote sensing (Carder and Steward, 1985; Tassan, 1994; Kahru and Mitchell, 1998; Stumpf et al., 1999). However, the complexity of optical properties still presents difficulties in the use of ocean color remote sensing in coastal waters.
   Oceanic waters have been commonly classified as Case 1 or Case 2 waters (Morel and Prieur, 1977; Gordon and Morel, 1983). Case 1 waters have been defined as those for which phytoplankton and their associated derived products are the main influence on the optical field, while in Case 2 waters, additional seawater constituents such as suspended sediments and dissolved organic matter that may not co-vary with phytoplankton, influences the optical field. Coastal and inland water bodies are optically complex and often referred to as Case 2 waters (Sathyendranath, 2000). Until recently, most studies have focused on Case 1 waters where generally reliable estimates of phytoplankton chlorophyll are being obtained. In contrast, estimates of various in-water constituents from ocean color remote sensing of Case 2 waters are often not accurate due to poor performance of standard algorithms (Darecki and Stramski, 2004). For example, in interpreting the CZCS data of the Orinoco River plume, Hochman et al., 1994 concluded that as much as 50% of the remotely sensed chlorophyll biomass within the plume could be an artifact due to colored dissolved organic matter (CDOM). Large variations in CDOM-to-chlorophyll ratios were reported for measurements conducted in

R.L. Miller et al. (eds.), Remote Sensing of Coastal Aquatic Environments, 129–155.
© 2005 *Springer.*

the northern Gulf of Mexico (Carder et al., 1989) with potential effects on both empirical and semi-analytic ocean color algorithms. In many coastal regions including those influenced by rivers, the often complex interactions of the physical, chemical and biological processes and lack of knowledge of the optical properties have hampered the routine use of ocean color remote sensing in these waters.

One of the keys to improving the accuracy of estimates of the seawater constituents from ocean color remote sensing is the need for a better understanding of the coastal bio-optical properties (Bukata et al., 1981; Carder et al., 1989,1991; Sathyendranath et al., 1989; Gallegos et al., 1990; Blough et al., 1993; Doerffer and Fisher, 1994; Bukata et al., 1995; Arrigo et al., 1998; Gould et al., 1999; Reynolds et al., 2001; Twardowski and Donaghay, 2001; Babin et al., 2003; Hamre et al., 2003). The optically significant seawater constituents such as chlorophyll (chlorophyll *a*), CDOM (generally denotes the colored material that passes through a 0.2 μm filter) and suspended particulate material (SPM) determine the inherent optical properties (IOPs) such as spectral absorption $a(\lambda)$, scattering $b(\lambda)$, and backscattering $b_b(\lambda)$ coefficients that are important for characterizing the marine optical field. In turn, these variables influence the apparent optical properties (AOPs) that define the in-water radiation field. The three most commonly studied AOPs of interest in remote sensing include the diffuse attenuation coefficient $K_d(\lambda)$, irradiance reflectance $R(\lambda)$, and the remote sensing reflectance $R_{rs}(\lambda)$. Empirical and semi-analytic bio-optical algorithms have been developed mainly for oceanic Case 1 waters to estimate both these IOPs as well the seawater constituents from the apparent optical properties (AOPs) such as remote sensing reflectance that is generally derived from ocean color satellite sensors (Gordon et al., 1988; O'Reilly et al., 1998; Carder et al., 1999). Empirical band-ratio algorithms, for example, are derived by statistical regression of radiance or reflectance ratios versus chlorophyll. The more complicated semi-analytic ocean color algorithms (Garver and Siegel, 1997; Carder et al., 1999) have resulted from a better understanding of the relationship between remote sensing reflectance, inherent optical properties such as backscattering and absorption, and in-water constituents. Bio-optical studies in various coastal waters have demonstrated the need for local empirical (Kahru and Mitchell, 2001; D'Sa et al., 2002a) and semi-analytic algorithms (Reynolds et al., 2001) in order to obtain better estimates of various in-water constituents.

Here we examine various relationships between the AOPs, IOPs and seawater constituents, generally limiting our discussion to coastal waters. Recent developments in instrumentation for *in situ* measurements of IOPs such as absorption and backscattering coefficients of seawater have provided better insight into the optical properties of coastal waters. We review some of the instrumentation and methods used in studies of coastal waters and examine linkages between the physical and bio-optical properties and the relationship between the various seawater constituents and the IOPs and AOPs. We interpret R_{rs} spectra and examine semi-analytic and empirical algorithms used for estimating the seawater constituents such as chlorophyll and CDOM (see also Twardowski et al., Chapter 4; Muller-Karger et al., Chapter 5; Del Castillo, Chapter 7). Finally, we identify potential areas of research that need to be addressed to obtain a better understanding of the bio-optical properties of coastal waters.

2. Background

The process of light absorption and scattering by the various seawater constituents such as particulate and dissolved organic matter are described by the inherent optical properties (IOPs). The apparent optical properties (AOPs) of seawater such as the vertical attenuation coefficient for downward irradiance or the irradiance reflectance are mainly determined by the inherent optical properties (Kirk, 1994). However, in comparison to the IOPs (e.g., absorption coefficient $a(\lambda,z)$, and backscattering coefficient $b(\lambda,z)$), routine radiometric measurements of the underwater light field have allowed for the easier calculation of the AOPs such as the diffuse attenuation coefficient for downwelling irradiance K_d, the irradiance reflectance R, and the remote sensing reflectance, R_{rs}.

2.1 SEMIANALYTIC BIO-OPTICAL MODELS

Various models have been developed to relate IOPs to or determine IOPs from AOPs (Gordon et al., 1975; Morel and Prieur, 1977; Gordon and Morel, 1983; Kirk, 1984, 1991; Gordon, 1991; Carder et al., 1999; Ciotti et al., 1999; Stramska et al., 2000). The AOPs commonly used in these models include $K_d(\lambda,z)$, the irradiance reflectance $R(\lambda,z)$ and the remote sensing reflectance R_{rs}. Some of these models are based on numerical simulations of the underwater radiative transfer and involve various types of assumptions (e.g., the spectral behavior of IOPs) or require knowledge of surface illumination or sea state as model inputs. Next we present some basic relationships that have been developed relating AOPs, IOPs and some seawater constituents.

2.1.1 *Remote sensing reflectance and IOPs*

The AOP that defines the color of the ocean and is of interest to remote sensing is the irradiance reflectance R (just beneath the sea surface) and is given by

$$R(\lambda,0^-) = E_u(\lambda,0^-)/E_d(\lambda,0^-), \tag{1}$$

where E_u is the upwelling irradiance (radiant flux per unit surface area) and E_d is the downwelling irradiance at null depth (denoted by 0^-).

Many studies have modeled the behavior of the underwater light field based on the absorption and scattering properties of the medium. One approach uses the Monte Carlo method to simulate the light field in waters with specified optical properties and environmental conditions such as the angle of the incident light field and sea conditions (Kirk, 1984). The results of these studies have shown that the reflectance at the sea surface is a function of the backscattering coefficient b_b and the absorption coefficient a (Gordon et al., 1975; Morel and Prieur, 1977; Gordon and Morel, 1983; Gordon et al., 1988), and can be approximated as

$$R(0^-) = fb_b/(a + b_b), \tag{2}$$

where f is a variable that depends upon the solar zenith angle, the optical properties of seawater, and the wavelength of light (Morel and Gentili, 1991; 1996). These IOPs and AOPs are generally functions of light wavelength λ and depth z and will be used when required for clarity.

However, since there is a directional component to the light observed by the remote sensor, it is common to deal with the term remote sensing reflectance, R_{rs}, which is defined just below the sea surface as

$$R_{rs}(\lambda,0^-) = L_u(\lambda,0^-)/E_d(\lambda,0^-), \, [\text{sr}^{-1}] \tag{3}$$

where L_u is the upwelling or water leaving radiance (radiant flux per unit surface area per steradian, $\text{Wm}^{-2}\text{nm}^{-1}\text{sr}^{-1}$) and is a function of both zenith and azimuth angles. For most oceanic waters the albedo for single scattering $(\varpi = b/(b+a))$ may not be sufficiently high for the upwelling radiances below the surface or the water-leaving radiances to form an isotropic field (Morel and Gentili, 1993, 1996). A proportionality factor Q, which relates a given upwelling radiance $L_u(\theta, \phi)$ to the upwelling irradiance E_u, expresses this nonisotropic character of the light field (here θ is the nadir angle, ϕ the azimuth angle, and $Q = E_u/L_u$). Based on the definition of Q and on Eqs. (1) and (3), Q relates R_{rs} to R (just beneath the surface) as

$$Q(\theta, \phi, \lambda) = R(\lambda, 0^-)/R_{rs}(\theta, \phi, \lambda, 0^-), \text{ [sr]}. \tag{4}$$

From Eqs. (2) and (4), we can express R_{rs} just below the sea surface as

$$R_{rs}(0^-) = (f/Q) * b_b/(a + b_b), \text{ [sr}^{-1}\text{]}. \tag{5}$$

The subsurface values of $R_{rs}(\lambda, 0^-)$ are extrapolated to above water (Mobley 1994) and generally given by the approximation

$$R_{rs} = 0.54 \, (f/Q) * b_b/(a + b_b), \text{ [sr}^{-1}\text{]}. \tag{6}$$

From Eq. (6) we observe the remote sensing reflectance as being related to the scattering and absorption properties of the water which are in turn determined by the seawater constituents such as phytoplankton biomass, CDOM and detrital or nonalgal particles. The value of the ratio f/Q has been shown to vary less than f and Q individually for oceanic waters (Morel and Gentili, 1993), and was estimated from a limited data set to have values between 0.09 and 0.12 (determined at the SeaWiFS satellite wavebands) for a river dominated coastal region (D'Sa and Miller 2003). The basic expression (Eq. (6)) is generally applicable to all waters including coastal waters.

The coefficients a and b_b in Eq. (6) are the sum due to contributions from the various individual seawater components including pure water. An example of absorption by phytoplankton, CDOM and pure water (Pope and Fry, 1997) in the spectral range 400 to 750 nm for a coastal location is shown in Fig. 1. The corresponding R_{rs} spectrum (Fig. 1, right panel) for the same location shows low reflectance in the blue wavebands that can be attributed to high absorption by phytoplankton and CDOM, while high reflectance in the green (around 555 nm) may be attributed to scattering by suspended material. A peak in the R_{rs} spectrum observed at around 683 nm can be attributed to phytoplankton chlorophyll fluorescence.

2.1.2 *Diffuse attenuation coefficient and IOPs*

The diffuse attenuation coefficient for downwelling irradiance $K_d(\lambda, z)$ is calculated from the spectral downwelling irradiance profiles $E_d(\lambda, z)$ and is defined as the slope of natural log transformation of downwelling irradiance as a function of depth. The vertical diffuse attenuation coefficient K_d has been related to IOPs (Gordon et al., 1975; Gordon, 1989) as

$$K_d = (a + b_b)/\mu_d, \text{ [m}^{-1}\text{]}, \tag{7}$$

where μ_d is the cosine of the mean downwelling light field. The value of μ_d has been shown to vary from 0.65 to 0.85 and depends on water depth, solar zenith angle, and the ratio of backscattering to absorption (Bannister, 1992). To a first approximation, K_d has been shown to be directly proportional to the absorption coefficient, and in phytoplankton dominated waters to chlorophyll concentrations. For example, a simple correlation analysis between $K_d(443)$ and absorption at 443 nm indicates that for some surface coastal waters, $K_d(443)$ is directly related to the absorption coefficient at the same waveband (Fig. 2).

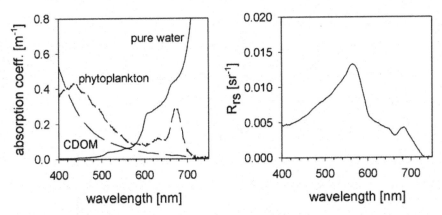

Figure 1. An example of phytoplankton and CDOM absorption spectra for surface water samples at a coastal location along with a pure water absorption spectrum (left panel). Corresponding remote sensing reflectance R_{rs} determined at the same location from above water radiance measurements (right panel).

Determination of the AOPs, K_d and R_{rs} from measurements of the underwater light field have been used to obtain estimates of absorption and backscattering coefficients. Furthermore, knowledge of the spectral characteristics of absorption, scattering and backscattering of the various seawater constituents are used in the parameterization of these IOPs in terms of these constituents. For oceanic waters, parameterization of the IOPs in terms of chlorophyll has enabled the retrieval of chlorophyll and other associated variables from the AOPs such as R_{rs}. However, these parameterizations may not be applicable in some coastal waters where the various seawater constituents do not

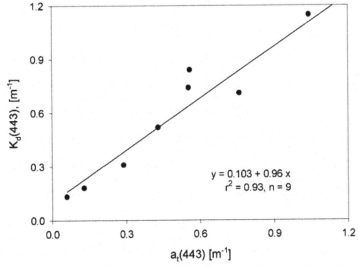

Figure 2. Diffuse attenuation coefficient $K_d(443)$ versus absorption $a_t(443)$ (particulate plus CDOM) for various locations in a coastal environment.

co-vary with chlorophyll. Nonetheless, we briefly review some of these absorption and scattering models and examine them in the context of their applicability to coastal waters.

2.1.3 *Parameterization of absorption*

The major light absorbing constituents in seawater are phytoplankton, detrital or nonalgal particles and CDOM. The total absorption a is the additive sum of contributions by pure water a_w (Pope and Fry, 1997) phytoplankton a_{phy}, detrital or nonalgal particles a_{nap} and CDOM absorption a_{cdom} and can be written as

$$a(\lambda) = a_w(\lambda) + a_{phy}(\lambda) + a_{nap}(\lambda) + a_{cdom}(\lambda), \ [m^{-1}] \tag{8}$$

where particle absorption $a_p(\lambda)$ is due to phytoplankton and nonalgal particles and can be written as

$$a_p(\lambda) = a_{phy}(\lambda) + a_{nap}(\lambda). \tag{9}$$

Phytoplankton absorption can account for a large fraction of the total absorption in oceanic waters and is usually defined in terms of its pigment-specific coefficients as

$$a_{phy}(\lambda) = [Chl] \ a^*_{phy}(\lambda), \tag{10}$$

where $a^*_{phy}(\lambda)$ is the chlorophyll-specific absorption coefficient of phytoplankton (m^2 mg Chl^{-1}) and is not a constant. The variability in the phytoplankton absorption coefficients per unit chlorophyll concentration or $a^*_{phy}(\lambda)$ can be attributed to various factors such as pigment composition, packaging, light history, cell size, nutrients, and temperature (Mitchell and Kiefer, 1988; Bricaud and Stramski, 1990). Many studies that have documented the variability in absorption properties of phytoplankton in natural populations have been mainly confined to oceanic waters (Mitchell and Kiefer, 1988; Bricaud and Stramski, 1990; Bricaud et al., 1995). Results from a large data set obtained from different regions of the world (covering the chlorophyll concentration range 0.02-25 mg m^{-3}) showed the a^*_{phy} values to decrease rather regularly (0.18 to 0.01 m^2 mg^{-1}) from oligotrophic to eutrophic waters (Bricaud et al., 1998).

CDOM absorption has been parameterized in the visible range (400 to 700 nm) with the spectrum typically fitting an exponential function (Bricaud et al., 1981) as

$$a_{cdom}(\lambda) = a_{cdom}(\lambda_0)e^{[S(\lambda - \lambda_0)]}. \tag{11}$$

where S (nm^{-1}) is the slope of the exponential decrease with wavelength and $a_{cdom}(\lambda_0)$ is the absorption coefficient for CDOM at a reference wavelength (e.g., 400 nm). While values between -0.017 and -0.014 nm^{-1} have been used in bio-optical models, the value of S can vary over a wider range (e.g., from -0.01 to -0.02) (Kirk, 1994; Carder et al., 1989; Miller et al., 2002), with values generally lower in coastal waters and increasing offshore (Vodacek et al., 1997).

Spectral absorption by detrital or nonalgal particles has been found to be similar to CDOM absorption and a value of 0.01 m^{-1} for the spectral slope S_{nap} has been reported for some inland waters (Roesler et al., 1989). Semianalytic bio-optical models (e.g., Carder et al., 1999; Ciotti et al., 1999) have combined the contributions by these two seawater constituents.

Based on field data and the assumption of CDOM and detrital particle absorption co-varying with chlorophyll, a well known bio-optical model for spectral absorption coefficient of Case 1 waters developed by Prieur and Sathyendranath (1981) and further modified by Morel (1991) is parameterized in terms of chlorophyll concentration and is given by

$$a(\lambda) = [a_w(\lambda) + 0.06a^*_{phy}(\lambda)[Chl]^{0.65}][1 + 0.2 \ exp^{(-0.014(\lambda - 440))}]. \tag{12}$$

The applicability of this model in coastal waters will depend on whether absorption by CDOM and nonalgal particles co-vary with chlorophyll, and on whether the other assumptions (e.g., spectral slope value of -0.014) are appropriate.

2.1.4 *Parameterization of scattering*

The total spectral scattering coefficient $b(\lambda)$ can be separated into contributions by pure water $b_w(\lambda)$ (Morel, 1974) and total particles $b_p(\lambda)$ as

$$b(\lambda) = b_w(\lambda) + b_p(\lambda), \ (m^{-1}). \tag{13}$$

Light transmission is also influenced in addition to the scattering coefficient by the angular distribution of scattering, and is defined by the normalized volume scattering function or the scattering phase function. It is obtained by dividing the volume scattering function $\beta(\psi)$ $(m^{-1}sr^{-1})$ by the total scattering function b as

$$\overline{\beta}(\psi) = \beta(\psi)/b. \tag{14}$$

Scattering and backscattering coefficients are obtained by integrating the normalized volume scattering function from 0 to 180 and from 90 to 180°, respectively. Although the backscattering coefficient is the important term in bio-optical models (e.g., Eq. (6)), it has, until recently, been a difficult parameter to measure directly. As such, particle scattering b_p has been related to particle backscattering b_{bp} through the backscattering ratio ($B = b_{bp}/b_p$), and the backscattering coefficient expressed as

$$b_b(\lambda) = \frac{1}{2} b_w(\lambda) + B\, b_p(\lambda), \text{ or} \tag{15}$$

$$b_b(\lambda) = b_{bw}(\lambda) + b_{bp}(\lambda). \tag{16}$$

where b_{bw} is the backscattering due to water molecules (Smith and Baker, 1981). In oceanic Case 1 waters with low chlorophyll content, backscattering contribution by water molecules to total backscattering is not negligible (Morel and Loisel, 1998), while in coastal waters scattering by suspended particles dominates. Based on a number of data points collected in oceanic and coastal waters (Petzold, 1972), a constant value (~0.0183) has generally been assumed for the backscattering ratio (Gould et al., 1999; Mobley et al., 2002). For oceanic waters, backscattering ratio has also been expressed as a decreasing function of increasing chlorophyll (Morel and Maritorena, 2001). The shapes of scattering and backscattering spectra have been described by functions of the form λ^n, with values between 0 and –2 being often used for the exponent n (Sathyendranath et al., 1989; Morel, 1988; Mobley, 1994). In eutrophic waters, the scattering coefficient was shown not to vary much spectrally (Gordon et al., 1988; Morel, 1988) due to the presence of larger particles.

A commonly used bio-optical model for scattering (Gordon and Morel, 1983) that is parameterized in terms of chlorophyll concentrations ([Chl], mg m^{-3}) is given by

$$b(\lambda) = 0.30\ [Chl]^{0.62}\ (550/\lambda). \tag{17}$$

A model for particle backscattering also expressed in terms of chlorophyll (Morel and Maritorena, 2001) is

$$b_{bp} = b_p(550)\{0.002 + 0.02[0.5 - 0.25\ log[Chl](550/\lambda)\}, \tag{18}$$

where $b_p(550)$ is given in terms of Chl (Loisel and Morel, 1998) and expressed as

$$b_p(550) = 0.416\ [Chl]^{0.76}. \tag{19}$$

These relationships are however based on the assumption that nonalgal particles (e.g., detrital and inorganic) co-vary with chlorophyll concentrations and may need to be evaluated for coastal waters.

2.2 EMPIRICAL REMOTE SENSING ALGORITHMS

Empirical algorithms on the other hand provide direct relationships between ratios of remote sensing reflectance or water leaving radiances and seawater constituents such as chlorophyll. They make use of the differential absorption of phytoplankton in the blue (maximum at 443 or 490 nm) and green (minimum at 555 nm) to obtain estimates of chlorophyll. Using simple linear regression of log-transformed data, the power equation has been widely used to relate water-leaving radiances to chlorophyll concentrations (Gordon et al., 1983; Mitchell and Holm-Hansen, 1991). Based on a large set of field data, empirical algorithms based on cubic polynomials have been developed for the SeaWiFS and MODIS satellite sensor wavebands. The two widely used algorithms are the ocean chlorophyll 2 (OC2) modified cubic polynomial and the ocean chlorophyll 4 (OC4) maximum band ratio algorithms (O'Reilly et al., 1998; O'Reilly et al., 2000). The two-band ratio or OC2 algorithm for example is given by the relation

$$[Chl] = 0.0929 + 10^{(0.2974-2.2429X+0.8358X2-0.0077X3)}, \tag{20}$$

where $X = log(R_{rs}490/R_{rs}555)$ and the remote sensing reflectance R_{rs} defined as nL_w/F_0 (F_0 is the solar constant), and nL_w is the normalized water leaving radiance. The OC4 is a four-band maximum band ratio formulation and uses a fourth order polynomial (five coefficients) to estimate [Chl] and is given by

$$[Chl] = 10^{(0.366-3.067X+1.930X2+0.649X3-1.532X4)}, \tag{21}$$

where $X = log(R_{rs}443/R_{rs}555 > R_{rs}490/R_{rs}555 > R_{rs}510/R_{rs}555)$. It uses one of the band ratios with the maximum value and is based on the shift of the maximum of R_{rs} spectra toward higher wavelengths with increasing chlorophyll. Although, the OC2 and OC4 empirical algorithms have been useful in estimating chlorophyll concentrations in both Case 1 and 2 waters, a few studies have shown some large errors in these estimates in some coastal waters (Darecki and Stramska 2004).

3. Methods and Instruments

Recent developments in optical instrumentation have demonstrated their utility in providing new insights into the bio-optical properties of various coastal waters (Twardowski et al., 1999; Sosik et al., 2001; D'Sa and Miller, 2003; Twardowski et al., Chapter 4). Optically complex coastal regions may include river-dominated coastal margins that are regions of high biogeochemical and physical activity. For example, in a SeaWiFS satellite derived chlorophyll image of the northern Gulf of Mexico (Fig. 3), we observe the Mississippi river plume associated with the brighter regions (high chlorophyll concentrations) and showing a highly variable surface chlorophyll distribution. Studies have revealed this region to be optically complex and mainly influenced by the physical and biogeochemical processes associated with river discharge (Sydor et al., 1998; Lohrenz et al., 1999; Miller and D'Sa, 2002).

Numerous instruments and methods are presently available for carrying out bio-optical measurements in oceanic waters (see Twardowski et al., Chapter 4). Protocols are available (Fargion and Mueller, 2000) for measuring bio-optical and radiometric data for satellite ocean color sensor validation. These include the calibration and characterization of field instruments, field methods and data analyses. Instruments used

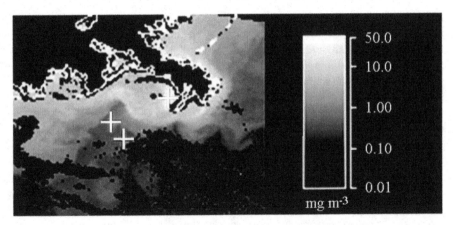

Figure 3. An example of an optically complex coastal region in the northern Gulf of Mexico influenced by the Mississippi River. The SeaWiFS (Sea-viewing Wide Field-of-view Sensor) derived chlorophyll (mg m^{-3}) image was processed using SEADAS software and the OC4 empirical algorithm. Pluses in the figure indicate one nearshore and two offshore locations (based on distance from the coast) as examples of waters with contrasting bio-optical properties. Land and clouds are masked to black.

to characterize the bio-optical properties in oceanic waters include radiometers and absorption and scattering meters. These instruments are often integrated along with a CTD instrument into a single bio-optical package (e.g., Miller and D'Sa, 2002) enabling simultaneous measurements of water column optical and physical variables. Discrete water samples are generally collected to calibrate the optical measurements. For example, spectrophotometric absorption measurements on filtered seawater samples are used to obtain the spectral absorption characteristics of particulate and dissolved components in seawater. Here we describe briefly some instruments that have been used to obtain vertical bio-optical profiles along with methods for discrete sampling and analyses.

3.1 WATER COLUMN MEASUREMENTS

Examples of water column bio-optical measurements include: absorption and attenuation (ac-9, WETLabs), backscattering (Hydroscat-2, HOBILabs; VSF3, WETLabs), spectral downwelling irradiance and upwelling radiance (OCR, OCI-200, Satlantic; PRR-800, Biospherical Instruments) at wavebands corresponding to ocean color satellite sensors, and chlorophyll fluorescence (Wetstar, WETLabs).

3.1.1 *IOP measurements*

A dual-path absorption and attenuation meter (ac-9, WETLabs) is used for measuring absorption and attenuation coefficients at nine 10-nm spectral bandwidths (412, 440, 488, 510, 532, 555, 650, 676 and 715 nm) in the visible and near infrared over a 25 cm path length (Moore et al., 1992; Zaneveld et al., 1992). Optically clean water (e.g., Barnstead NANOpure) is used as a reference to calibrate the ac-9, and thus the estimated absorption and attenuation coefficients does not include contributions by pure water. Measured absorption and attenuation are corrected for temperature, salinity and scattering to give the corrected absorption coefficients $a_t(\lambda,z)$ (i.e. absorption due to particulate and dissolved components) and the attenuation coefficients $c(\lambda)$ (Zaneveld et al., 1994; Pegau et al., 1997; Twardowski et al., 1999) Particle scattering coefficients

$b_p(\lambda,z)$ at the eight visible wavebands are calculated as the difference between the attenuation and absorption coefficients. Vertical profiles of absorption coefficients due to CDOM $a_{cdom}(\lambda,z)$ are obtained by attaching a 0.2 μm pleated maxicapsule filter (Gelman) to the intake tubing of the ac-9. Particle absorption $a_p(\lambda,z)$ is then computed as

$$a_p(\lambda,z) = a_t(\lambda,z) - a_{cdom}(\lambda,z). \tag{22}$$

Instruments for measuring scattering have been constructed with either the detector or source/projector being able to rotate relative to each other (Petzold, 1972; Kirk, 1994) enabling the calculation of the volume scattering function $\beta(\omega)$ (per meter per steradian) over the instrument measuring range. The backscattering coefficient b_b can be estimated by integrating the measured $\beta(\omega)$ from 90 to 180° in scattering angle. A method based on Mie computations has shown that there exists an approximately constant ratio between the backscattering coefficients and $\beta(\omega)$ at 120° (Oishi, 1990). Two commercial field instruments have been developed based on this concept; the Hydroscat-2 and 6 (HOBI Labs), and the ECO BB (WETLabs). The Hydroscat-2 measures backscattering at two wavebands (442 and 532 nm) while the Hydroscat-6 at six wavebands, and is based on a calibration method to convert the optical backscattering signal to a measurement of $\beta(\omega)$ at a single backscattering angle of 140° (Maffione and Dana, 1997). The 117° angle of scattering measured by the ECO BB meter was based on the determination of this angle as a minimum convergence point for variations in the volume scattering function $\beta(\omega)$ induced by suspended materials and water itself (Boss and Pegau, 2001). Estimated $\beta(\omega)$ at this angle is converted to backscattering and further corrected to account for losses due to attenuation of some backscattered light in the path between the sampling volume and detector. The ECO-VSF (WETLabs) optical backscattering sensor measures the optical scattering at three distinct angles of 100, 125, and 150°, thus allowing the determination of the backscattering coefficient through interpolation and extrapolation from 90 to 180°.

3.1.2 Radiometric measurements

Spectral radiometric measurements of the marine light field are now routinely collected with commercial radiometers. Important in-water radiometric profile measurements of spectral downwelling irradiance $E_d(\lambda,z)$, upwelling irradiance $E_u(\lambda,z)$, and upwelling radiance $L_u(\lambda,z)$ allow the determination of AOPs and are used in the calibration and validation of ocean color sensors, bio-optical modeling and primary production studies (Morel, 1988; Behrenfeld and Falkowski, 1997; Fargion and Mueller, 2000; Hooker and Maritorena, 2000; Stramska et al., 2000).

Radiometric measurements of downward irradiance (E_d) and upward radiance (L_u) at wavebands corresponding to satellite ocean color sensors (e.g., 412, 443, 490, 510, 555, 665, and 683 nm) made using radiometers (e.g., OCI and OCR-200, Satlantic; PRR-800 and PRR-2600, Biospherical Instruments) allow the determination of AOPs such as K_d and R_{rs}. Depth-dependent diffuse attenuation coefficients for downwelling irradiance K_d are computed for each wavelength as $d\{ln[Ed(z)]\}/dz$. Similarly, the diffuse attenuation coefficients for upwelling radiance K_L are computed using L_u measurements. Values of $K_d(\lambda,z)$ and $K_L(\lambda,z)$ at the shallowest depth are then used to estimate $E_d(\lambda,0^-)$ and $L_u(\lambda,0^-)$ just beneath the sea surface. Based on known values of Fresnel reflectance at the water-air interface for upwelling radiance and downwelling irradiance, the sub-surface values $E_d(\lambda,0^-)$ and $L_u(\lambda,0^-)$ are extrapolated to values just above the sea surface (Mueller and Austin, 1995). Remote sensing reflectance $R_{rs}(\lambda,0^+)$ is then calculated as

$$R_{rs}(\lambda,0^+) = L_u(\lambda,0^+)/E_d(\lambda,0^+), \tag{23}$$

where $L_u(\lambda,0^+)$ is the water-leaving radiance measured in the nadir direction just above the sea surface and $E_d(\lambda,0^+)$ is the downwelling irradiance incident on the sea surface. Primary sources of uncertainty in the radiometric measurements and the AOP determination include ship perturbation (Voss et al., 1986), instrument shading (Gordon and Ding, 1992) and variability due to sky conditions (e.g., clouds) during the in-water measurements.

3.1.3 *Hydrographic and fluorescence measurements*

Hydrographic data consisting of temperature and salinity are derived from measurements of conductivity-temperature-depth (CTD) using a profiler (e.g., SBE 49, SeaBird). These measurements are important in determining the properties of water masses associated with different physical processes (e.g., fronts, upwelling, and plumes) that often have characteristic optical properties. Temperature and salinity measurements are also required for correcting the absorption measurements made with the ac-9 absorption and attenuation meter.

Continuous profiles of *in situ* chlorophyll fluorescence intensity are generally obtained in conjunction with CTD measurements using commercial fluorometers (e.g., WETStar, WETLabs; Aquatraka II, Chelsea Instruments). Integration of these instruments with the optical instruments is desirable in coastal waters since it allows for simultaneous measurements of multiple variables that often change at small spatial and temporal scales. Chlorophyll fluorescence profiles are generally calibrated against chlorophyll pigments obtained from discrete samples using either the High Performance Liquid Chromatography (HPLC) or those derived from fluorometric analysis. As observed (Fig. 4) high correlations between chlorophyll fluorescence (volts) and chlorophyll concentrations (mg m^{-3}) enables reliable estimates of chlorophyll profiles to be obtained along with optical measurements.

3.2 ABOVE WATER REMOTE SENSING REFLECTANCE

An alternative to deriving R_{rs} from in-water radiometric measurements described previously is to derive R_{rs} from radiance measurements made above the water (Carder and Steward, 1985; Carder et al., 1993; Mobley, 1999). These measurements are generally made from a ship deck using a radiometer to measure radiance signals that are proportional to that emanating from i) the sea surface L_{sea} at nadir angle θ (usually chosen between 30 and 50°) and azimuth angle ϕ (usually with an orientation of 90°) to the plane of the sun, ii) the sky radiance L_{sky} measured with the radiometer looking upward (at same angles but in the zenith direction), and iii) the radiance reflected from a horizontal plaque (L_{pl}) with a known bi-directional reflectance. Measurements of the radiance reflected from the calibrated plaque is used as a proxy for $E_d(\lambda,0^+)$ (Eq. (23)). Also, due to skylight reflected off the sea surface contaminating the L_{sea} measurements, L_{sky} values are used to correct for it. Remote sensing reflectance is calculated as (Mueller and Austin, 1995; Toole et al., 2000)

$$R_{rs} = (L_{sea} - \rho(\theta)L_{sky})/(\pi L_{pl}/\rho_{pl}) - residual(750), \tag{24}$$

where $\rho(\theta)$ is the Fresnel reflectance of the sea surface at a viewing angle θ, and ρ_{pl} is the reflectance of the plaque. Any residual reflected sky radiance is removed by subtracting the signal at 750 nm (Toole et al., 2000).

High spectral resolution radiometers generally used for shipboard determinations of above-water radiance include i) Analytical Spectral Devices (ASD) (Boulder, Colorado)

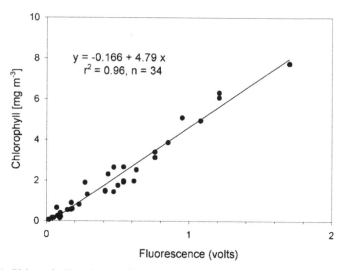

Figure 4. Chlorophyll estimates from discrete samples obtained using the HPLC method plotted as a function of chlorophyll fluorescence from the profiling fluorometer. Fluorescence values correspond to depth of discrete samples.

512-channel, fiber-optic spectrometer, and ii) the GER 1500 (Geophysical and Environmental Research), a 512-channel, 1-nm spectral resolution fiber-optic system. A radiometer (GER 1500) was adapted with optical fibers for shipboard operations and used in the determination of R_{rs} in a river-dominated coastal region (Miller and D'Sa, 2002). A Spectralon plaque (Labsphere) with a nominal reflectance of 98% was used as a reference. Following SeaWiFS protocols (Mueller and Austin, 1995), a typical measurement sequence with the fiber optic radiometer was to measure radiance spectra of the sea surface at 30° nadir viewing angle, followed by the sky radiance at 30° zenith angle, and the radiance reflected off the Spectralon plaque. These measurements are usually repeated three or more times. Following Eq. (24), the data are processed to estimate the above water $R_{rs}(\lambda)$. However, uncertainties in the above-water determination of R_{rs} due to sea-state effects (Toole et al., 2000) such as rough sea conditions and presence of white caps and foam could possibly render some of these measurements impractical (Kirk, 1994).

3.3 DISCRETE MEASUREMENTS

Water samples at discrete depths are generally collected using a CTD rosette multi-bottle array system immediately prior to or following optical profiles for better correlation with optical measurements. These samples are processed at sea or taken to the laboratory for various types of analyses such as pigment concentrations, particulate and CDOM absorption using SeaWiFS protocol (Mueller and Austin, 1995) and will be briefly discussed in the following section.

3.3.1 *Phytoplankton pigment concentrations*

Phytoplankton pigment concentrations have been determined using the fluorometric measuring technique (Parsons et al., 1984) and in more recent times the application of High-Performance Liquid Chromatography (HPLC) method (Wright et al., 1991) has been recommended (Fargion and Mueller, 2000).

Depending on phytoplankton biomass, a variable volume of water (50-500 ml) is filtered through Whatman GF/F filters for fluorometric analysis of chlorophyll concentration (Parsons et al., 1984). The filters are homogenized and kept refrigerated in the dark while pigments are extracted in 90% acetone for approximately 6 hours or longer. Fluorescence is measured following extraction in a Turner Designs fluorometer previously calibrated with pure chlorophyll *a* (Sigma). The fluorometric method however, has been shown to significantly under- or over-estimate chlorophyll concentrations due to the presence of chlorophyll degradation products and accessory pigments (Trees et al., 1985; Bianchi et al., 1995).

HPLC determination of phytoplankton pigments such as chlorophyll *a* and phaeopigments and other accessory pigments have been recommended for validation of bio-optical algorithms for ocean color sensors (Fargion and Mueller, 2000) and include protocols and standards that closely follow those adopted by the Joint Global Ocean Flux Study (JGOFS) (UNESCO, 1994). Briefly, seawater samples are filtered onto a 25 cm glass fiber filter (Whatman, GF/F) under low vacuum. These are then stored in cryotubes kept in liquid nitrogen and analyzed in the laboratory with high-performance liquid chromatography (HPLC). Frozen sample filters are homogenized in 1.5 ml 90% acetone, centrifuged, and diluted with 0.5 M aqueous ammonium acetate before injection. Peak identifications were made using standards for the various pigments followed by the determination of chlorophyll and other pigments.

3.3.2 Particle absorption

Light absorption of particles suspended in seawater are obtained by filtering the particles following Mitchell and Kiefer (1988). Depending on the particle load, different volumes of seawater samples are filtered onto a 25 mm glass fiber filter (Whatman GF/F) at low vacuum and immediately analyzed or frozen in liquid nitrogen for latter spectroscopic analysis. The optical density of the filters (OD_f) are measured in a dual-beam scanning spectrophotometer (e.g., Lambda-18, Perkin-Elmer) generally between 380 and 750 nm. A filtrate saturated Whatman GF/F filter is used as a blank. Particle absorption $a_p(\lambda)$ is calculated as

$$a_p(\lambda) = 2.3\ OD_f(\lambda)\ A/V\beta, \tag{25}$$

where A is the clearance area of the filter (m^2), V is the volume of seawater filtered (m^3), and multiplication by 2.3 converts base-10 logarithm to natural logarithm. The value for the pathlength amplification factor of the filters β is taken to be either a constant (Roesler, 1998) or to be wavelength dependent (Bricaud and Stramski, 1990). After particle measurements, the sample filters are extracted in methanol (Kishino et al., 1985) or bleached (Tassan and Ferrari, 1995) to remove phytoplankton pigments, and the optical density of the filter is determined again in the spectrophotometer. The absorption coefficient of nonpigmented particulates or $a_{nap}(\lambda)$ is also determined using Eq. (25). Phytoplankton or algal absorption $a_{phy}(\lambda)$ is then calculated as the difference between $a_p(\lambda)$ and $a_{nap}(\lambda)$.

3.3.3 CDOM absorption

Measurements of spectral absorption of the soluble material in seawater or the colored dissolved organic matter (CDOM) have been obtained in diverse oceanic and coastal waters using spectrophotometers with 10 cm pathlength cells (Bricaud et al., 1981; Blough et al., 1993). Seawater samples are filtered through pre-rinsed 0.2 μm membrane filters (e.g., Nuclepore nylon), and the optical density spectra $OD_{cdom}(\lambda)$ of the filtrate relative to purified water are obtained using a spectrophotometer. Necessary care should be taken during handling of the seawater samples to avoid contamination and photo-degradation. CDOM absorption coefficients $a_{cdom}(\lambda)$ are calculated as

$$a_{cdom}(\lambda) = 2.303\ OD_{cdom}(\lambda)/l,\ [m^{-1}] \tag{26}$$

where l is the pathlength of the sample cell or cuvette (usually 0.1 m), and OD_{cdom} is the optical density relative to pure water.

An alternate method for determining CDOM absorption is the use of a long pathlength aqueous capillary waveguide based spectrophotometer that effectively increases the sensitivity of measurements (D'Sa et al., 1999; D'Sa and Steward, 2001). This technology was further developed into a multiple pathlength aqueous waveguide system (Fig. 5) that enabled the determination of CDOM absorption spectra in diverse inland, coastal and oceanic waters (Miller et al., 2002).

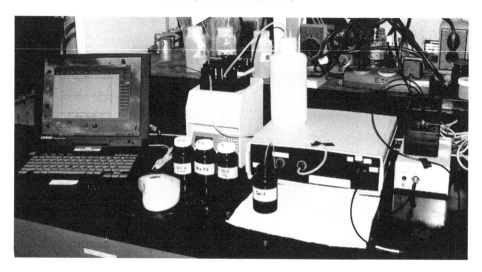

Figure 5. Multiple pathlength capillary waveguide system used for measuring CDOM absorption at sea.

4. Coastal Bio-optical Properties

Bio-optical properties of optically complex Case 2 waters have been studied in numerous regions that include: coastal, river-dominated, upwelling regions, fronts and estuaries. Here we review some studies that examined the physical influences on optical properties and the distribution of seawater constituents. We also examine light variability in relation to IOPs, AOPs and present examples of bio-optical data in a river-dominated coastal region.

4.1 LINKAGES BETWEEN PHYSICAL AND BIO-OPTICAL PROPERTIES

Physical influences associated with advective currents, tides, resuspension, stratification, mixing and flow of river waters often determine the bio-optical properties and carbon fluxes in coastal waters. In studies conducted on the New England continental shelf, physical processes associated with characteristic seasonal patterns in stratification and mixing contributed to the optical variability mostly through effects on phytoplankton (Sosik et al., 2001). In coastal waters of the Antarctic Peninsula, elevated values of beam attenuation coefficients were associated with phytoplankton blooms in stratified surface waters and resuspension in bottom waters (Mitchell and Holm-Hansen, 1991). Over longer time scales, variations in chlorophyll fluorescence on the inner shelf off Duck, North Carolina were related to changes in water mass properties that could be attributed to alternating events of upwelling and downwelling (D'Sa et al., 2001). In examining the specific absorption for total particulates, detritus and phytoplankton in the California Current System, the variability in specific absorption due to phytoplankton $a_{ph}*(\lambda)$ was found to be related to the hydrographic and chemical environment (Sosik and Mitchell, 1995).

Optical properties of the U.S. Middle Atlantic Bight were examined during four cruises along a cruise track extending from Delaware Bay to the Sargasso Sea (DeGrandpre et al., 1996). The overall seasonal and spatial ranges for CDOM and particulate absorption (at 442 nm) and chlorophyll concentrations were 0.02-0.41 m^{-1}, 0.01-0.49 m^{-1}, and 0.07-9.4 mg m^{-3}, respectively. Largest values in these variables were near the coast, with many of the peaks along the transect corresponding to frontal structures. Additionally, CDOM absorption was found to be highest in April, most likely due to increased freshwater inputs from spring runoff. Studies of CDOM absorption in another location in the Mid Atlantic Bight (Boss et al., 2001) revealed that on short time scales, variability in the vertical distribution of CDOM absorption was mostly due to high frequency internal waves, while over longer periods and episodically, CDOM variability was dominated by storms.

In river-dominated coastal regions, CDOM absorption and salinity are often well correlated (Blough et al., 1993). Freshwater input from surrounding watersheds and new CDOM production from *in situ* biologic activity were factors found contributing to the temporal and spatial variability of CDOM absorption in Narragansett Bay and Block Island Sound (Rhode Island) (Keith et al., 2002). The seasonal flow of river discharge by the Mississippi river strongly influenced the spatial and temporal variability in the bio-optical properties in the Northern Gulf of Mexico (Miller and D'Sa, 2002). In the Mississippi River Bight, stratification due to lower salinity surface waters overlying more saline oceanic waters strongly influenced the spatial and vertical structure of the bio-optical field (D'Sa and Miller, 2003).

An example of the large variability in the bio-optical properties associated with changes in the physical properties (e.g., temperature and salinity) in a river dominated coastal system can be observed in Fig. 6. At the nearshore location, surface waters with lower salinity and temperature were associated with high phytoplankton biomass (~ 8 mg m^{-3} of chlorophyll) that decreased with depth and increasing salinity and temperature. Optical properties of absorption ($a_t = a_p + a_{cdom}$) at 440 nm also decreased with depth as the chlorophyll profile. The backscattering profile b_b however showed lower values in surface waters that increased with depth, indicating that phytoplankton was not the main contributor to the backscattering signal. At the offshore location, more oceanic conditions prevailed (~35.8 psu) with low chlorophyll concentrations (~ 0.5 mg m^{-3}), and the optical properties an order of magnitude lower than the nearshore location. This data clearly indicates the strong linkages between the physical and bio-optical properties and provides the opportunity to evaluate bio-optical models for water with stratified (Gordon and Clark, 1980) or non-uniform pigment profiles (Sathyendranath and Platt, 1989).

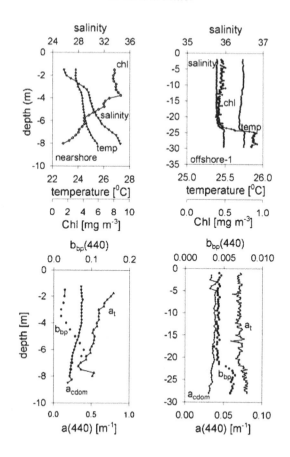

Figure 6. Vertical profiles of temperature, salinity and chlorophyll at two coastal locations (nearshore and offshore) in the northern Gulf of Mexico. Corresponding water column profiles of absorption, scattering and backscattering (440 nm) at the same locations (lower panel).

4.2 ABSORPTION AND SCATTERING CHARACTERISTICS

For any given water body, the total spectral absorption coefficient is the sum of the known absorption coefficients of pure water and that due to dissolved and particulate matter. In the case of scattering and backscattering coefficients, the contributions are from pure water and particulate matter. In oceanic waters, phytoplankton and its derived products are the main contributors to absorption and scattering. In these waters, the various components (except that due to pure water) are often assumed to covary with chlorophyll concentration (Morel and Prieur, 1977). Based on this assumption, and knowledge of the spectral characteristics of these constituents, reflectance models and ocean color algorithms have been developed for estimating seawater constituents from remote sensing (Gordon et al., 1988; Roesler and Perry, 1995; Garver and Siegel, 1997; Bricaud et al. 1998; Carder et al., 1999; Reynolds et al., 2001). The application of these models to coastal waters may require a better understanding of the absorption and scattering characteristics of the various seawater constituents. Next we review studies that have documented the optical properties in some coastal waters.

4.2.1 *Absorption by seawater constituents*

In a comprehensive study in coastal waters around Europe (Babin et al., 2003), the relationship between phytoplankton absorption and chlorophyll concentration was found to be generally similar to those for open ocean waters (Bricaud et al., 1995). However, the study found significant departures from the general trend due to unusual cell size and high phaeopigment concentrations.

A few studies on the absorption properties of detrital or nonalgal particles in coastal waters have been described (Roesler et al., 1989; Duarte et al., 1998; Babin et al., 2003). In most cases, the absorption spectra of these particles were well described by an exponential function. Average values for the slope of the exponential function were about 0.011 nm^{-1} in inland marine waters (Roesler et al., 1989) while an average value of 0.0123 nm^{-1} was obtained in coastal waters around Europe (Babin et al., 2003). Light absorption by marine mineral particles in coastal waters, showed a similar behavior as that of nonalgal particles and were shown to have an average spectral slope of about 0.011 nm^{-1} (Bowers et al., 1996).

Optical properties of Case 2 waters can be substantially affected by CDOM absorption and numerous studies have shown their effect on ocean color algorithms. Blough et al. (1993) determined that the discharge of colored organic matter by the Orinoco River strongly influenced the optical quality of the Caribbean waters with major implications for the carbon cycle.

Spectral absorption of seawater constituents (phytoplankton, detrital or nonalgal particles, and CDOM) determined from samples taken at different locations in a river dominated coastal environment (Fig. 7) demonstrates the large variability in particulate ($\sim$0.08 to 0.6 m^{-1}) and CDOM ($\sim$0.1 to 0.8 m^{-1}) absorption at 400 nm. For nearshore locations, detrital and CDOM absorption were greater than phytoplankton absorption, while for the offshore location, phytoplankton absorption was dominant. Variability in spectral absorption (particulate plus CDOM absorption) at six spectral bands at a nearshore coastal location (Fig. 8, left panel) shows absorption decreasing with increasing wavelength. Such datasets obtained from various coastal waters would be essential in evaluating or developing new absorption parameterizations (e.g., Eq. (18)) for the seawater constituents.

4.2.2 *Scattering and backscattering*

Scattering in the marine environment is dependent on the amount and composition of the suspended particles through their size, shapes and refractive indices (Stramski and Kiefer, 1991). Uncertainties remain on the contributions of various particle size fractions to light scattering (Beardsley et al., 1970; Brown and Gordon, 1973), or on the relative importance to scattering by organic or inorganic particles (Brown and Gordon, 1973; Zaneveld et al., 1974). In coastal waters the larger range and size distribution of particles present greater challenges, with backscattering by suspended sediments often contributing significantly to the ocean color signal.

Petzold (1972) determined b and b_b from measurements of volume scattering function obtained at numerous locations in Case 1 and Case 2 waters that included the turbid San Diego Harbor. The values of the backscattering ratios obtained from the above dataset have been widely used in determining backscattering from scattering. While b_b and b was shown to be linearly related for the Petzold data set (Gould et al., 1999), field measurements in the Mississippi River Bight indicated non-linear relation-

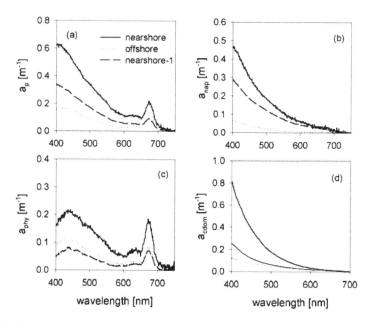

Figure 7. Spectral absorption corresponding to (a) total particulate a_p, (b) nonalgal particles a_{nap}, (c) phytoplankton a_{phy}, and (d) CDOM a_{cdom} for surface waters in a coastal river dominated environment.

ship between the two variables (D'Sa and Miller, 2003). In presenting the results of field measurements in turbid waters, Whitlock et al. (1981) observed the ratio b_b/b to have only small spectral variations, while the magnitude changed with increased water turbidity. Spectrally, the backscattering ratio in coastal waters was found to be only weakly sensitive to changes in wavelength (Risovic, 2002). Bulk refractive index for understanding particle composition in Case 1 and Case 2 waters were described in terms of the backscattering ratio and the hyperbolic slope of the particle size distribution (Twardowski et al., 2001). Estimates of the refractive index agreed with expected values for algal cells and inorganic minerals.

An example of the water-column structure of spectral scattering (412, 440, 488, 510, 555, 650 nm) and backscattering (442 and 589 nm) for a nearshore location (Fig. 8, right panel) shows little correlation with corresponding chlorophyll profile (Fig. 5, left panel). We observe both scattering and backscattering increasing with depth while chlorophyll decreased, indicating the greater influence of suspended sediments on scattering. Spectrally, both scattering and backscattering decreased with increasing wavelength. The data presented here indicates that characterizing scattering and backscattering in terms of chlorophyll may not be appropriate in some coastal waters and more field data may be needed to better understand the relationships (Hamre et al., 2003).

4.3 LIGHT FIELD CHARACTERISTICS AND OPTICAL PROPERTIES

The amount of visible radiant energy and its spectral content emerging from the sea depends on the downwelling incident light, the sun elevation, sky state, sea state, and the optical properties of absorption and backscattering of the seawater itself. Numerous numerical models have been developed to simulate the underwater light field (Mobley et

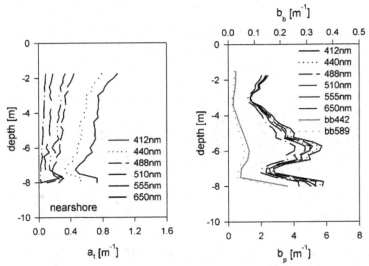

Figure 8. Water column profiles of absorption a_t (left panel) and scattering b_p (right panel) spectra (412, 440, 488, 510, 555, 650 nm) for a coastal nearshore location. Backscattering, $b_b(442)$ and $b_b(589)$ are also shown (—— and ·····) in the right panel.

al., 1993). The underwater light field is most frequently defined in terms of irradiance (e.g., downwelling irradiance) since it is important in photosynthesis and plays a central role in the theory of radiation transfer in water (Kirk, 1994). The spectral composition and intensity of the downwelling light flux or the downwelling irradiance $E_d(\lambda,z)$ changes progressively with increasing depth z as a result of its interaction with seawater and its constituents through the process of absorption and scattering. $E_d(\lambda,z)$ diminishes in an approximately exponential manner with depth and is given by

$$E_d(\lambda,z) = E_d(\lambda,0^-)\, e^{-K_d(\lambda)\, dz},\qquad (27)$$

where $E_d(\lambda,z)$ and $E_d(\lambda,0^-)$ are the values of downward irradiance at depth z meters and just below the surface respectively.

Kirk (1984) determined a relationship between the diffuse attenuation coefficient K_d and absorption a and scattering b using a Monte Carlo model of the propagation of photons underwater. This model was used to extract the IOPs of absorption and scattering in a turbid estuary from *in situ* radiometric measurements (Gallegos et al., 1990). A similar study using modeled derived IOPs (a and b) from radiometric estimates of AOPs (R_{rs} and K_d) (Morel and Gentili, 1993; Gordon, 1989) showed good agreement with field measurements of a and b in the blue-green region of the spectrum (McKee et al., 2003).

The behavior of the light field E_d, K_d and IOPs for two coastal locations is illustrated in Figure 9. The top panel shows the changes in the spectral downwelling light field with depth at a nearshore and offshore locations. The spectra of the attenuation coefficient for downwelling irradiance $K_d(\lambda)$ for surface waters are shown in the mid row, while the corresponding profiles of total absorption (at 440 nm) and chlorophyll are shown in the bottom panel. The light level diminishes rapidly at the nearshore location in the blue waveband ($\sim 1\%$ at 5 m depth) in comparison to the offshore location where the 1% level in the blue is at ~ 30 meters. At the nearshore coastal location, light attenuation in the blue due to absorption by high levels of CDOM and phytoplankton is

comparable to that in the red. In contrast, at the offshore location, the light level
diminishes much more slowly in the blue than in the red. Significant differences are
observed in the spectral diffuse attenuation coefficients $K_d(\lambda)$ at the two coastal
locations that can be attributed primarily to absorption by particulate and dissolved
components in seawater (D'Sa et al., 2002b).

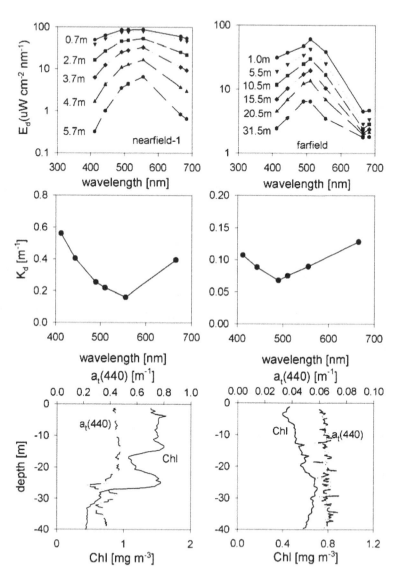

Figure 9. Spectral downwelling light field at various depths at a nearshore (left panel)
and an offshore location (right panel). Spectral diffuse attenuation coefficient $K_d(\lambda)$ in
surface waters (mid row) and (c) vertical profiles of absorption (particulate plus CDOM)
at 440 nm and chlorophyll (bottom row) are shown for the same locations.

Remote sensing reflectance or water-leaving radiance defines the color of the ocean that contains information on the inherent optical properties and the seawater constituents. In analyzing the spectral reflectance curves obtained from Case 1 and 2 waters (Morel and Prieur, 1977), it was observed that the spectra were modified in different ways by the seawater constituents through their effects on absorption and scattering properties. Observations from a 3-year field program in the Santa Barbara Channel, California (Toole and Siegel, 2001) showed that the remote sensing reflectance variability was tightly coupled to biologically and terrestrially derived particles, with the largest variability in the spectra determined by the backscattering processes.

The basic relationship between remote sensing reflectance and the IOPs (e.g., Eq. (6)) or its approximations (Morel and Prieur, 1977; Gordon and Morel, 1983) derived from numerical simulations of the radiative transfer equation has been used to interpret remote sensing reflectance or ocean color data in terms of seawater optical properties and constituents (Carder et al., 1999; Lee et al., 2002). Volume absorption coefficients of turbid coastal waters determined from R_{rs} were in agreement with field observations for a certain absorption range and were based on assumptions related to sea surface roughness, spectral scattering characteristics and absorption in the near-infrared (Sydor et al., 1998). Bio-optical studies have also considered the need to account for the effects of bottom reflectance, CDOM fluorescence, and water Raman scattering on the remote sensing reflectance signals in coastal waters (Maritorena et al., 1994; Lee et al., 1994, 1998; Ackleson, 2003).

Examples of remote sensing reflectance spectra (shown at wavebands corresponding to the SeaWiFS satellite sensor) in a river dominated coastal environment (Fig. 10) are seen to exhibit two general characteristics. At locations with low surface chlorophyll and CDOM concentrations, the reflectance values were higher in the blue spectral region and exhibited high blue to green ratios (Group 2, Fig. 10). At nearshore locations, higher chlorophyll and CDOM concentrations resulted in low reflectance in the blue wavebands and thus low values in the blue to green ratios (Group 1, Fig. 10). Reflectance at 555 nm was also higher due to backscattering contributions by suspended sediments. This example broadly demonstrates the complexity of the R_{rs} spectra observed in coastal waters.

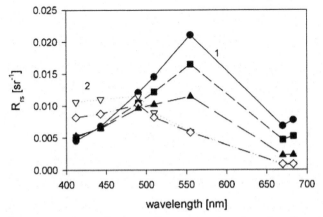

Figure 10. Remote sensing reflectance R_{rs} spectra at various locations in a coastal river dominated region shown at SeaWiFS wavebands. Group 1 (closed symbols) correspond to nearshore locations, while Group 2 (open symbols) correspond to offshore locations.

Empirical algorithms such as the OC2 and OC4 algorithms (O'Reilly et al., 1998) use ratios of remote sensing reflectances (e.g., 412/555, 443/555 or 490/555 nm) to estimate chlorophyll concentrations from atmospherically corrected satellite reflectance signals. Various studies have indicated that the coefficients in the empirical algorithms for chlorophyll may have to be optimized for regional and seasonal conditions (Kahru and Mitchell, 1999; D'Sa and Miller, 2003). Similar empirical algorithms have been developed for CDOM absorption (Kahru and Mitchell, 2001; D'Sa and Miller, 2003). Simple empirical relationships relating water leaving radiance or reflectance and suspended sediments have also been presented for coastal waters (Munday and Alfoldi, 1979; Stumpf, 1988). These studies have demonstrated that empirical algorithms may need to be optimized for regional coastal waters to obtain better estimates of concentrations of seawater constituents.

5. Summary

The application of ocean color remote sensing in coastal waters includes the ability to retrieve various seawater constituents such as chlorophyll, colored dissolved organic matter and suspended sediment concentrations. However, many of the bio-optical and empirical models have been developed mainly for oceanic or the relatively simple Case 1 waters and their application to the optically complex Case 2 waters present many challenges. In this study, we described some of these bio-optical models and assessed their use in coastal waters. A review of bio-optical studies conducted in some coastal waters indicated close linkages between physical and bio-optical properties. As expected, absorption and scattering properties of seawater constituents in coastal waters were observed to be more complex than oceanic Case 1 waters. Scattering and backscattering of particles have been the least studied and new optical instruments provide the potential to obtain a greater understanding of this optical variable and its role in determining ocean color in coastal waters. The spectral composition and intensity of the light field is strongly influenced by the seawater constituents and are reflected in the derived AOPs of diffuse attenuation coefficients and remote sensing reflectance. With increasing number of operational ocean color satellites and improvements in atmospheric correction algorithms, better satellite estimates of water-leaving radiance or remote sensing reflectance are now being obtained. The ability to invert these signals to obtain reliable estimates of the IOPs and seawater constituents is still a major challenge for optically complex coastal waters. While standard empirical ocean color algorithms are being used to obtain estimates of chlorophyll concentrations, a few studies have demonstrated the need for regional empirical algorithms to locally optimize such estimates. Increasing such bio-optical studies is the key to better utilization of ocean color satellite data for coastal waters.

6. References

Ackleson, S.G. 2003. Light in shallow waters: a brief research review. Limnology and Oceanography, 48:323-328.

Arrigo, K.R., D.H. Robinson, D.L. Worthen, B. Schieber and M.P. Lizotte. 1998. Bio-optical properties of the southwestern Ross Sea. Journal of Geophysical Research, 103:21683-21695.

Babin, M., D. Stramski, G.M. Ferrari, H. Claustre, A. Bricaud, G. Obolenski and N. Hoepffner. 2003. Variations in the light absorption coefficients of phytoplankton, nonalgal particles, and dissolved organic matter in coastal waters around Europe. Journal of Geophysical Research, 108(C7):3211-doi:10.1029/2001JC000882.

Bannister, T.T. 1992. Model of the mean cosine of underwater radiance and estimation of underwater scalar irradiance. Limnology and Oceanography, 37:773-780.

Beardsley, G.F.Jr., H. Pak, K.L. Carder and B. Lundgren. 1970. Light scattering and suspended particles in the eastern equatorial Pacific Ocean. Journal of Geophysical Research, 75:2837-2845.

Behrenfeld, M.J. and P.G. Falkowski. 1997. A consumer's guide to phytoplankton primary productivity models. Limnology and Oceanography, 42:1479-1491.

Berwald, J., D. Stramski, C.D. Mobley and D.A. Kiefer. 1995. Influences of absorption and scattering on vertical changes in the average cosine of the underwater light field. Limnology and Oceanography, 40:1347-1357.

Bianchi, T.S., C. Lambert and D.C. Biggs. 1995. Distribution of chlorophyll *a* and phaeopigments in the northwestern Gulf of Mexico: a comparison between fluorometric and high-performance liquid chromatography measurements. Bulletin of Marine Science, 56:25-32.

Blough, N.V., O.C. Zafiriou and J. Bonilla. 1993. Optical absorption spectra of waters from the Orinoco River outflow: Terrestrial input of colored organic matter to the Caribbean. Journal of Geophysical Research, 98:2271-2278.

Boss, E. and W.S. Pegau. 2001. Relationship of light scattering at an angle in the backward direction to the backscattering coefficient. Applied Optics, 40:5503-5507.

Boss, E., W.S. Pegau, J.R.V. Zaneveld and A.H. Barnard. 2001. Spatial and temporal variability of absorption by dissolved material at a continental shelf. Journal of Geophysical Research, 106:9499-9507.

Bowers, D.G., E.L. Harker and B. Stephan. 1996. Absorption spectra of inorganic particles in the Irish Sea and their relevance to remote sensing of chlorophyll. International Journal of Remote Sensing, 17:2449-2460.

Bricaud, A., A. Morel and L. Prieur. 1981. Absorption by dissolved organic matter of the sea (yellow substance) in the UV and visible domains. Limnology and Oceanography, 26:43-53.

Bricaud, A. and D. Stramski. 1990. Spectral absorption coefficients of living phytoplankton and nonalgal biogenous matter: A comparison between the Peru upwelling area and Sargasso Sea. Limnology and Oceanography, 35:562-582.

Bricaud, A., M. Babin, A. Morel and H. Claustre. 1995. Variability in the chlorophyll-specific absorption coefficients of natural phytoplankton: Analysis and parameterization. Journal of Geophysical Research, 103:13321-13332.

Bricaud, A., A. Morel, M. Babin, K. Allali and H. Claustre. 1998. Variations of light absorption by suspended particles with the chlorophyll *a* concentration in oceanic (case 1) waters: Analysis and implications for bio-optical models. Journal of Geophysical Research, 103:31033-31044.

Brown, O.B. and H.R. Gordon. 1973. Two component Mie scattering models of Sargasso Sea particles. Applied Optics, 12:2461-2465.

Bukata, R.P., J.H. Jerome, J.E. Burton, S.C. Jain and H.H. Zwick. 1981. Optical water quality model of Lake Ontario, I. Determination of the optical cross sections of organic and inorganic particulates in Lake Ontario. Applied Optics, 20:1696-1703.

Bukata, R.P., J.H. Jerome, K.Y. Kondratyev and D.V. Pozdnyakov. Optical properties and remote sensing of inland and coastal waters, C.R.C. Press, Boca Raton (1995) 362 pp.

Carder, K.L. and R.G. Steward. 1985. A remote-sensing reflectance model of a red tide dinoflagellate off West Florida. Limnology and Oceanography, 30:286-298.

Carder, K.L., R.G. Steward, G.R. Harvey and P.B. Ortner. 1989. Marine humic and fulvic acids: Their effects on remote sensing of ocean chlorophyll. Limnology and Oceanography, 34:68-81.

Carder, K.L., S.K. Hawes, K.A. Baker, R.C. Smith, R.G. Steward and B.G. Mitchell. 1991. Reflectance model for quantifying chlorophyll *a* in the presence of productivity degradation products. Journal of Geophysical Research, 96:20599-20611.

Carder, K.L., P. Reinersman, R.F. Chen, F. Muller-Karger, C.O. Davis and M. Hamilton. 1993. AVIRIS calibration and application in coastal oceanic environments. Remote Sensing of Environment, 44:205-216.

Carder, K.L, R.F. Chen, Z.P. Lee and S.K. Hawes. 1999. Semianalytic moderate-resolution imaging spectrometer algorithms for chlorophyll *a* and absorption with bio-optical domains based on nitrate-depletion temperatures. Journal of Geophysical Research, 104:5403-5421.

Ciotti, A.M., J.J. Cullen and M.R. Lewis. 1999. A semi-analytical model of the influence of phytoplankton community structure on the relationship between light attenuation and ocean color. Journal of Geophysical Research, 104:1559-1578.

Darecki, M. and D. Stramski. 2004. An evaluation of MODIS and SeaWiFS bio-optical algorithms in the Baltic Sea. Remote Sensing of Environment, 89:326-350.

DeGrandpre, M.D., A. Vodacek, R.K. Nelson, E.J. Bruce and N.V. Blough. 1996. Seasonal seawater optical properties of the U.S. Middle Atlantic Bight. Journal of Geophysical Research, 101:22727-22736.

Doerffer, R. and J. Fischer. 1994. Concentrations of chlorophyll, suspended matter, and gelbstoff in case II waters derived from satellite coastal zone color scanner data with inverse modeling methods. Journal of Geophysical Research, 99:7457-7466.

D'Sa E.J., R.G. Steward, A. Vodacek, N.V. Blough and D. Phinney. 1999. Optical absorption of seawater colored dissolved organic matter determined using a liquid capillary waveguide. Limnology and Oceanography, 44:1142-1148.

D'Sa E.J., S.E. Lohrenz, J.H. Churchill, J. Largier, V.L. Asper and A.J. Williams. 2001. Chloropigment distribution and transport on the inner shelf off Duck, North Carolina. Journal of Geophysical Research, 106:11581-11596.

D'Sa E.J., and R.G. Steward. 2001. Liquid capillary waveguide application in absorbance spectroscopy (reply to comment). Limnology and Oceanography, 46:742-745.

D'Sa, E.J., C. Hu, F.E. Muller-Karger and K.L. Carder. 2002a. Estimation of colored dissolved organic matter and salinity fields in case 2 waters using SeaWiFS: Examples from Florida Bay and Florida Shelf. Earth and Planetary Science (Indian Academy of Sciences), 111:197-207.

D'Sa, E.J., R.L. Miller, B.A. McKee and R. Trzaska. 2002b. Apparent optical properties in waters influenced by the Mississippi River. In: Proceedings 7th International Conference on Remote Sensing for Marine and Coastal Environments. Miami, FL.

D'Sa, E.J., and R.L. Miller. 2003. Bio-optical properties in waters influenced by the Mississippi River during low flow conditions. Remote Sensing of Environment, 84:538-549.

Duarte, C.M., S. Augusti, M.P. Satta and D. Vaque. 1998. Partitioning particulate light absorption: a budget for a Mediterranean bay. Limnology and Oceanography, 43:236-244.

Fargion, G.S. and J.L. Mueller. 2000. Ocean optics protocols for satellite ocean color sensor validation, revision 2. NASA Technical Memo. 209966, NASA Goddard Space Flight Center, Greenbelt, MD., 184 pp.

Gallegos, C.L., D.L. Correll and J.W. Pierce. 1990. Modeling spectral diffuse attenuation, absorption, and scattering coefficients in a turbid estuary. Limnology and Oceanography, 35:1486-1502.

Garver, S.A. and D.A. Siegel. 1997. Inherent optical property inversion of ocean color spectra and its biogeochemical interpretation. I. Time series from the Sargasso Sea. Journal of Geophysical Research, 102:18607-18625.

Gordon, H.R., O.B. Brown and M.M. Jacobs. 1975. Computed relationships between the inherent and apparent optical properties of a flat, homogeneous ocean. Applied Optics, 14:417-427.

Gordon, H.R. and D.K. Clark. 1980. Remote sensing optical properties of a stratified ocean: an improved interpretation. Applied Optics, 19:3428-3430.

Gordon, H.R. and A. Morel. Remote assessment of ocean color for interpretation of satellite visible imagery: A review. Springer-Verlag, New York (1983) 114 pp.

Gordon, H.R., D.K. Clark, J.W. Brown, O.B. Brown, R.H. Evans and W.W. Brockow. 1983. Phytoplankton pigment concentrations in the Middle Atlantic Bight: Comparison of ship determinations and CZCS estimates. Applied Optics, 22:20-36.

Gordon, H.R., O.B. Brown, R.H. Evans, J.W. Brown, R.C. Smith, K.S. Baker and D.K. Clark. 1988. A semianalytic radiance model of ocean color. Journal of Geophysical Research, 93:10909-10924.

Gordon, H. R. 1989. Can the Lambert-Beer law be applied to the diffuse attenuation coefficient of ocean water?. Limnology and Oceanography, 34:1389-1409.

Gordon, H.R. 1991. Absorption and scattering estimates from irradiance measurements: Monte Carlo simulations. Limnology and Oceanography, 36:769-777.

Gordon, H.R. and K. Ding. 1992. Self shading of in-water optical instruments. Limnology and Oceanography, 37:491-500.

Gould, R.W., R.A. Arnone and P.M. Martinolich. 1999. Spectral dependence of the scattering coefficient in case 1 and case 2 waters. Applied Optics, 38:2377-2383.

Hamre, B., O. Frette, S.R. Erga, J.J. Stamnes and K. Stamnes. 2003. Parameterization and analysis of the optical absorption and scattering coefficients in a western Norwegian fjord: a case II water study. Applied Optics, 42:883-892.

Hedges, J.I. and R.G. Keil. 1995. Sedimentary organic-matter preservation-an assessment and speculative synthesis. Marine Chemistry, 49:81-115.

Hochman, H.T., F.E. Muller-Karger and J.J. Walsh. 1994. Interpretation of the coastal zone color scanner signature of the Orinoco River plume. Journal of Geophysical Research, 99:7443-7455.

Hooker, S.B. and S. Maritorena. 2000. An evaluation of oceanographic radiometers and deployment methodologies. Journal of Atmospheric and Oceanic Technology, 17:811-830.

Kahru, M. and B.G. Mitchell. 1998. Spectral reflectance and absorption of a massive red tide off southern California. Journal of Geophysical Research, 103:21601-21609.

Kahru, M. and B.G. Mitchell. 1999. Empirical chlorophyll algorithm and preliminary SeaWiFS validation for the California Current. International Journal of Remote Sensing, 20:3423-3429.

Kahru, M. and B.G. Mitchell. 2001. Seasonal and nonseasonal variability of satellite-derived chlorophyll and colored dissolved organic matter concentration in the California Current. Journal of Geophysical Research, 106:2517-2529.

Keith, D.J., J.A. Yoder and S.A. Freeman. 2002. Spatial and temporal distribution of coloured dissolved organic matter (CDOM) in Narragansett Bay, Rhode Island: Implications for phytoplankton in coastal waters. Estuarine, Coastal and Shelf Science, 55:705-717.

Kirk, J.T.O. 1984. Dependence of relationship between inherent and apparent optical properties of water on solar altitude. Limnology and Oceanography, 29:350-356.

Kirk, J.T.O. 1991. Volume scattering function, average cosines, and the underwater light field. Limnology and Oceanography. 36:455-467.

Kirk, J. T. O. Light and photosynthesis in aquatic ecosystems, 2nd edition, Cambridge University Press (1994) 509 pp.

Kishino, M., M. Takahasi, N. Okami and S. Ichimura. 1985. Estimation of the spectral absorption coefficients of phytoplankton in the sea. Bulletin of Marine Science, 37:634-642.

Lee, Z.P., K.L. Carder, S.K. Hawes, R.G. Steward, T.G. Peacock, and C.O. Davis. 1994. Model for the interpretation of hyperspectral remote-sensing reflectance. Applied Optics. 33:5721-5732.

Lee, Z.P., K.L. Carder, R.G. Steward, T.G. Peacock, C.O. Davis and J.S. Patch. 1998. An empirical algorithm for light absorption by ocean water based on color. Journal of Geophysical Research, 103:27967-27978.

Lee, Z.P., K.L. Carder and R.A. Arnone. 2002. Deriving inherent optical properties from water color: a multiband quasi-analytical algorithm for optically deep waters. Applied Optics, 41:5755-5772.

Lohrenz, S.E., G.L. Fahnenstiel, D.G. Redalje, G.A. Lang, X. Chen and M.J. Dagg, T.E. Whitledge and Q. Dortch. 1999. Nutrients, irradiance, and mixing as factors regulating primary production in coastal waters impacted by the Mississippi River. Continental Shelf Research, 19:1113-1141.

Loisel, H. and A. Morel. 1998. Light scattering and chlorophyll concentration in case 1 waters: a reexamination. Limnology and Oceanography, 43:847-858.

Maffione, R.A. and D.R. Dana. 1997. Instruments and methods for measuring the backward-scattering coefficient of ocean waters. Applied Optics, 36:6057-6067.

Maritorena, S., A. Morel and B. Gentili. 1994. Diffuse reflectance of oceanic shallow waters: influence of water depth and bottom albedo. Limnology and Oceanography, 39:1689-1703.

McKee, D., A. Cunningham, J. Slater, K.J. Jones and C.R. Griffiths. 2003. Inherent and apparent optical properties in coastal waters: a study of the Clyde Sea in early summer. Estuarine, Coastal and Shelf Science, 56:369-376.

Miller, R.L., M. Belz, C.E. Del Castillo and R. Trzaska. 2002. Determining CDOM absorption spectra in diverse aquatic environments using a multiple pathlength, liquid core waveguide system. Continental Shelf Research, 22:1301-1310.

Miller, R.L. and E.J. D'Sa. 2002. Evaluating the influence of CDOM on the remote sensing signal in the Mississippi River Bight. In: EOS Transactions AGU, Honolulu, HI, pp. 171.

Mitchell, B.G. and D.A. Kiefer. 1988. Variability in the pigment specific fluorescence and absorption spectra in the northeastern Pacific Ocean. Deep Sea Research, Part A, 35:665-689.

Mitchell, B.G. and O. Holm-Hansen. 1991. Bio-optical properties of Antarctic Peninsula waters: Differentiation from temperate ocean models. Deep Sea Research, 38:1009-1028.

Mobley, C. D., B. Gentili, H.R. Gordon, Z. Jin, G.W. Kattawar, A. Morel, P. Reinersmann, K. Stamnes, and R.H. Stavn. 1993. Comparison of numerical models for computing underwater light fields. Applied Optics, 32:7484-7504.

Mobley, C.D. Light and Water: Radiative transfer in natural waters. San Diego: Academic Press (1994) 592 pp.

Mobley, C.D. 1999. Estimation of the remote-sensing reflectance from above-surface measurements. Applied Optics, 38:7442-7455.

Mobley, D.D., L.K. Sundman and E. boss. 2002. Phase function effects on oceanic light fields. Applied Optics, 41:1035-1050.

Moore, C.M., J.R.V. Zaneveld and J.C. Kitchen. 1992. Preliminary results from an in situ spectral absorption meter. Ocean Optics XI, Proc. SPIE Int. Soc. Opt. Eng., 1750:330-337.

Morel, A. 1974. Optical properties of pure water and sea water 1. In: Optical Aspects of Oceanography. N.G. Gerlov and E. Steemann-Nielsen (eds.), Academic Press, New York, 1-24 pp.

Morel, A. and L. Prieur. 1977. Analysis of variations in ocean color. Limnololgy and Oceanography, 22:709-722.

Morel, A. 1988. Optical modeling of the upper ocean in relation to its biogenous matter content (case 1 waters). Journal of Geophysical Research, 93:10749-10768.

Morel, A. 1991. Light and marine photosynthesis: A spectral model with geochemical and climatological implications. Progress in Oceanography, 26:263-306.

Morel, A. and B. Gentili. 1991. Diffuse reflectance of oceanic waters: its dependence on Sun angle as influenced by the molecular scattering contribution. Applied Optics, 30:4427-4438.

Morel, A. and B. Gentili. 1993. Diffuse reflectance of oceanic waters. II. Bidirectional aspects. Applied Optics, 32:6864-6879.

Morel, A. and B. Gentili. 1996. Diffuse reflectance of oceanic waters. III. Implication of bidirectionality for the remote-sensing problem. Applied Optics, 35:4850-4862.

Morel, A. and H. Loisel. 1998. Apparent optical properties of oceanic water: dependence on the molecular scattering contribution. Applied Optics, 37:4765-4776.

Morel, A. and S. Maritorena. 2001. Bio-optical properties of oceanic waters: A reappraisal. Journal of Geophysical Research, 106:7163-7180.

Mueller, J.L. and R.W. Austin. 1995. Ocean optics protocols for SeaWiFS validation, revision 1. In: S. B. Hooker & E. R. Firestone (Eds.), NASA Technical Memo. 104566, vol. 25, Greenbelt, MD: NASA Goddard Space Flight Center, Greenbelt, MD., 67 pp.

Muller-Karger, F.E., J. Walsh, R.H. Evans and M.B. Meyers. 1991. On the seasonal phytoplankton concentration and sea surface temperature cycles of the Gulf of Mexico as determined by satellites. Journal of Geophysical Research, 96:12645-12665.

Munday, J.C. and T.T. Alfoldi. 1979. LANDSAT test of diffuse reflectance models for aquatic suspended solids measurements. Remote Sensing of the Environment. 8:169-183.

Oishi, T. 1990. Significant relationship between the backward scattering coefficient of sea water and the scatterance at 120°. Applied Optics, 29:4658-4665.

O'Reilly, J.E., S. Maritorena, B.G. Mitchell, D.A. Siegel, K.L. Carder, S.A. Garver, M. Kahru and C. McClain. 1998. Ocean color algorithms for SeaWiFS. Journal of Geophysical Research, 103:24937-24953.

O'Reilly, J.E. et al. 2000. Ocean color chlorophyll *a* algorithms for SeaWiFS, OC2 and OC4: Version 4, In: SeaWiFS Postlaunch Calibration and Validation Analyses. Part 3, S. B. Hooker and E. R. Firestone (Eds). NASA Technical Memo. 206892, vol. 11, NASA Goddard Space Flight Center, Greenbelt, MD., 8-22 pp.

Parsons, T.R., Y. Maita and C.M. Lalli. A manual of chemical and biological methods for seawater analysis. 1st edn. Pergamon, Oxford (1984) 173 pp.

Petzold, T.J. 1972. Volume scattering functions for selected ocean waters, Scripps Institute of Oceanography, San Diego. Ref. 72-78, 79 pp.

Pegau, W.S., D. Gray and J.R.V. Zaneveld. 1997. Absorption and attenuation of visible and near-infrared light in water: Dependence on temperature and salinity. Applied Optics, 36:6035-6046.

Platt, T., S. Sathyendranath, C.M. Caverhill and M.R. Lewis. 1988. Ocean primary production and available light: further algorithms for remote sensing. Deep Sea Research, 35:855-879.

Pope, R.M. and E.S. Fry. 1997. Absorption spectrum (380-700 nm) of pure water. II. Integrating cavity measurements. Applied Optics, 36:8710-8723.

Prieur, L. and S. Sathyendranath. 1981. An optical classification of coastal and oceanic waters based on the specific spectral absorption of phytoplankton pigments, dissolved organic matter and particulate materials. Limnology and Oceanography, 26:671-689.

Reynolds, R.A., D. Stramski and B.G. Mitchell. 2001. A chlorophyll-dependent semi analytical reflectance model derived from field measurements of absorption and backscattering coefficients within the Southern Ocean. Journal of Geophysical Research, 106:7125-7138.

Risovic, D. 2002. Effect of suspended particulate-size distribution on the backscattering ratio in the remote sensing of seawater. Applied Optics, 41:7092-7101.

Roesler, C.S. 1998. Theoretical and experimental approaches to improve the accuracy of particulate absorption coefficients derived from the quantitative filter technique. Limnology and Oceanography, 43:1649-1660.

Roesler, C.S. and M.J. Perry. 1995. In situ phytoplankton absorption, fluorescence emission, and particulate backscattering spectra determined from reflectance. Journal of Geophysical Research, 100:13279-13294.

Roesler, S.R., M.J. Perry and K.L. Carder. 1989. Modeling in situ phytoplankton absorption from total absorption spectra in productive inland marine waters. Limnology and Oceanography, 34:1510-1523.

Sathyendranath, S., and T. Platt. 1989. Remote sensing of ocean chlorophyll: consequence of nonuniform pigment profile. Applied Optics, 28:490-495.

Sathyendranath, S., L. Prieur and A. Morel. 1989. A three-component model of ocean color and its application to remote sensing of phytoplankton pigments in coastal waters. International Journal of Remote Sensing, 10:1373-1394.

Satyendranath S. 2000. Remote sensing of ocean color in coastal, and other optically complex waters. In: *Report of the International Ocean Color Coordinating Group.* Darmouth,Canada, Mooers, M. J. Bowman and B. Zeitschel (eds.), Springer-Verlag, New York 145 pp.

Smith, R.C. and K.S. Baker. 1981. Optical properties of the clearest natural waters (200-800 nm). Applied Optics, 20:177-184.

Sosik, H.M. and B.G. Mitchell. 1995. Light absorption by phytoplankton, photosynthetic pigments and detritus in the California Current. Deep_Sea Research I, 42:1717-1748.

Sosik, H.M., R.E. Green, W.S. Pegau and C.S. Roesler. 2001. Temporal and vertical variability in optical properties of New England shelf waters during late summer and spring. Journal of Geophysical Research, 106:9455-9472.

Stramska, M., D. Stramski, B.G. Mitchell and C.D. Mobley. 2000. Estimation of the absorption and backscattering coefficients from in-water radiometric measurements. Limnology and Oceanography, 45: 628-641.

Stramski, D. and D. A. Kiefer. 1991. Light scattering by microorganisms in the open ocean. Progress in Oceanography, 28: 343-383.

Stumpf, R.P. 1988. Sediment transport in Chesapeake Bay during floods: Analysis using satellite surface observations. Journal of Coastal Research, 4:1-15.

Stumpf, R.R., M.L. Frayer, M.J. Durako and J.C. Brock. 1999. Variations in water clarity and bottom albedo in Florida Bay from 1985 to 1997. Estuaries, 22:431-444.

Sydor, M., R.A. Arnone, R.W. Gould, G.E. Terrie, S.D. Ladner, and C.G. Wood. 1998. Remote sensing technique for determination of volume absorption coefficient of turbid water. Applied Optics, 37:4944-4950.

Tassan, S. 1994. Local algorithms using SeaWiFS data for the retrieval of phytoplankton, pigments, suspended sediment, and yellow substance in coastal waters. Applied Optics, 33:2369-2377.

Tassan, S. and G.M. Ferrari. 1995. An alternative approach to absorption measurements of aquatic particles retained on filters. Limnology and Oceanography, 40:1358-1368.

Toole, D.A., D.A. Siegel, D.W. Menzies, M.J. Neumann and R.C. Smith. 2000. Remote-sensing reflectance determinations in the coastal ocean environment: impact of instrumental characteristics and environmental variability. Applied Optics, 39:456-468.

Toole, D.A. and D.A. Siegel. 2001. Modes and mechanisms of ocean color variability in the Santa Barbara Channel. Journal of Geophysical Research, 26985-27000.

Trees, C.C., M.C. Kennicutt II and J.M. Brooks. 1985. Errors associated with the standard fluorometric determination of chlorophylls and phaeopigments. Marine Chemistry, 17:1-12.

Twardowski, M.S., J.M. Sullivan, P.L. Donaghay and J.R.V. Zaneveld. 1999. Microscale quantification of the absorption by dissolved and particulate material in coastal waters with an ac-9. Journal of Atmospheric and Oceanic Technology, 16:691-707.

Twardowski, M.S., E. Boss, J.B. MacDonald, W.S. Pegau, A.H. Barnard, and R.V. Zaneveld. 2001. A model for estimating bulk refractive index from the optical backscattering ratio and the implications for understanding particle composition in Case I and II waters. Journal of Geophysical Research, 106:14129-14142.

Twardowski, M.S. and P.L. Donaghay. 2001. Separating in situ and terrigenous sources of absorption by dissolved materials in coastal waters. Journal of Geophysical Research. 106:2545-3560.

UNESCO, 1994. Protocols for the Joint Global Ocean Flux Study (JGOFS) Core measurements. Manual and Guides. 29: 170 pp.

Vodacek, A., N.V. Blough, M.D. DeGrandpre, E.T. Peltzer and R.K. Nelson. 1997. Seasonal variation of CDOM and DOC in the Middle Atlantic Bight: Terrestrial inputs and photooxidation. Limnology and Oceanography, 42:674-686.

Voss, K.J., J.W. Nolten and G. D. Edwards. 1986. Ship shadow effects on apparent optical properties. In: Proceedings of Ocean Optics VIII. M Blizard, (Ed.), 637:186-190.

Whitlock, C.H., L.R. Poole, J.W. Usry, W.M. Houghton, W.G. Witte, W.D. Morris and E.A. Gurganus. 1981. Comparison of reflectance with backscatter and absorption parameters for turbid waters. Applied Optics, 20:517-522.

Wollast, R. 1991. The coastal organic carbon cycle: Fluxes, sources and sinks. In: Ocean margin processes in global change. R.F. Mantoura, J.-M. Martin and R. Wollast, (eds). Wiley and Sons, Chichester, England, 469 pp.

Wright, S.W., S.W. Jeffrey, R.F.C. Mantoura, C.A. Llewellyn, T. Bjornland, D. Repeta and N. Welschmeyer. 1991. Improved HPLC method for the analysis of chlorophylls and carotenoids from marine phytoplankton. Marine Ecology Progress Series, 77:183-196.

Zaneveld, J.R.V., D.M. Roach and H. Pak. 1974. The determination of the index of refraction distribution of oceanic particulates. Journal of Geophysical Research, 79:4091-4095.

Zaneveld, J.R.V., J.C. Kitchen, A. Bricaud and C.M. Moore. 1992. Analysis of in situ spectral absorption meter. In: Proceedings of Ocean Optics XI, 1750:187-200.

Zaneveld, J.R.V., J.C. Kitchen and C. Moore. 1994. The scattering error correction of reflecting-tube absorption meters. In: Proceedings of Ocean Optics XII, J. S. Jaffe (Ed.), 44-55.

Chapter 7

REMOTE SENSING OF ORGANIC MATTER IN COASTAL WATERS

CARLOS E. DEL CASTILLO

National Aeronautics and Space Administration, Earth Science Applications Directorate, Stennis Space Center, MS, 39529 USA

1. Introduction

The atmosphere, land, oceans, and sediments interact in the coastal environment through sometimes contrasting processes that result in the burial, destruction of old and formation of new organic carbon. Colored dissolved organic matter (CDOM), a large fraction of the organic matter pool in rivers, interferes with remote-sensing determinations of chlorophyll concentrations that are needed to estimate primary productivity (Müller-Karger et al., 1989) and, in many instances, is the most important factor controlling light penetration in coastal waters (Del Castillo et al., 1999). Kalle (1966) originally coined the term Gelbstoff from the German yellow substance. Others have suggested names like gilvin (Kirk, 1994) and chromophoric dissolved organic matter. Here it will be referred to as CDOM.

The influence of CDOM is not limited to coastal areas adjacent to a river mouth. Large rivers can broadcast their influence into waters that are normally considered oligotrophic. Muller-Karger et al. (1989) documented how the Orinoco River plume enters the eastern Caribbean, almost reaching the southern coast of Puerto Rico. Del Castillo et al. (2001) documented an intrusion of the Mississippi River plume into normally oligotrophic waters of the Gulf of Mexico (Fig. 1). In both cases, large amounts of terrigenous organic carbon were injected thousands of miles into the ocean, and the presence of CDOM contributed to erroneous remote-sensing estimates of chlorophyll by ocean color satellite sensors.

Processes that alter the concentration and chemical properties of CDOM result in changes in the optical properties of the water column. Understanding these processes is important for the development of better remote-sensing techniques in coastal waters and for a better quantification of carbon budgets. This chapter discusses the relationship between chemical and optical properties of CDOM, how biogeochemical processes in coastal environments affect these properties, and the resulting implications for the remote sensing of organic matter in coastal waters.

2. The Chemistry of Color

The optical properties of organic matter (its color and all photochemical processes) are determined by interactions between electromagnetic radiation and the organic molecule's outer shell of electrons. The color of a substance results from the preferential absorption of certain wavelengths and the scattering of others. A good example is the leaf of a tree. Chlorophyll, the pigment responsible for the color of plants, has a strong

R.L. Miller et al. (eds.), Remote Sensing of Coastal Aquatic Environments, 157–180.
© 2005 *Springer.*

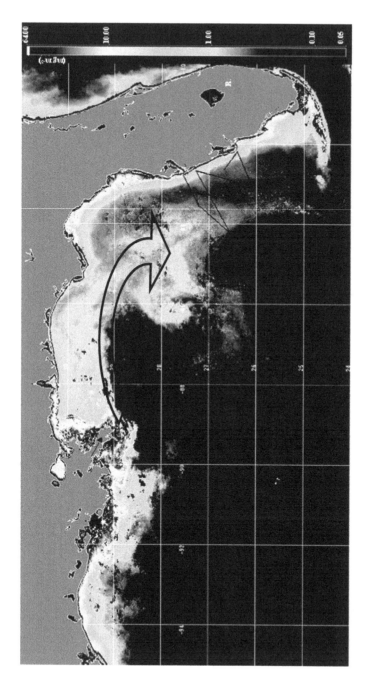

Figure 1. SeaWiFS image obtained in 1998 showing the Mississippi River plume intruding into normally oligotrophic waters of the Gulf of Mexico. The arrow shows the river flow and the line off Tampa Bay represents the cruise track of the R/V Suncoaster (Del Castillo et al., 2001). The image was processed by Frank Müller-Karger (University of South Florida). Chlorophyll concentrations shown in the image were retrieved using the OC4 standard chlorophyll algorithm. In general, chlorophyll concentrations were overestimated. This was caused in part by large CDOM absorption in the Mississippi River Plume.

absorption bands in the blue and in the red. Green light is not strongly absorbed but scattered (reflected or transmitted) resulting in the color of foliage (Fig. 2). The color of an organic molecule is a manifestation of its chemical structure and its chemical environment. This relationship is the basis for the use of several spectroscopic techniques (IR, UV-Vis, and fluorescence spectroscopy) to study the composition and structure of organic compounds. This relationship becomes more relevant to the application of remote sensing in areas where biogeochemical processes change the composition and concentration of organic matter and the color of the water – the water leaving radiance that is measured by a remote sensor.

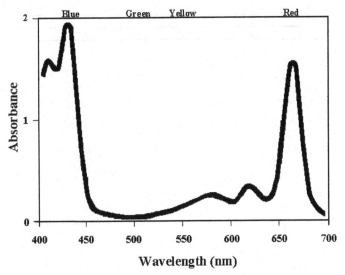

Figure 2. Absorption spectrum of chlorophyll *a* showing the correspondence between color and wavelength. Note the low absorption of green wavelengths.

The phenomenon of light absorption and subsequent photochemical events are best explained using the Jablònski diagram (Fig. 3). Light absorption occurs through interactions between photons and electrons in the ground state (S_0). Upon absorption of a photon, an electron at the ground state is promoted to an excited (antibonding) energy level (S_1 or S_2). These states are unstable, therefore, electrons relax to the lowest vibrational level within S_1. This transition is fast ($<10^{-12}$ seconds) and involves the release of energy through vibrational relaxation and internal conversions. From the S_1 state the electron can return to the ground state releasing energy through collisions with other molecules, intersystem crossing, or by emitting a photon as fluorescence. In some molecules, an electron in the S_1 level can change its spin resulting in a metastable triplet state. From the triplet state (T_1) relaxation can occur through emission of a photon in a process called phosphorescence.

The Jablònski diagram shows that the energy of the emitted photons is lower than that of the absorbed light. Therefore, the emitted photons have longer wavelengths (less energetic) than the excitation photon. This phenomenon is known as Stokes's shift (after George Stokes -1819-1903, Lucasian Professor at Cambridge). In most cases, the emission spectrum of a pure fluorophore is the mirror images of the S_0 to S_1 absorption peak. There are few exceptions to this rule usually caused by multiple geometrical arrangements of the excited state.

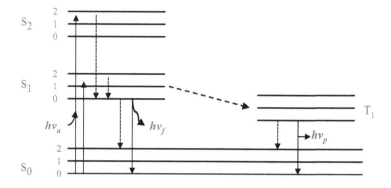

Figure 3. A version of the Jablònski diagram showing electronic transitions resulting in the absorption of a photon (hv_a –solid upward lines). The diagram shows the singlet ground, first, and second electronic states (S_0, S_1, and S_2). Within each of these levels electrons can be at several vibrational levels (showing only 0, 1, and 2 for simplicity). T_1 denotes the triplet state. Dashed lines indicate internal transitions that result in losses of energy without emission of a photon. Solid downward lines indicate releases of energy by emission of a photon as fluorescence (hv_f) or phosphorescence (hv_p). The terms singlet and triplet come from the number of possible orientations of the spin angular momentum corresponding to the electron total spin quantum number. Calculated as 2S + 1, where S is the summation of the spins. For the singlet state, $2(-\frac{1}{2} + \frac{1}{2}) +1 = 1$; for the triplet state $2(\frac{1}{2} + \frac{1}{2}) +1 = 3$.

In organic molecules, light absorption in the near UV and visible wavelengths is mainly caused by double bonds between atoms of carbon (π bonds) and by non-bonding (n) to σ^* or n to π^* transitions. The n to σ^* transition corresponds to absorption in the far UV, like the non-bonding pairs of oxygen electrons in methanol (CH_3OH). Other interactions between non-bonding electrons and π systems can have significant effects upon the absorption spectra.

Figure 4 shows the electronic transition between S_0 and excited S_n for a π bond, and the consequent change in electronic density distribution from π (bonding) to π^* (anti-bonding).

The simplest way to explain the relationship between chemical characteristics and optical properties in an organic molecule is to show how optical properties change as we modify the structure of a molecule. We will start with the simple organic compound Ethane,

$$CH_3 - CH_3.$$

The carbon-carbon single bonds (σ bonds) are very strong and require high energie to produce a σ to σ^* transition. The wavelength associated with such transition energy can be calculated as

$$\lambda = hc/\Delta E, \tag{1}$$

where λ is the absorption wavelength, h is Planck's constant (6.63×10^{-34} J s), c is the speed of light (3×10^8 m s^{-1}), and ΔE is the energy needed to excite an electron from the bonding to the antibonding level. For a single carbon bond, as in Ethane the σ to σ^* (C-C) transition energy is about 921 kJ mol^{-1}. The corresponding wavelength is ~130 nm,

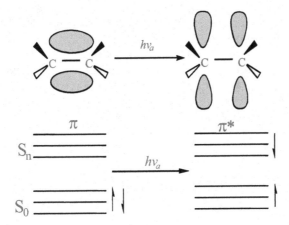

Figure 4. Electronic transition and change in geometry of the π bond in Ethene.

beyond the UV range and cannot be measured with conventional UV-Vis spectrophotometers. A similar hydrocarbon, Ethene, ($CH_2=CH_2$) contains σ and a π bonds. The π electrons require less energy to be excited to π^* and the absorption wavelength is ~ 160 nm.

Increasing the number of carbon double bonds can result in conjugation. Electrons in a conjugated π system are delocalized, shared by all the atoms in the system. As a consequence, the area of probability where the electrons can be found is larger (more separation between electrons and nuclei) and the energy needed to promote these electrons is lower. The result is that a molecule like β-carotene (Fig. 5), with 11 conjugated π bonds, has a broad absorption spectrum, with two λ max at ~452 and 478 nm, depending on the solvent.

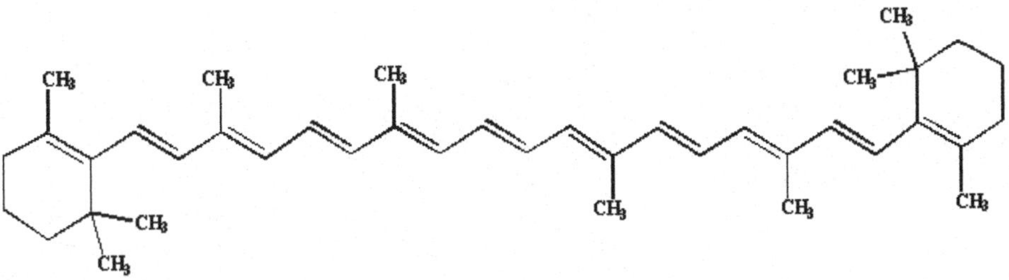

Figure 5. Molecular structure of β-carotene showing conjugation of π bonds.

Another effect of the delocalized electrons is that there is a larger number of possible vibrational energy sub-levels within the ground state (within S_0, Fig. 3). The Bohr condition states that the energy required to force a transition from any S_0 sub-level to an excited state must correspond to the energy of the photon that is absorbed, and the energy of the photon is related to its wavelength by $\lambda = hc/\Delta E$. Therefore, as the number of possible energy transitions increases, so does the range of possible wavelengths that can be absorbed. This causes broadening of the absorption spectrum. Conversely, if one reduces the possible number of vibrational states by shortening the π system (or by cooling the sample, which reduces the vibration of the electrons), the spectrum gets narrower. As a general rule, reactions that increase the length of the conjugated system result in stronger absorption at longer wavelengths. Addition of groups that extend the π system, for example non-bonding electrons in carbonyl groups, also result in absorption at longer wavelengths. Conversely, changes that shorten the length of the conjugated system result in light absorption at shorter wavelength. The relationship between chemical structure and optical properties of organic matter is a fascinating and extensive subject albeit away from the scope of this book. The treatise by Calvert and Pitts (1966) *Photochemistry* and the introductory text *Introduction to Spectroscopy* (Pavia et al., 2000) are good sources of information. In following sections, I will discuss how biogeochemical processes may result in changes in concentration and optical properties of CDOM.

3. Organic Matter in Natural Waters

CDOM is defined operationally by the method used to separate suspended and dissolved material. The two most common methods are filtration through glass-fiber filters (GF/F 0.7 μm) and polycarbonate or polysulfone membranes (0.2 μm). Both methods have their advantages and disadvantages. GF/F filters allow passage of some bacteria, viruses, and colloids, which are not considered dissolved materials. However, GF/F filters can be ashed (500 °C), reducing the possibility of sample contamination. Membrane filters exclude more suspended particles, but are difficult to clean. In blue waters, the concentration of suspended particles is so low that filtration for dissolved organic carbon (DOC) and CDOM measurements may be unnecessary and risky as samples could get contaminated during the procedure. However, in coastal environments, filtration is necessary and the differences in CDOM content between GF/F and membrane filtrates can be significant (Del Castillo, unpublished data). There is a final consideration. Bio-optics research strives to account for the contribution of all light-absorbing species in water. The method most commonly used for measuring absorption by particles (alive and dead) involves capture of the particles unto GF/F filters. If membrane filters are used to obtain CDOM samples, a fraction of organic material will not be accounted for. The significance of this fraction should be evaluated before making a decision on which CDOM filtration method will be used.

The organic matter in seawater is not composed of a few types of organic molecules but of countless uncharacterized organic compounds. Humic and Fulvic acids, important components of CDOM, were originally discovered and studied in soils. For this reason, the study of these substances in the ocean inherited the jargon and operational definitions from Soil Chemistry. Humic and Fulvic acids are separated by their solubility at different pH. Their source and mechanisms of formation have been a very active area of research since their discovery in the 17th century. There are two main schools of thought in soil chemistry explaining the formation of these organic acids: (1) that they are derivatives of lignin and formed by decomposition of plant detritus; and, (2) that they are derivatives of sugars. The presence of phenol in Humic and Fulvic acid and the susceptibility of sugars to biodegradation has resulted in the prevalence of the first theory. However, the presence of Humic and Fulvic acids (but relative absence of lignin) in the oceans, and the discovery of new formation mechanisms, have gained the sugar hypothesis new followers.

The source of marine CDOM is more obscure. Clearly, the sources of carbon have to be primary productivity and respiration by-products in the ocean; however, most of these products are carbohydrates and lipids that are not colored. The mechanisms that result in the modification of these compounds to form CDOM are in dispute. Harvey et al. (1983,1984; Harvery and Boran, 1985) proposed a mechanism involving UV-induced cross-linking reactions between lipids. This mechanism was challenged by Laane (1984), who proposed polymerization reactions between sugars and amino acids. Both mechanisms have their merits and pitfalls. Others have proposed that marine organic matter could be the result of exudation from algae, and recently Azam and Worden (2004) stressed the importance of bacterial processes in the formation and destruction of organic matter in the oceans. In a recent study, Bontempi et al. (2003) found that CDOM from Laguna Madre, a small lagoon connecting to the Sinaruco River, Venezuela, has excitation-emission spectra identical to those of water leachates from tree leaves collected around the lagoon. This shows that plant material can be a source of CDOM in some rivers. Regardless of the mechanisms and sources of material responsible for the formation of CDOM, one must remember that in river-dominated coastal environments, the organic matter is a mixture of compounds of terrestrial and marine origin. The practical result is that there are strong gradients in chemical and optical properties of CDOM.

The chemical differences between marine and terrestrial CDOM explain the differences in their optical properties. The carbon skeleton of terrestrial organic matter is larger, more complex, and contains more aromatic groups than marine organic matter. The marine organic matter skeleton is smaller, simpler, more aliphatic, and contains more carboxylic groups and sugars than terrestrial organic matter. The higher degree of aromaticity and molecular complexity in terrestrial CDOM, when compared to marine CDOM, result in stronger absorption coefficients and more absorption towards longer wavelengths – red-shift. Fluorescence spectra are equally red-shifted, indicating a higher degree of π conjugation and molecular complexity. Figure 6 shows the difference between CDOM fluorescence spectra from two seawater samples with different riverine CDOM contributions.

Optical properties of CDOM change along river plumes (Fig. 7) reflecting the differences in chemical composition between riverine and marine end-members. Longer emission wavelengths (river water) indicate complex chemical structures, including extended aromatic rings, whereas shorter emission wavelengths (seawater) indicate simpler structures with localized aromatic rings. Changes along the salinity gradient are caused by the mixing between end-members, and probably by photodegradation of CDOM in clear, high-salinity waters.

The absorption spectra of CDOM, unlike that of a pure compound, show a featureless increase in absorption with decrease in wavelength (Fig. 8). Such absorption spectra could be explained by the overlapping of the absorption spectra of the different chromophores found in natural waters. However, recent work shows that part of the absorption, particularly that at wavelengths longer than 350 nm, cannot be explained by a simple superposition of the absorption spectra of non-interacting chromophores, and has been instead proposed to result from intramolecular electron transfer between hydroxyaromatic donors and quinoid acceptors (Del Vecchio and Blough, 2002, 2004).

The absorption curve of CDOM can be fitted to an exponential equation of the form $a(\lambda) = a(\lambda_0)e^{-S(\lambda-\lambda_0)}$, where $a(\lambda)$ is the absorption coefficient at any λ, $a(\lambda_0)$ is the absorption measured at a reference wavelength λ_0, typically 400 or 440 nm. The absorption coefficient is calculated as 2.023A/l, where A is the absorbance (Log I_0/I), l is the pathlength (m), and 2.303 is the conversion factor between Log_{10} and ln. S is the spectral slope calculated from the least square regression of λ vs. ln $a(\lambda)$, which results in a linear plot. Non-linear fitting techniques have also been used to calculate S (Blough and Del Vecchio, 2002). The method used to calculate S should always be reported for

Del Castillo

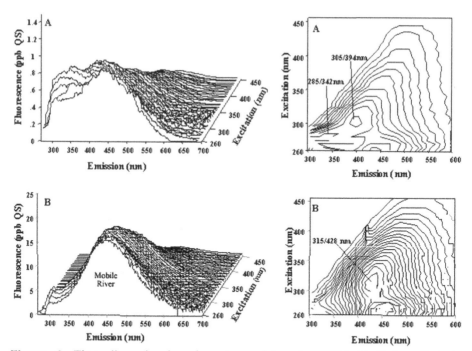

Figure 6. Three-dimensional and contour plots of CDOM Excitation Emission fluorescence matrices (Coble, 1996) for samples collected along the plume of the Mobile River, USA. Sample A had low riverine CDOM concentrations (salinity ~ 35). Sample B (salinity ~26) was collected near the mouth of the river and had higher quantities of riverine CDOM than sample A. Fluorescence units are in equivalents of quinine sulfate (QS).

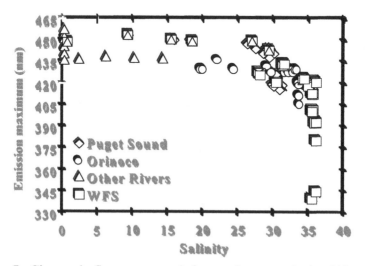

Figure 7. Changes in fluorescence emission maxima at excitation 310 nm for several river plumes in the world. WFS denotes several rivers from the West Florida Shelf. Data for Puget Sound and Other Rivers are from Coble (1996).

the use of different wavelength intervals, and curve-fitting methods can produce different results. Blough and Del Vecchio (2002) present a very detailed discussion on the subject.

The spectral slope, *S*, can be used to study changes in chemical properties of organic matter. Carder et al. (1989) related changes in *S* to variations in the proportion of Humic and Fulvic acids in CDOM. Along a river plume, *S* and fluorescence maxima (Figs. 7 and 9) do not change in low salinity waters due to the preponderance of CDOM from the fresh-water end-member. However, changes in *S* are observed when the riverine CDOM has been diluted enough by seawater to allow for the detection of marine CDOM. The salinity at which changes in optical properties can be observed is determined by the concentration of CDOM in the riverine end-member.

The changes in optical properties shown in Figures 7 and 9 are mainly caused by mixing between riverine and seawater end-members. These changes can be modeled easily using a simple linear mixing model (Del Castillo et al., 2000). However, processes other than mixing can also be of importance. The following sections discuss some of these processes and their consequences to optical properties and concentrations of CDOM, and how they affect remote sensing of organic matter.

4. Biogeochemical Processes Responsible for Changes in CDOM Properties in Coastal Environments

The processes controlling CDOM concentration and optical properties are mixing, flocculation and particle aggregation, photodegradation, bacterial degradation, resuspension, and biological production. These processes could change the optical

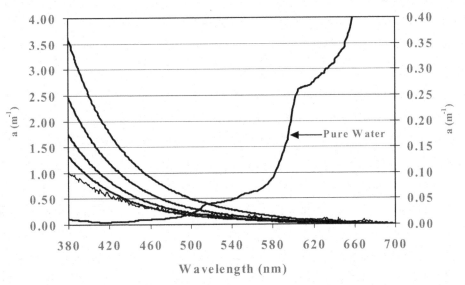

Figure 8. Absorption spectra of CDOM and pure water. CDOM spectra correspond to samples collected in the Mississippi River plume and Gulf of Mexico. From top to bottom, the salinities of the samples were 0, 5, 15, 19, and 35. The absorption scale for pure-water and for the blue-water sample (salinty = 35, bottom curve) are in the right Y-axis. Pure-water values are from Pope and Fry (1997).

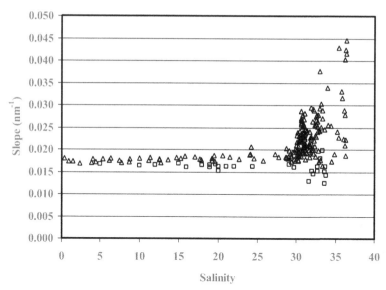

Figure 9. Spectral slopes vs. salinity for samples collected along the Mississippi and Orinoco River plumes (□), and Delaware and Chesapeake bays, and the Mid Atlantic Bight (DCB-MAB, Δ). The slopes for the river samples were calculated from 300 nm to the λ, at which absorption values reached the accuracy limit of the instrument (~0.046 m^{-1} for a 10 cm cell) using linear regression analysis. Spectral slopes for the DCB-MAB were calculated from data between 290 and 700 nm using non-linear regression methods. DCB-MAB data are from Del Vecchio and Blough, (2004).

properties of CDOM by destroying chromophores or by changing the distribution of dissolved compounds. This section presents an overview of these processes and the effect that they could have on CDOM.

Flocculation has been reported in low-salinity regions of estuaries and river plumes where the fresh water end-member starts mixing with seawater. As a result, variable amounts of organic matter can be removed from the dissolved pool, incorporated into the particulate organic carbon fraction, removed from the water column, and potentially buried. In areas with high sedimentation rates, this process could represent a significant sink of river organic matter. Sholkovitz (1976) estimated that 3 to 11% of the dissolved organic matter (DOM) is removed depending on the river. Flocculation, if present, should preferentially eliminate from solution large molecules (see De Souza Sierra et al., 1997). Such change in size distribution of chromophores could result in changes in optical properties (see Carder et al., 1989). The assumption is that large compounds can have larger conjugated systems and, therefore, should have absorption maxima at longer wavelengths; eliminating them from the CDOM pool will decrease absorption at longer wavelengths, resulting in a change in the fluorescence and absorption spectra (for example, increasing *S*). However, Del Castillo et al. (2000) performed a series of flocculation experiments to assess the effect of these processes upon concentration and optical properties of CDOM during estuarine mixing. In their experiments, river water was mixed at different proportions with seawater, filtered and analyzed for changes in CDOM concentration and optical properties. Simple mixing models were used to predict changes in the concentrations and optical properties of CDOM based exclusively on conservative behavior (see also De Sousa-Sierra et al., 1997). Results showed no difference in the behavior of the mixtures and models, suggesting absence of flocculation.

Particle adsorption may also be an important mechanism for the removal of CDOM. Recent experiments by Shank et al. (2004) showed that high-molecular-weight CDOM preferentially sorbs to particles, resulting in an increase in the spectral slope of CDOM. However, they also reported that in natural conditions less than 1% of the CDOM will be extracted from the water column by particle adsorption. Although flocculation and adsorption apparently have little effect upon optical properties of CDOM, their contribution to carbon burial in estuarine areas may be significant.

Photodegradation is a very active process in coastal environments, probably responsible for the destruction of most of the riverine CDOM that enters the oceans. Morell and Corredor (2001), for example, showed that exposure of CDOM-rich waters from the Gulf of Paria to solar radiation resulted in exponential decay of CDOM fluorescence and that in the Orinoco River plume (as detected by fluorescence) was depleted through photodegradation as the plume traverses the eastern Caribbean. Photodegradation changes the optical properties of CDOM by preferentially degrading chromophores that absorb light at mid-to-long wavelengths through direct and secondary photoreactions. However, in estuaries with high CDOM inputs and low residence times, the effects of photodegradation upon optical properties are often negligible. Blough and Del Vecchio (2002) showed that for photodegradation to have an effect on optical properties of the water column (natural samples, not incubations), the first optical depth should be similar to the mixed layer depth. This allows for the cumulative effect of photodegradation on the thin surface layer to propagate through the water column. Similarly, Del Castillo et al. (1999) only found evidence for CDOM photodegradation in areas of the Orinoco River Plume where light penetration was high. However, in estuaries with low flow and high residence times, the effects of photodegradation may be significant. For example, Twardowski and Donaghay (2002) observed changes in absorption spectra of CDOM during incubation experiments in waters of East Sound (Washington, USA). They concluded that cumulative photodegradation was responsible for the high spectral slopes observed in oceanic waters. Other studies have shown that photodegradation processes occur on time scales of weeks to months (Vodacek et al., 1997; Opsahl and Benner, 1998), explaining why conservative behavior is most often observed in river plumes. After all, photodegradation can only occur during day-time, whereas mixing of river and seawater is a continuous process.

Biological activity releases organic matter into water through exudation by organisms, viral lysis, and sloppy feeding by zooplankton. Consisting mainly of lipids, carbohydrates, and proteins, these compounds can increase the concentration of DOC, but not the concentration of CDOM as they are mostly colorless. They are also labile, so their contribution to the dissolved carbon pool can be brief. However, there is evidence of formation of CDOM under bloom conditions, or in estuaries with high productivity and long residence times. For example, Twardowski and Donaghay (2001) reported increases in CDOM absorption at 412 nm that were well-correlated with chlorophyll concentrations. Kahru and Mitchell (2001) reported good correlation between remote sensing-estimates of CDOM and chlorophyll in waters off the California current. Cannizzaro et al. (2002) and Carder et al. (2004) also reported increases in CDOM associated with a bloom of the toxic dinoflagellate *Karenia brevis* in waters of the West Florida Shelf. Clearly, blooms can release CDOM. However, there is no evidence that these colored compounds are as recalcitrant as riverine or old marine CDOM, so their contribution to ocean color could be ephemeral. Changes in CDOM due to bacterial degradation are more difficult to observe due to the recalcitrant nature of CDOM and the effect on CDOM absorption is usually very small (Moran and Zepp., 1997; Miller and Moran., 1997). However, Kieber et al. (1989, 1990) showed that photodegradation of CDOM results in the release of low-molecular weight carbonyl compounds that are labile, so bacterial degradation of CDOM could follow partial photodegradation.

Mixing is the dominating factor controlling concentrations and optical properties of CDOM along a river plume. During mixing between fresh and seawater, the high concentration of river CDOM masks the optical signal of the marine end-member. As a result, changes in optical properties are usually negligible at salinities below ~30 (See Figs. 7 and 8 in previous section – Blough et al., 1993; Coble, 1996; De Souza Sierra et al., 1997; Del Castillo et al., 1999). Other researchers observed conservative behavior of DOC in estuaries (e.g. Laane, 1981; Mantoura and Woodward, 1983; Ferrari, 2000), and CDOM (Blough et al., 1993, Del Castillo et al., 1999, 2000; Chen et al., 1999; Ferrari, 2000). Blough et al. (1993) observed conservative behavior in CDOM within the Gulf of Paria (Orinoco River). Minor deviations from conservative behavior outside the Gulf were attributed to temporal variability (probably also to inputs from rivers in Trinidad). Del Castillo et al. (1999) reported conservative behavior in both absorption and fluorescence of CDOM, within and outside the Gulf of Paria. Chen et al. (1999), working in four estuaries of North America (Boston Harbor, Chesapeake Bay, San Diego Harbor, and San Francisco Bay), reported that DOC is conservative in some, non-conservative in others, and that CDOM is mostly conservative. Del Castillo et al. (2000a) observed conservative behavior in both CDOM and DOC in rivers from Alabama and the West Florida coast (Mobile, Apalachicola, Suwannee, Hillsborough, Peace, and Cloosahatche rivers).

During conditions of low flow and long residence times, some of the non-conservative processes listed above could increase in importance, particularly photodegradation. However, mixing is a continuous process, whereas photodegradation and related processes can only occur under strong solar irradiance.

The molecular structure of CDOM is unknown. Therefore we do not know with certainty how the processes described above will affect its chemical structure and optical properties. However, CDOM is composed of photoreactive organic molecules so the general principles of photophysics described here apply. It is clear that more research is needed to elucidate the structure of CDOM. This information will provide valuable insight into the sources of CDOM and better understanding on the effect of biogeochemical and physical processes upon its optical properties.

5. CDOM and Ocean Color

The changes in concentration and optical properties of CDOM produced by the processes described above have profound effects upon the optical properties of the water column and remote-sensing measurements. Figure 10 shows remote-sensing reflectance curves modeled from inherent optical properties from waters along the Mississippi River plume. Remote-sensing reflectance at several wavelengths ($R_{rs}(\lambda)$) was calculated as

$$R_{rs}(\lambda) = C \frac{b_b(\lambda)}{a_T(\lambda) + b_b(\lambda)}, \tag{2}$$

where b_b is total backscattering, a_T is total absorption by water, CDOM, phytoplankton, and detritus, and C is a constant. To show how remote sensing reflectance changes dramatically due to changes only in CDOM absorption, R_{rs} was calculated taking only into consideration absorption and scattering by pure water, and CDOM absorption from several natural samples. The S parameters were nearly identical in these samples except for the one with salinity of 35. Therefore, the changes that we observed in R_{rs} are mainly the result of changes in CDOM absorption.

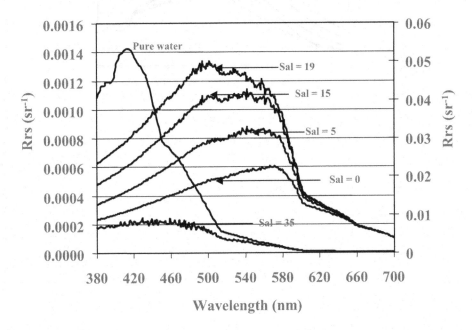

Figure 10. R_{rs} calculated from absorption and backscattering from pure water and CDOM absorption spectra collected in the Gulf of Mexico. The curves correspond the absorption spectra shown in Figure 8. Pure water and seawater R_{rs} scales are on the right axis.

It is well established that CDOM interferes with remote-sensing determinations of chlorophyll, especially with algorithms that use blue to green ratios. Figure 10 shows how changes in CDOM concentration can affect R_{rs} when the only absorbing components are water and CDOM. To present a more realistic picture of how CDOM can affect determinations of chlorophyll, we modeled R_{rs} for CDOM concentrations as in Figure 11, but added absorption of chlorophyll and b_b due to water and phytoplankton (Fig. 12) based on the model presented by Morel (1988). This model was developed for Case 1 water, however, it is used here because it offers a solution for b_b based on the contributions of pure water and a known concentration of chlorophyll. The absorption spectrum of phytoplankton used corresponded to a concentration of Chl of 6.6 mg m^{-3} and was left unchanged to show only the effects of changing CDOM. From these modeled R_{rs} curves we calculated chlorophyll concentrations using the default turbid water algorithm of Carder et al. (1999). Figure 10 shows the sensitivity of the algorithms to changes in CDOM. The results from top to bottom are 28, 53, 87, 125 mg m^{-3}. This clearly shows the need to account for the CDOM contribution to R_{rs}.

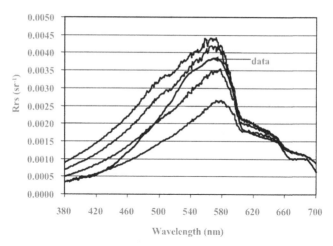

Figure 11. R_{rs} modeled from b_{bw}, $b_{b\phi}$ (Morel, 1988), a_w, a_g, and a_ϕ. The concentration of chlorophyll was 6.6 mg m^{-3} (HPLC). For simplicity, we ignored a_d and b_{bd}. The curve marked as data was generated from measurements done in the study area with a GER spectroradiometer.

River plumes can be divided in two regions based on the optical properties of surface waters. These are: 1) waters where absorption by CDOM changes due to dilution, but where optical properties (S and fluorescence maxima) do not change due to the preponderance of the freshwater end-member. These correspond to samples with salinities < 30 in Figs. 7 and 9; and 2) waters where both CDOM absorption and optical properties change with salinity. These correspond to samples with salinities > 30 in Figs. 7 and 9. *Rrs* is influenced by both CDOM absorption and optical properties, so changes in *Rrs* can be very dramatic in areas of the river plume where both CDOM absorption and optical properties change rapidly (Fig. 12).

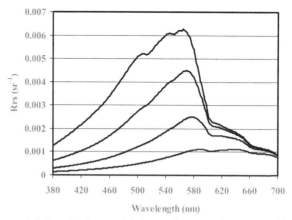

Figure 12. R_{rs} modeled as in Figure 10, but using samples with different S and CDOM concentrations. The CDOM slopes from top to bottom are 0.20, 0.018, 0.016, and 0.014 nm^{-1}.

6. CDOM and DOC

Remote-sensing measurements of organic carbon concentrations can only be done if there is a good relationship between CDOM absorption and DOC concentration. This good relationship does not exist over most of the ocean, including coastal waters away from the influence of rivers (Siegel and Michaels, 1996; Nelson et al., 1998; Nelson and Siegel, 2002; Blough and Del Vecchio, 2002). The relationship can even be inverse when analyzed over large areas of the ocean (Hansell et al., 1997; Nelson and Siegel, 2002). Data available from BATS and cruises to the Sargasso Sea show that the concentrations of colored and non-colored components of DOM are controlled by local, independent processes. For example, photodegradation can reduce absorption coefficients of CDOM without a concomitant reduction in DOC concentration (Vodacek et al., 1997). Inversely, biological activity can increase concentrations of DOC without changing CDOM. Vertical transport of deep-water CDOM by Ekman pumping, upwelling, and turbulent mixing can increase concentrations of CDOM in surface waters (Coble et al., 1998; Del Castillo and Coble, 2000; Nelson and Siegel, 2002).

Coastal regions influenced by river discharge exhibit good relationships between DOC and CDOM. These are also regions of the ocean that receive large inputs of organic material and, consequently, where remote-sensing tools are useful to quantify transport of terrigenous organic carbon to the oceans. The same processes that decouple the concentrations of DOC and CDOM in the open ocean act in river-dominated coastal areas, although ameliorated by the preponderance of riverine organic matter and low light penetration. Nevertheless, deviations from conservative behavior and no covariance between CDOM and DOC are, however, still possible. To be able to measure DOC remotely, conservative behavior of DOC is neccesary, and the relationship between CDOM and DOC in the river end-member should not have a strong seasonal variability. This will allow for the development of robust empirical algorithms. The following section discusses how these conditions are met in river-dominated waters using in most cases data collected in the Orinoco and Mississippi River Plumes.

River-ocean mixing processes overshadow the effects of non-conservative processes on CDOM and DOC concentrations. Figure 12A shows a good example of conservative behavior of CDOM along the Mississippi River plume using samples collected over three years, from the surface (< 1 m) and covering a wide salinity range. The slopes of the regression lines for the three cruises were similar. Differences in y-intercept were caused by seasonal changes in the concentrations of CDOM in the river end-member. These seasonal changes should not affect the relationship between DOC and CDOM in the river plume if conservative processes dominate. However, if the relationship between DOC and CDOM in the end-member changes seasonally, it will be necessary to change regional algorithms accordingly. Models used to derive DOC concentrations from CDOM can be modified based upon measurements of the CDOM/DOC relationship in the end-member or, in the best of cases, by using an empirical relationship between river flow and the relationship between DOC and CDOM. Figure 13 shows seasonal changes in DOC and CDOM in waters of the Mississippi River (sal=0) within a year. There are seasonal changes in both CDOM and DOC but they co-vary. Changes in the DOC to CDOM relationship however, are possible particularly during events of high water flow, when organic material accumulated in terrestrial systems during low flow is flushed by the river. The relationship between CDOM and DOC in the end-member should be monitored over periods longer than one year using the highest possible sampling frequency to ensure the robustness of the empirical relationships.

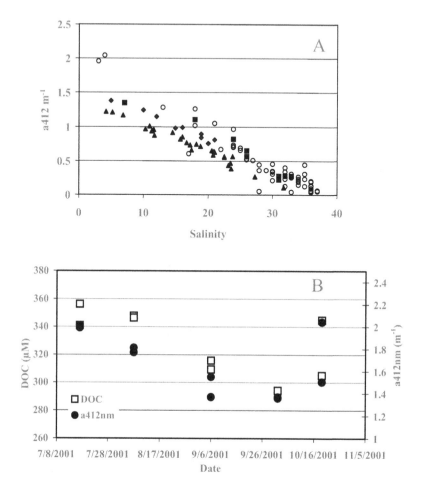

Figure 13. A: Relationship between CDOM and salinity in coastal waters of the Gulf of Mexico influenced by the Mississippi River plume. Samples were taken December 2001 (circles), July, 2002 (squares), and August, 2003 (triangles). B: Seasonal changes in DOC and CDOM in waters of the Mississippi River (salinity=0). Samples were collected at Fort Jackson, Louisiana in 2001.

Several studies have shown good correlations between CDOM absorption coefficients and DOC concentrations in river-ocean mixing zones. Vodacek et al. (1997), working along a transect between the mouth of Delaware Bay and the Sargasso Sea (salinities from 0 to > 36), found good correlations between DOC and CDOM for most of the year ($r^2 = 0.86$). Only during August, when high irradiance and stratified water column favored photodegradation, did they find lack of correlation between these parameters. Ferrari et al. (1996), working in the Baltic Sea found good ($r^2 > 0.90$) correlations between DOC and CDOM during periods of high river flow. Under conditions of low river flow and stratified water column, which are favorable for photobleaching, the correlation between CDOM and DOC concentrations fell to 0.70. Only during August, when flow was at the lowest, and stratification and irradiance were highest, did the correlation between CDOM and DOC disappear. Nevertheless, these samples were within the regression line obtained for the data collected during other periods of the year. They concluded that the relationship between CDOM and DOC is stable. Del Castillo et al. (1999) found a good correlation (r^2 = 0.71) in samples collected along the Orinoco River plume. These samples included waters from the Gulf of Paria and eastern Caribbean, and their salinities ranged from 19 to 35. Del Castillo et al. (2000) working in the Gulf of Mexico and the West Florida Shelf found correlations of 0.80 in samples that included coastal waters influenced by effluents from Tampa Bay, the Apalachicola, Suwannee, and Mobile River. Salinities fluctuated between 27 and 36. Data along each individual river plumes yielded correlation coefficients higher than 0.9. Del Vecchio and Blough (personal communication), working off the west Florida Shelf in waters influenced by the Shark River, found correlations better than 0.9. These data included samples with salinities that fluctuated between 33 and 35, collected during a heavy drought and very sunny conditions. Clearly, it is common to find a good correlation between CDOM and DOC in coastal waters influenced by river plumes. These relationships loose statistical significance only when conditions favor photodegradation or high productivity.

Figure 14 shows a good relationship between DOC and CDOM in the Mississippi and Orinoco Rivers. The samples from the Orinoco River included very high-salinity surface and sub-surface oceanic waters, and were possibly influenced by small rivers from Trinidad. Others have reported that re-suspension may affect organic matter concentrations in the region (Blough et al., 1993). This could explain the lower correlation between DOC and CDOM in these waters. The samples from the Mississippi River were collected during two short cruises to the river plume. These were all surface samples, and there was no influence from other rivers. In this simpler case, the correlation between CDOM and DOC is excellent.

Coastal regions that are not influenced by river discharge do not behave as described above. The relationship between CDOM and DOC is more similar to those observed in the open ocean. In many coastal areas, CDOM concentrations can be affected by other processes like upwelling and re-suspension. Even areas with riverine influence can show deviations from conservative behavior due to these processes. For example, Sosik et al. (2001) reported that CDOM contribution to optical properties of the water column in the New England Continental Shelf remains constant during the year, and that changes in optical properties are caused by variability in particle size distribution. Boss et al. (2001), working in the Mid-Atlantic Bight, reported that some variability in CDOM concentrations is caused by re-suspension of sediments during storms. In some cases, like a re-suspension event caused by a storm, effects upon CDOM distribution can be ephemeral and should not affect long-term studies of riverine carbon transport using remote sensing.

7. Remote Sensing and CDOM

Empirical and semi-analytical algorithms have been used with various degrees of success to derive absorption coefficients of colored organic matter. The absorption due to CDOM cannot be separated from that of colored detrital particulate organic matter using remote-sensing techniques, because they have similar spectral shapes. For this

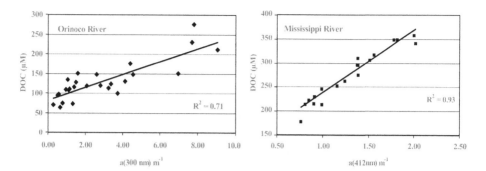

Figure 14. Relationship between DOC and CDOM in waters of the Mississippi and Orinoco River plumes.

reason, their influence on the inherent optical properties (IOP) of water are considered together. Here I will call the combination of the dissolved and particulate organic material *Colored Detrital Material* (CDM) in the nomenclature of Siegel and Michaels, (1996), and will refer to the remote-sensing product as a_{cdm}. The differences between semi-analytical and empirical algorithms, and details about radiative transfer theory are discussed in other chapters so they will not be discussed here. Below, I discuss a few (of many) examples of both types of algorithms and their use to estimate a_{cdm} from remotely sensed data in coastal environments.

Siegel et al. (2002) developed a semi-analytical algorithms that was used to retrieve a_{cdm} values from SeaWiFS (Sea-viewing Wide Field-of-view Sensor) satellite imagery. The data were used to study global distribution and dynamics of colored organic matter. The model used in this work was based on the model presented by Garver and Siegel (1997) and Maritorena et al. (2002). This algorithm uses fixed values of S, and performs well in blue waters. However, the estimates of CDOM and chlorophyll are very sensitive to the S value, particularly in areas where pigments concentrations are very low and CDOM controls light absorption in the blue. Therefore, it cannot return good values in Case 2 waters.

Carder et al. (1999) developed a semi-analytical model that computes chlorophyll concentrations and a_{cdm} for MODIS (Moderate Resolution Imaging Spectroradiometer). The model requires inputs of normalized water leaving radiance (L_{wn}) at 412, 443, and 551 nm. The algorithm requires several components to be derived empirically. For example b_b is defined after Lee et al. (1994) as

$$b_b(\lambda) = X \bullet \left[\frac{551}{\lambda} \right]^Y ,$$

$$(3)$$

where X and Y were determined empirically. These semi-analytical algorithms use fixed values that have been found to work well in most waters. However, Carder et al. (1999) recommends that these values should be determined regionally, particularly in areas dominated by riverine discharge. Absorption by organic matter is defined by the classic relationship (Bricaud et al., 1981):

$$a_g(\lambda) = a_g(400nm)\exp^{-S(\lambda-400)} .$$

$$(4)$$

The algorithm uses a default value of $S = 0.017$ nm^{-1}; however, Carder et al.(1999) emphasizes that this and other parameters (i.e. X and Y) are not absolute and should be changed according to the regions. In very turbid waters, with high concentrations of CDOM and chlorophyll, Carder's algorithm defaults to an empirical formulation for a_{cdm} based on the work of Lee et al. (1994) and expressed as

$$a_{cdm}(400nm) = 1.5 \bullet 10^{-1.147-1.963p15+0.056p25+1.702p25^2}, \qquad (5)$$

where $p15 = \dfrac{R_{rs}410nm}{R_{rs}555nm}$ and $p25 = \dfrac{R_{rs}440nm}{R_{rs}555nm}$.

In this case, the pitfalls of using a fixed S value are avoided, but the coefficients in the relationship also need to be determined regionally.

Hoge et al. (1995) used the 443 nm band from the Coastal Zone Color scanner (CZCS) to retrieve values of $a_{cdm}(443$ nm) in waters with low influence of chlorophyll and particulate detrital material. Later, Hoge et al. (2001) used the 412, 490, and 555nm bands from SeaWiFS to retrieve values of $a_{cdm}(412nm)$. They used a linear matrix inversion technique and for the formulation of a_{cdm} a fixed spectral slope of 0.018 nm^{-1}. These and other excellent semi-analytical formulations share the same characteristic. One or various parameters are fixed based on the empirical relationships determined for a particular region of the world.

Kahru and Mitchell (2001) worked in the California Current and developed an empirical algorithm to estimate $a_{cdm}(300$ nm). Their algorithm used the band ratios L_{wn} 443/520 nm for OCTS (Ocean Color and Temperature Scanner) data and L_{wn} 443/510 nm for SeaWiFS data. Their formulation was a power function of the form

$$a_{cdm}(300nm) = 10^{(a_0+a_1 \bullet R)}, \qquad (6)$$

where R is the band ratio of L_{wn}, and a_0 and a_1 are empirically derived coefficients.

Their results showed good agreements ($r^2 = 0.77$) between measured CDOM and modeled $a_{cdm}(300$ nm). They also compared their results with the semi-analytical algorithm of Carder et al. (1999) using an S parameter optimized for the California Current. They found that the general semi-analytical algorithm overestimates CDOM with respect to their regional model.

The empirical algorithm by Kahru and Mitchell (2001) takes advantage of the high contribution of CDOM to total absorbance at 443 nm relative to chlorophyll, and the low contribution of CDOM in the 500 to 520 nm range.

Johannessen et al. (2002) used in-situ remote sensing reflectance data to derive empirical relationships for diffuse attenuation coefficient and absorption due to CDOM from coastal to off shelf waters from the Mid Atlantic Bight and the Bering Sea. The relationships found for CDOM are:

$$a_{CDOM}(323) = 0.904\ K_d(323) - 0.00714, \qquad r^2=0.93$$

$$a_{CDOM}(338) = 0.858\ K_d(338) - 0.0190, \qquad r^2=0.92$$

$$a_{CDOM}(380) = 0.972\ K_d(380) - 0.0171. \qquad r^2=0.66$$

They also used these relationships to determine the diffuse attenuation coefficient slope coefficient (S_{kd}), similar to S, the CDOM absorption slope coefficient. They were able to map S using SeaWiFS data for the Georgia Bight and discuss seasonal and spatial changes in terms of terrestrial runoff and photobleaching.

An empirical relationship was found by Del Castillo (2004) when working in turbid waters of the Mississippi River plume. The ratio $R_{rs}510/R_{rs}670$ was found to correlate well with a412 nm (Fig. 15). The relationship found by Del Castillo (2004) takes further advantage of the high CDOM absorption in waters of the Mississippi River Plume, where the absorption by CDOM at 510 nm is still very significant, whereas that of chlorophyll and accessory pigments is low. As in Kahru and Michaels (2001) the longer wavelengths used in the ratios are affected by absorption by pigments and particles. However, they effectively derive concentrations of CDOM in their respective regions. In this case study, the empirical relationship between DOC and CDOM (from Fig. 14) was used to produce a relationship between $R_{rs}510/R_{rs}670$ and DOC concentrations and to derive DOC concentrations from field measurements of R_{rs}. Figure 16 shows a comparison between modeled and measured DOC concentrations in the Mississippi River plume waters.

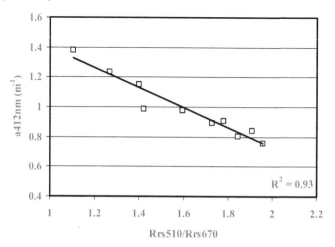

Figure 15. Relationship between CDOM absorption coefficient at 412 nm and a ratio of R_{rs} from the 510 and 670 nm bands. Measurements of CDOM were made from filtered (GF/F) water samples collected in the river plume using a Perking-Elmer dual-beam spectrophotometer. R_{rs} were derived from measurements made with a GER spectroradiometer.

The effect of changing optical properties (*S*) and absorption of CDOM upon the performance of a_{cdm} retrieval algorithms should be considered carefully. As shown above, their effects can be easily modeled using good-quality field samples. Empirical relationships are often based on measured values of CDOM absorbance at one wavelength (i.e. 412 nm), and the effect of changing *S* are seldom considered. This is justified by the strong conservative behavior of CDOM and the good relationships between CDOM and DOC.

An important obstacle for the implementation of any of these algorithms is that low L_{wn} in colored Case 2 waters results in low signal to noise ratios in the satellite signal. Therefore, algorithms based in solid empirical relationships may still produce incorrect CDOM retrievals from satellite data. New generations of geostationary, hyperspectral, high-resolution satellite sensors (under design) will allow for longer data collections with improvements in the quality of the signal.

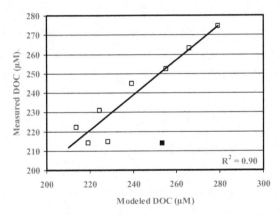

Figure 16. Relationship between modeled and measured DOC based on an empirical relationship between DOC and CDOM in the Mississippi River plume. The models were tested in low-salinity waters (<30). The closed square corresponds to the outlier in Figure 15. It was not included in the regression.

8. Conclusions

It is now possible to obtain estimates of CDOM concentrations from remote sensors. Future improvements in atmospheric correction methods in coastal waters, the establishment of calibration and validation sites, and the development of better regional algorithms will allow accurate measurements of CDOM using ocean color sensors. The implementation of a nested system of regional algorithms can yield inputs of terrestrial carbon for all coastal regions of the world impacted by large rivers. This is an important development to help quantify carbon fluxes from the land to oceans. The crux of the problem remains in the relationship between CDOM and DOC. It is clear that, in some areas, particularly those with strong riverine discharge, the relationship is robust and stable. In other areas, particularly where productivity introduces significant amounts of non-colored DOC, the remote-sensing estimates of DOC based on measurements of CDOM will represent carbon inputs from land, not changes in carbon concentration due to local processes.

9. Acknowledgements

I want to thank Mayra I. Berríos, Neil Blough, Jorge E. Corredor, Rossana Del Vecchio, and Ajit Subramaniam, who provided invaluable technical and editorial comments.

10. References

Azam, F. and A.Z. Worden. 2004. Microbes, molecules, and marine ecosystems. Science, 303:1622-1624.

Blough, N.V., O.C. Zafiriou, and J. Bonilla. 1993. Optical absorption spectra of waters from the Orinoco River outflow: Terrestrial input of colored organic matter to the Caribbean. Journal of Geophysical Research, 98:2271-2278.

Blough, N.V. and R. Del Vecchio. 2002. Chromophoric DOM in the Coastal Environment. In: Biogeochemistry of Marine Dissolved Organic Matter. D.A. Hansell and C.A. Carlson eds. Academic Press. 774 pp.

Bontempi, P., S. Davis, C.E. Del Castillo, D. Roelke, and K. Winemiller. 2003. Transformation of allochthonous dissolved organic carbon in a tropical blackwater river as measured by fluorescence analysis. Application to foodweb ecology. Ocean Optics XVI, Santa Fe, NM.

Boss, E., S.W. Pegau, R. Zaneveld, and A. H. Barnard. 2001. Spatial and temporal variability of absorption by dissolved material at a continental shelf. Journal of Geophysical Research, 106:9499-9507.

Bricaud, A., A. Morel, and L. Prieur 1981. Absorption by dissolved organic matter of the sea (Yellow Substance) in the UV and visible domains. Limnology and Oceanography, 26(1): 43-53.

Calvert, J.G., and J.N. Pitts 1966. Photochemistry, Wiley, New York, 748 pp.

Cannizzaro, J.P., K.L. Carder, F.R. Chen, J.J. Walsh, Z.P. Lee, and C. Heil. 2002. A novel optical classification technique for detection of red-tides in the Gulf of Mexico: Application to the 2001-2002 bloom event. In Proceedings of the 10th International Conference on Harmful Algae, St. Pete Beach, Florida.

Carder, K.L., R.G. Steward, G. Harvey, and P. Ortner. 1989. Marine humic and fulvic acids: Their effects on remote sensing of ocean chlorophyll. Limnology and Oceanography, 34(1): 68-81.

Carder, K.L., F.R. Chen, Z.P. Lee, S.K. Hawes, and D. Kamykowski. 1999. Semianalytic moderate-resolution imaging spectrometer algorithm for chlorophyll-a and absorption with bio-optical domains based on nitrate-depletion temperatures. Journal of Geophysical Research, 104:5403-5421.

Carder, K.L., J.P. Cannizzaro, F.R. Chen, C. Heil, and G.A. Vargo. 2004. Karenia brevis blooms on the West Florida Shelf: A bridge between optics and physiology. ASLO/TOS Meeting. Honolulu, Hawaii.

Chen, R.F., G.B Gardner, Y Zhang, P Vlahos., X.Wang, and S.M.Rudnick. 1999. Chromophoric dissolved organic matter (CDOM) in four US estuaries. EOS 80:92.

Coble, P.G. 1996. Characterization of marine and terrestrial DOM in seawater using excitation-emission matrix spectroscopy. Marine Chemistry, 51(4):325-346.

Coble, P. G., C. E. Del Castillo, and B. Avril. 1998. Distribution of DOM in the Arabian Sea during the SW Monsoon. Deep-Sea Research II, 45:2195-2223

Del Castillo, C.E., P.G. Coble, J.M. Morel, J.M. Lopez, and J.E. Corredor. 1999. Analysis of the optical properties of the Orinoco River Plume by absorption and fluorescence spectroscopy. Marine Chemistry, 66:35-51.

Del Castillo, C.E., F. Gilbes, P.G. Coble, and F.E. Muller-Karger. 2000. On the dispersal of riverine colored dissolved organic matter over the West Florida Shelf. Limnology and Oceanography, 45:1425-1432.

Del Castillo, C.E., and P.G. Coble. 2000. Seasonal variability of the colored dissolved organic matter during the 1994-95 NE and SW Monsoons in the Arabian Sea. Deep-Sea Research II, 47:1563-1579.

Del Castillo, C.E., P.G. Coble, R.N. Conmy, F.E. Muller-Karger, L. Vanderbloemen, and G. Vargo. 2001. Multispectral in situ measurements of organic matter and chlorophyll fluorescence in seawater: Documenting the intrusion of the Mississippi River Plume in the West Florida Shelf. Limnology and Oceanography, 46:1836-1843.

Del Castillo, C.E. 2004. Optical consequences of changes in organic matter composition in the Mississippi River Plume. ASLO/TOS meeting. Honolulu, Hawaii.

De Souza-Sierra, M.M., O.F.X. Donard, and M. Lamotte. 1997. Spectral identification and behavior of dissolved organic fluorescent material during estuarine mixing processes. Marine Chemistry, 58: 51-58.

Del Vecchio, R. and N.V. Blough. 2002. Photobleaching of chromophoric dissolved organic matter in natural waters: kinetics and modeling Marine Chemistry, 78:231-253

Del Vecchio R. and N.V. Blough. 1994 On the origin of optical properties of organic matter. In Press.

Del Vecchio R. and N.V. Blough. 2004. Spatial and seasonal distribution of chromophoric dissolved organic matter (CDOM) and dissolved organic carbon (DOC) in the middle Atlantic Bight. Marine Chemistry, In Press.

Ferrari, G.M., M.D. Dowell, S. Grossi, and C. Targa. 1996. Relationship between the optical properties of chromophoric dissolved organic matter and total concentration of dissolved organic carbon in the southern Baltic Sea Region. Marine Chemistry, 55:299-316.

Ferrari, G.M. 2000. The relationship between chromophoric dissolved organic matter and dissolved organic carbon in the European Atlantic coastal area and in the West Mediterranean Sea (Gulf of Lions). Marine Chemistry, 70:339-357.

Gagosian, R.B. and D.H. Stuermer. 1977. The Cycling of Biogenic Compounds and Their Diagenetically Transformed Products in Seawater. Marine Chemistry, 5:605-632.

Garver, S.A., and D.A. Siegel. 1997. Inherent optical property inversion of ocean color spectra and its biogeochemical interpretation: 1. Time series from the Sargasso Sea. Journal of Geophysical Research, 102:18607-18625.

Hansell, D.A., C.A. Carlson, N.R. Bates, and A. Poisoon. 1997. Horizontal and vertical removal of organic carbon in the equatorial Pacific Ocean: A mass balance assessment. Deep-Sea Research I, 44:2115-2130.

Harvey, G.R., D.A. Boran, L.A. Chesal, and J.M. Tokar 1983. The Structure of Marine Fulvic and Humic Acids. Marine Chemistry, 12:119-132.

Harvey, G.R., D.A. Boran, S.R. Piotrowicz, and C.P. Weisel 1984. Synthesis of Marine Humic Substances from Unsaturated Lipids. Nature, 309:244-246.

Harvey, G.R., and D.A. Boran. 1985. Geochemistry of humic substances in seawater. In Humic Substances in Soils, Sediments, and Water. Geochemistry, Isolation, and Characterization. (G.R. Aiken, D. McKnight, R.L. Wershaw, and P. MacCarthy, eds.) Wiley-Interscience, New York.

Hoge, F.E., M.E. Williams, R.N. Swift, J.K. Yungel, and A. Vodacek. 1995. Satellite retrieval of the absorption coefficient of chromophoric dissolved organic matter in continental margins. Journal of Geophysical Research, 100:28,847-24,854.

Hoge, F.E., C.W. Wright, P.E. Lyon, R.N. Swift, and J.K. Yungel. 2001. Inherent optical properties imagery of the western North Atlantic Ocean: Horizontal spatial variability of the upper mixed layer. Journal of Geophysical Research, 106:31,129-31,140.

Johannessen S.C., W.L. Miller, and J.J. Cullen. 2003. Calculation of UV attenuation and colored dissolved organic matter absorption spectra from measurements of ocean color. Journal of Geophysical Research, 108.C9 3301, doi:10.1029/2000JC000514.

Kalle, K. 1966. The problem of Gelbstoff in the sea. Oceanography Marine Biology Annals Review, 4:91-104.

Kahru, M. and B.G., Mitchell. 2001 Seasonal and non-seasonal of satellite-derived chlorophyll and dissolved organic matter concentration in the California Current. Journal of Geophysical Research, 106:2517-2529.

Kieber, R.J., J. McDaniel, and K. Mopper. 1989. Photochemical source of biological substrates in seawater: Implications for Carbon Cycling. Nature, 341:637-639.

Kieber, R.J., X. Zhou, and K. Mopper. 1990. Formation of carbonyl compounds from UV-induced photodegradation of humic substances in natural waters: Fate of riverine carbon in the sea. Limnology and Oceanography, 35(7):1503-1515.

Kirk, J.T.O. 1994. Light and Photosynthesis in the Aquatic Ecosystems. Second edition. Cambridge University Press. 509 pp.

Laane, R.W.P.M. 1981. Composition and distribution of dissolved fluorescent substances in the Ems-Dollart Estuary. Netherlands Journal of Sea Research, 15(1): 88-99.

Laane, R.W.P.M. 1984. Comment on the structure of marine fulvic and humic acids. Marine Chemistry, 15:85-87.

Lee, Z.P., K.L. Carder, S.H. Hawes, R.G. Steward, T.G. Peacock, and C.O. Davis. 1994. A model for interpretation of hyperspectral remote-sensing reflectance. Applied Optics, 33:5721-5732.

Mantoura, R.F.C. and E.M.S. Woodward 1983. Conservative behavior of riverine dissolved organic carbon in the Severn Estuary: Chemical and Geochemical implications. Geochimica et Cosmochimica Acta, 47:1923-1309.

Maritorena, S., D.A. Siegel, amd A.R. Paterson. 2002. Optimization of a semianalytical ocean color model for global-scale applications. Applied Optics, 41:2705-2714.

Miller, W.L and M.A. Moran. 1997. Interaction of photochemical and microbial processes in the degradation of refractory dissolved organic matter from a coastal marine environment. Limnology and Oceanography, 42:1317-1324.

Moran, M.A. and R.G. Zepp. 1997. Role of Photoreactions in the formation of biologically labile compounds from dissolved organic matter. Limnology and Oceanography, 42:1307-1316

Morel, A. 1988. Optical modeling of the upper ocean in relation to its biogenous matters content (Case 1 waters). . Journal of Geophysical Research, 93:10749-10768.

Morell J.M. and J. E. Corredor. 2001. Photomineralization of fluorescent dissolved organic matter in the Orinoco River plume: Estimation of ammonium release. Journal of Geophysical Research, 106:16,807-16,813.

Müller-Karger, F. E., C. R. McClain, T. R. Fisher, W. E. Esaias, and R. Varela. 1989. Pigment distribution in the Caribbean Sea: Observation from space. Progress Oceanography, 23:23-64.

Nelson, N.B., D.A. Siegel, and A.F. Michaels. 1998. Seasonal dynamics of colored dissolved material in the Sargasso Sea. Deep-Seas Research I, 45:931-957.

Nelson, N.B. and D.A. Siegel. 2002. Chromophoric DOM in the open ocean. In: Biogeochemistry of Marine Dissolved Organic Matter. D.A. Hansell and C.A. Carlson eds. Academic Press. 774 pp.

Opsahl, S, and R. Benner.1998. Photochemical reactivity of dissolved lignin in river and ocean waters. Limnology and Oceanography, 43:1297-1304.

Pavia, D.L., G.M. Lampman, and G.S. Kriz. 2000. Introduction to Spectroscopy. 3rd ed. Brooks Cole. 515 pp.

Pope, R.M., and E.S. Fry. 1997. Absorption spectrum (380-700 nm) of pure water. II. Integrated cavity measurements. Applied Optics, 36:8710-8723.

Shank, G.C., R.G. Zeep, and M.L. Smith. 2004. Variations in the spectral properties of estuarine waters caused by CDOM partitioning onto river and estuarine sediments. ASLO/TOS Ocean Research Conference. Honolulu, Hawaii.

Sholkovitz, E.R. 1976. Flocculation of dissolved organic and inorganic matter during mixing of river and seawater. Geochimica et Cosmochimica Acta, 40:831-845.

Siegel, D.A. and A.F. Michaels. 1996. Quantification of non-algal light attenuation in the Sargasso Sea: Implications for biogeochemistry and remote sensing. Deep-Sea Research II, 43:321-345.

Siegel, D.A., S. Maritorena, N.B. Nelson, D.A. Hansell, and M. Lorenzi-Kayser. 2002. Global ocean distribution and dynamics of colored dissolved and detrital organic materials. . Journal of Geophysical Research, 107: c12, 3228,doi:10.1029/2001JC000965.

Sosik, H.M., R.E. Green, W.S. Pegau, and C.S. Roesler. 2001. Temporal and vertical variability in optical properties of New England Shelf Waters during late summer and spring. . Journal of Geophysical Research, 106:9455-9472.

Stevenson, F.J., 1982. Humus Chemistry, genesis, composition, reactions. A. Wiley International Publisher, New York 443 pp.

Twardowsky, M.S. and P.L. Donaghay. 2001. Separating in situ and terrigenous sources of absorption by dissolved materials in coastal waters. Journal of Geophysical Research, 106:2545-2560.

Twardowsky, M.S. and P.L. Donaghay. 2002. Photobleaching of aquatic dissolved materials: Absorption removal, spectral alterations, and their interrelationships. Journal of Geophysical Research, 107:C8, 3091.

Vodaceck, A., N.V. Blough, M.D. DeGramdpre, E.T. Peltzer, and R.K. Nelson. 1997. Seasonal variation of CDOM and DOC in the Middle Atlantic Bight: Terrestrial inputs and photooxidation. Limnology and Oceanography, 42:674-678.

Chapter 8

HYPERSPECTRAL REMOTE SENSING

[1]ZHONGPING LEE AND [2]KENDALL L. CARDER

[1]*Naval Research Laboratory, Code 7333, Stennis Space Center, MS, 39529 USA*
[2]*College of Marine Science, University of South Florida, St. Petersburg, FL, 33701 USA*

1. Introduction

Increasing scientific interest in the environment and ocean-atmosphere interactions has created a need to map the constituents of the world's oceans such as phytoplankton, dissolved organic matter, and suspended sediments (Gordon and Morel, 1983; Jerlov, 1976; Kirk, 1994; Mobley, 1994), and to study the ocean dynamics associated with environmental variations. For the world ocean, it is impractical to use traditional, single point water-sampling methods to obtain high-resolution global measurements. Instead, satellite sensors provide the most effective means for frequent, synoptic water-quality observations over large areas. Contrary to field methods, which physically or chemically separate the components of water, the satellite method resolves the components based on spectral appearance - a process called spectral remote sensing (Austin, 1974; Gordon and Morel, 1983). This idea is based on the early observations made by Austin (1974) and Jerlov (1976) of color shifts from blue to green when the amount of dissolved or suspended constituents in the water increases.

Until recently, spectral remote sensing has been limited by sensor configurations and processing algorithms to a few pre-defined spectral bands (Gordon and Morel, 1983; Morel and Prieur, 1977), or data from multi-spectral sensors. Multi-spectral data, however, may miss spectral regions with important spectral information, hindering the retrieval of the desired property. For example, without a spectral band in the blue, the Coastal Zone Color Scanner (CZCS) sensor could not effectively separate the absorption effects of Colored Dissolved Organic Matter (CDOM) from that of phytoplankton (Carder et al., 1991; Carder et al., 1986; Gordon et al., 1980a). With advanced sensor technologies and algorithms, a hyperspectral system may resolve this problem.

A precise definition of hyperspectral is taken as a sensor or dataset that has continuous spectral observations. In practice, however, hyperspectral is a conceptual term used for a sensor that has more than a few spectral bands. For instance, the AVIRIS and HYPERION sensors, which have spectral bands spaced about every 10 nm, are commonly called "hyperspectral" sensors. MERIS of ESA (Medium Resolution Imaging Spectrometer, European Space Agency), however, which has 12 selected bands within the 400 – 800 nm range, is called a multi-spectral sensor.

Spectral remote sensing uses water color as input to statistically or analytically derive water properties of interest, such as the attenuation coefficient (Austin and Petzold, 1981; Stumpf and Pennock, 1991), chlorophyll concentration (Gordon et al., 1983; Morel and Prieur, 1977; Sathyendranath et al., 2001), mass concentration of suspended sediments (Bukata et al., 1991b; Bukata et al., 1995; Dekker et al., 2001; Doerffer and Fisher, 1994), and bottom depth (when the sea bottom is shallow enough to be optically detected) (Lee et al., 2001; Lee et al., 1999). These properties are important

R.L. Miller et al. (eds.), Remote Sensing of Coastal Aquatic Environments, 181–204.
© 2005 *Springer.*

for evaluating the status of the water environment (Jerlov, 1976; Kirk, 1994), a
provide critical inputs for studies of oceanic photosynthesis (Kirk, 1994; Lee et a
1996b; Marra et al., 1992; Platt and Sathyendranath, 1988; Sathyendranath et al., 1989
and heat transfer (Lewis et al., 1990; Morel and Antoine, 1994).

Water color is measured by the quantity water-leaving radiance, $L_w(0^+)$ (Gordon a
Morel, 1983; Gordon et al., 1980b; Gordon and Wang, 1994), which is the downwar
irradiance from the sun and sky that penetrates the water surface, interacts wi
molecules and particulates in the water, then emerges from below the water surfac
Since the intensity of downward irradiance varies with solar altitude and sky conditio
(Gregg and Carder, 1990; Mobley, 1994), the variation of water-leaving radian
includes the variations of solar and sky inputs and water properties.

To better describe and quantify the water effects on water color, a quantity remot
sensing reflectance (R_{rs}) is defined (Carder and Steward, 1985; Gordon et al., 1980l
which is the ratio of water-leaving radiance to the downwelling irradiance above tl
surface, $E_d(0^+)$, a measure of solar and sky inputs. Therefore R_{rs} (with units of sr
avoids the intensity variation associated with solar inputs. Many models a
applications use instead subsurface irradiance reflectance (ratio of subsurface upwelli
irradiance to downwelling irradiance) (Gordon et al., 1975). The description a
discussion regarding this property is not presented here, but can be found in numero
references (Kirk, 1991; Morel and Prieur, 1977; Sathyendranath and Platt, 1997). Sin
R_{rs} is the basic quantity for spectral remote sensing, the following discussions will foc
on how R_{rs} relates to water properties and how to use R_{rs} as an input to process spectr
derivations.

Figure 1 presents examples of hyperspectral $R_{rs}(\lambda)$ curves, collected fro
representative parts of the ocean. As shown in these examples, $R_{rs}(\lambda)$ varies in bo
spectral shape and magnitude. These variations provide the fundamental basis f
remotely estimating properties of the water column and the visible bottom.

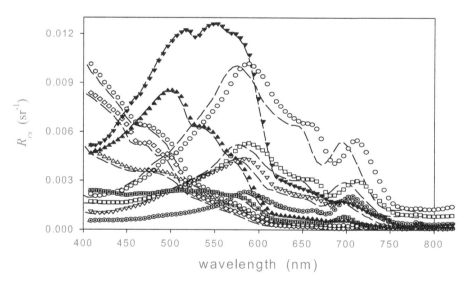

Figure 1. Examples of hyperspectral remote-sensing reflectance (R_{rs}) obtained fro
different ocean regions (Lee and Carder, 2002). Reprinted with permission of the auth
and publisher (Optical Society of America).

2. Relationships between R_{rs} and Inherent Optical Properties (IOPs)

Until recently, the focus within the field of ocean color remote sensing was on the retrieval of the concentration ($[C]$) of chlorophyll a and other pigments (Gordon et al., 1983; Gordon and Morel, 1983), with simple, empirical algorithms comparing pigment concentration to the ratios of R_{rs} (or $L_w(0^+)$) at two or more wavelengths (Clark, 1981; Gordon and Morel, 1983; Kahru and Mitchell, 1999; Lee et al., 1998b; O'Reilly et al., 1998; Sathyendranath et al., 2001). For such an approach, there was no strong need to know the physics relating R_{rs} and the inherent optical properties (IOPs) (Preisendorfer, 1976), as long as a large volume of data containing both R_{rs} and $[C]$ are collected to build the required statistical regression (Mitchell and Kahru, 1998; O'Reilly et al., 1998).

It has been found that such an approach performs reasonably well in obtaining the qualitative $[C]$ patterns for the world ocean, but is inadequate to accurately retrieve the $[C]$ quantity (Mitchell and Holm-Hansen, 1990; Smyth et al., 2002). The errors can easily be a factor of 2 or more (Carder et al., 1991; Carder et al., 1986; Gordon et al., 1983; Gordon and Morel, 1983; Mitchell and Holm-Hansen, 1990), especially for nearshore or coastal waters, areas important to the quality of human life.

Recently, more attention and efforts have been given to using R_{rs} as input to analytically or semi-analytically derive important environmental properties from collected spectral information (IOCCG, 2000), a process requiring knowledge of the fundamental relationships of radiative transfer. For this approach, the first requirement is to know how R_{rs} is related to the IOPs of the water (Morel and Prieur, 1977; Sathyendranath et al., 1989b).

R_{rs} is a quantity measured above the surface with r_{rs}, its counterpart measured below the surface. r_{rs} is simply the ratio of $L_w(0^-)$ to $E_d(0^-)$. Due to the refractive effects of the water surface (Austin, 1974) R_{rs} is not equal to r_{rs}. For R_{rs}, and its counterpart r_{rs}, their relationship can be described as (Austin, 1980; Gordon et al., 1988; Lee et al., 1998a; Mobley, 1994)

$$R_{rs} = \frac{\zeta \, r_{rs}}{1 - \Gamma \, r_{rs}}, \tag{1}$$

with ζ representing the divergence effect (Austin, 1974) and Γ representing the internal-reflection effect (Gordon et al., 1988) of the air-sea interface. For nadir-viewed R_{rs} and r_{rs}, $\zeta \approx 0.52$ and $\Gamma \approx 1.7$ (Lee et al., 1999), and the relationship of Eq. (1) is accurate within a few percent, therefore R_{rs} and r_{rs} are, in general, interchangeable.

2.1 OPTICALLY DEEP WATERS

Optically deep waters are those for which the sea bottom has negligible effects on the magnitude or appearance of water color. Such cases include waters of the ocean over 500 m deep, where effectively no photons from the sun and sky reach the bottom even for clearest natural waters. It also includes geometrically-shallow coastal waters that have large attenuation coefficients, where the sea bottom may only be a few meters deep, but the solar and sky photons are quickly lost to strong absorption effects before reaching the bottom. This is why only when the geometrical depth is a meter or so can red photons reach the bottom and back to surface again.

For optically deep waters, it is the spectral IOPs of the water column that determine the spectral R_{rs} (Gordon et al., 1975; Mobley, 1994). These IOPs include the absorption and backscattering coefficients and the scattering phase function of the particles (Mobley, 1994), which are determined by water and constituents dissolved and suspended in the water. From analytical derivations (Zaneveld, 1982; Zaneveld, 1995) and numerical simulations (Gordon et al., 1988) of the radiative transfer equation, it is found that, in general, r_{rs} can be expressed as

$$r_{rs} = g \frac{b_b}{a + b_b}, \qquad (2)$$

with a and b_b the total absorption and backscattering coefficients of the water medium, and g (with units of sr^{-1}) a model parameter to link IOP with r_{rs}. Due to the complexity of the light field, however, values and variations of g cannot be directly derived from the radiative transfer equation, unless it is simplified to single-scattering scenarios (Hojerslev, 2001; Zaneveld, 1995). For natural waters where the light field is determined by multiple scattering, values of g have been derived numerically.

Through Monte Carlo simulations, Gordon et al. (1988) found that g of nadir-viewed r_{rs} can be modeled as

$$g = g_0 + g_1 \frac{b_b}{a + b_b}, \qquad (3)$$

with $g_0 = 0.0979$ and $g_1 = 0.0794$. For realistic scattering phase functions, Gordon et al. (1988) pointed out that Eq. (3) is accurate within 20% for sun angles greater than 20° from zenith.

Similarly, Morel and Gentili (1993) expressed r_{rs} as

$$r_{rs} = \frac{f}{Q} \frac{b_b}{a}, \qquad (4)$$

with f a model parameter for the subsurface irradiance reflectance, and Q the ratio of upwelling irradiance to upwelling radiance. Both f and Q vary with the solar zenith angle and the IOPs of the water (Morel and Gentili, 1993). For selected $[C]$, wavelengths, and solar and sensor-observing angles, values of f/Q have been derived numerically based on a Case 1 bio-optical model and tabulated (Morel et al., 2002; Morel and Gentili, 1993).

Jerome et al. (1996), based on results from Monte Carlo simulations, empirically expressed nadir-viewed r_{rs} as

$$r_{rs} = -0.00042 + 0.112 \frac{b_b}{a} - 0.0455 \left(\frac{b_b}{a} \right)^2. \qquad (5)$$

Values for g, by Eq. (3) (or Eq. (5)), will be the same for different b_b and a values as long as their b_b/a ratios are the same. However, as indicated in theoretical (Zaneveld, 1995) and numerical studies (Mobley et al., 2002), different g values should be expected between one case that b_b is made of molecular scattering and the other case that b_b is made of particle scattering, though both cases may have the same b_b/a ratios. This is simply because molecule and particle scatterings have significantly different scattering phase functions (Mobley, 1994; Petzold, 1972) and that g value depends on the shape of phase function (Mobley et al., 2002). With the increase of particle scattering, the phase function of the medium shifts from that of molecular scattering to that of particle scattering. To explicitly include this phase-function shift into the r_{rs} model, Lee et al. (2002b) expanded the above expressions based on a solution (Zaneveld, 1995) to the radiative transfer equation to yield

$$r_{rs} = g_w \frac{b_{bw}}{a+b_b} + g_p \frac{b_{bp}}{a+b_b} . \tag{6}$$

Here b_{bw} and b_{bp} are the backscattering coefficients of water molecules and suspended particles (a collective term for all other scattering components), while g_w and g_p are the model parameters determined by the two different phase functions. Molecule-particle inter-scatterings are inexplicitly embedded into the two terms (Lee et al., 2002b).

Numerical simulations made using the commercially available *Hydrolight* (Mobley, 1995) radiative transfer model for nadir-viewed r_{rs} and particles that follow the Petzold averaged scattering function (Mobley, 1994) suggest that $g_w = 0.113$ and

$$g_p = 0.197 \left[1 - 0.636 \exp\left(-2.552 \frac{b_{bp}}{a+b_b} \right) \right] . \tag{7}$$

Eq. (6) basically indicates that the traditional model-parameter g of Eq. (2) is

$$g = g_w \frac{b_{bw}}{b_{bw} + b_{bp}} + g_p \frac{b_{bp}}{b_{bw} + b_{bp}} . \tag{8}$$

Therefore, when the backscattering is dominated by molecular scattering (open ocean at blue wavelengths, for instance), g values approach that determined by the phase function of molecular scattering; on the other hand, when the backscattering is dominated by that of particle scattering (red wavelengths, for instance), g values approach that determined by the phase function of particle scattering. Such separation produces results more consistent with the scattering phenomena and radiative transfer.

Interestingly, for an "extreme experimental" case where $b_b/(a+b_b)$ equals 0.95, r_{rs} from Eqs.2&3 and Eqs.6&7 are about 0.164 and 0.179 sr^{-1}, respectively. This is in reasonable agreement with *HydroLight* predicted value of 0.233 sr^{-1}. This simple comparison suggests that Eqs.2&3 and Eqs.6&7 could be applicable to waters with a quite wide range of $b_b/(a+b_b)$.

2.2 OPTICALLY SHALLOW WATERS

When it becomes necessary to consider sea bottom effects, the treatment of r_{rs} is more complex than for optically deep waters. Since contributions of the bottom also affect water color, and this effect has to be removed if we want to derive the properties of the water column. Conversely, for derivation of bottom depth or bottom features, the effects of the water column have to be accounted for. Note that due to multiple scattering, the remotely measured signal is an integration of that from the water column and that from the bottom. Sophisticated models and methods are required to adequately separate the individual signals.

Single or quasi-single scattering theory (Gordon, 1994) and numerical simulations (Lee et al., 1998a) suggest that subsurface remote-sensing reflectance can be approximated as a sum of contributions from the water column and from the bottom (Lyzenga, 1978; Lyzenga, 1981; Maritorena et al., 1994; Philpot, 1989):

$$r_{rs} = r_{rs}^{dp} \left[1 - e^{-2KH} \right] + \frac{\rho}{\pi} e^{-2KH} . \tag{9}$$

The first term on the right side of Eq. (9) is the water-column contribution, and the second term is the bottom contribution. r_{rs}^{dp} is the r_{rs} for optically deep waters (Eq. (2) or Eq. (6)), while ρ is the bottom reflectance. H is the depth of the bottom. K is usually categorized as an "effective" attenuation coefficient (Maritorena et al., 1994).

Because the term "effective" attenuation coefficient is ambiguous and not known from remotely measured data, and the diffuse-attenuation coefficient for downwelling light is not generally equal to the diffuse attenuation coefficients for upwelling light (Gordon et al., 1975), Lee et al. (1998a; 1999) expanded the nadir-viewed r_{rs} to an explicit expression after *HydroLight* simulations,

$$r_{rs} = r_{rs}^{dp}\left[1 - \exp(-(D_0 + D_1(1 + D'_1 u)^{0.5})\kappa H)\right] + \frac{\rho}{\pi}\exp(-(D_0 + D_2(1 + D'_2 u)^{0.5})\kappa H). \quad (10)$$

Here $\kappa = a + b_b$, $u = b_b/(a+b_b)$, and $D_0 = 1/\cos(\theta_w)$ with θ_w the subsurface solar zenith angle, $D_1 = 1.03$, $D'_1 = 2.4$, $D_2 = 1.04$, and $D'_2 = 5.4$. This semi-analytical model for r_{rs} is supported by recent field measurements of Voss et al. (2003).

2.3 CONTRIBUTIONS OF INELASTIC SCATTERING

The above expressions provide explicit descriptions for contributions resulting from elastic scattering to r_{rs}. In the natural light field, there are also contributions from inelastic scattering such as the fluorescence from phytoplankton pigments (Yentsch and Phinney, 1985; Yentsch and Yentsch, 1979) and CDOM (Hawes et al., 1992; Peacock et al., 1990) as well as Raman scattering (Marshall and Smith, 1990; Stavn and Weidemann, 1990). For clear oceanic waters, the contribution of Raman scattering could be more than 10% in the longer wavelengths (Morel et al., 2002; Stavn and Weidemann, 1990). But the most distinctive component of these inelastic effects in the measured $r_{rs}(\lambda)$ spectrum is chlorophyll fluorescence (Gordon, 1979), which is strong and centered at 683 nm within a well defined band (see Fig. 1). The contributions of CDOM fluorescence and Raman scattering to $r_{rs}(\lambda)$, however, are not as distinctive as that of chlorophyll fluorescence. This is due to the fact that CDOM fluorescence and Raman scattering in natural waters are weak and broad band (Lee et al., 1994), making their contributions to $r_{rs}(\lambda)$ difficult to resolve. Since their contributions to $r_{rs}(\lambda)$ is generally small (less than the errors resulted from the current atmosphere-correction process), their effects are practically ignored in the current satellite algorithms. However, semi-analytical models have been developed for evaluating their effects when the water IOP's are known (Lee et al., 1994; Marshall and Smith, 1990; Sathyendranath and Platt, 1998).

3. Hyperspectral Models of IOPs

3.1 THE MATHEMATICAL PROBLEM OF SPECTRAL REMOTE SENSING

The primary objective of ocean-color remote sensing is to quantitatively derive properties below the surface, such as concentrations of chlorophyll and suspended sediments that reside within the water column, and/or bottom depth for optically shallow waters.

For waters with vertical homogeneity, and ignoring the inelastic scattering contributions, R_{rs} can be generally expressed as

$$R_{rs}(\lambda) = Fun\{a(\lambda), b_b(\lambda), \rho(\lambda), H\}. \quad (11)$$

Fun in Eq. (11) represents a function of listed variables, and for the purpose of illustrating mathematical approaches in solving R_{rs}, the angular variation of R_{rs} is not explicitly presented here.

In ocean optics, the total absorption coefficient can be generally separated as (Carder et al., 1991; Gordon et al., 1980b)

$$a(\lambda) = a_w(\lambda) + a_{ph}(\lambda) + a_g(\lambda) + a_d(\lambda), \tag{12}$$

where $a_w(\lambda)$ represents the contribution of nothing but water molecules, or the so-called absorption coefficient of pure water. Values of $a_w(\lambda)$ have been reported since 1930's (Smith and Baker, 1981). Due to sample preparation and techniques used in its measurement, there are differences among those reported values (Smith and Baker, 1981). Currently, the laboratory measured values by Pope and Fry (1997) are widely accepted and used by the ocean-optics community. The subscript ph, g, and d represent the contributions from phytoplankton pigments, Gelbstoff, and detritus, respectively. All components vary with wavelength.

The backscattering coefficient of the water medium ($b_b(\lambda)$) is commonly separated into two components (Gordon et al., 1980b; Kirk, 1991; Morel and Gentili, 1991; Smith and Baker, 1981):

$$b_b(\lambda) = b_{bw}(\lambda) + b_{bp}(\lambda), \tag{13}$$

with $b_{bw}(\lambda)$ the contribution of water molecules, and $b_{bp}(\lambda)$ the contribution of other scatterers (collectively called particles). Values of $b_{bw}(\lambda)$ reported by Morel (1974) are widely used.

After the separation of these major components, for a sensor with n spectral channels, there is

$$R_{rs}(\lambda_1) = Fun\{a_w(\lambda_1), b_{bw}(\lambda_1), \boldsymbol{a}_{ph}(\lambda_1), \boldsymbol{a}_g(\lambda_1), \boldsymbol{a}_d(\lambda_1), \boldsymbol{b}_{bp}(\lambda_1), \boldsymbol{\rho}(\lambda_1), \boldsymbol{H}\}$$

$$R_{rs}(\lambda_2) = Fun\{a_w(\lambda_2), b_{bw}(\lambda_2), \boldsymbol{a}_{ph}(\lambda_2), \boldsymbol{a}_g(\lambda_2), \boldsymbol{a}_d(\lambda_2), \boldsymbol{b}_{bp}(\lambda_2), \boldsymbol{\rho}(\lambda_2), \boldsymbol{H}\}$$

$$\vdots \tag{14}$$

$$R_{rs}(\lambda_n) = Fun\{a_w(\lambda_n), b_{bw}(\lambda_n), \boldsymbol{a}_{ph}(\lambda_n), \boldsymbol{a}_g(\lambda_n), \boldsymbol{a}_d(\lambda_n), \boldsymbol{b}_{bp}(\lambda_n), \boldsymbol{\rho}(\lambda_n), \boldsymbol{H}\}$$

Eq. (14) provides the fundamental basis for deriving water properties analytically or semi-analytically. It also points out the intrinsic limitations in spectral remote sensing: an underdetermined system. For the n equations, the knowns are $R_{rs}(\lambda_{1-n})$ and other ancillary information such as solar zenith angle, sensor viewing angle, water temperature, and geographic location, etc. However, there are at least $4n$ (or $5n+1$) unknowns that need to be derived for optically deep (or shallow) waters. Mathematically there is no solution for such a problem, unless additional relationships are established to reduce the number of unknowns (or increase the number of equations). Fortunately field measurements have provided evidence that the five individual spectra do show some co-varying spectral dependences, so that $a_{ph,g,d}(\lambda_{1-n})$ and $b_{bp}(\lambda_{1-n})$ as well as $\rho(\lambda_{1-n})$ are not exactly independent variables among those spectral bands. Therefore each spectrum could be mathematically modeled with a few parameters.

3.2 SPECTRAL MODELS OF $a_g(\lambda)$ AND $a_d(\lambda)$

Based on extensive measurements from field samples, it has been found that both $a_g(\lambda)$ and $a_d(\lambda)$ can be well modeled by (Bricaud et al., 1981; Carder et al., 1989; Nelson and Robertson, 1993; Roesler et al., 1989)

$$a_x(\lambda) = a_x(\lambda_0)\exp(-S_x(\lambda - \lambda_0)), \tag{15}$$

with subscript x representing either g or d. S_x is the spectral slope of both absorption coefficients, with S_g generally in a range of 0.01 – 0.02 nm^{-1} (Kirk, 1994) and S_d in a range of 0.005 – 0.015 nm^{-1} (Mobley, 1994). From global measurements, mean values of S_g have been mostly reported in a range of 0.014 – 0.018 nm^{-1} and S_d in a range of 0.008 - 0.013 nm^{-1}.

Since $a_g(\lambda)$ and $a_d(\lambda)$ are very similar in spectral shape, it is almost impossible to analytically separate them remotely from $R_{rs}(\lambda)$. So in practice, the contributions of $a_g(\lambda)$ and $a_d(\lambda)$ are summed together to give (Carder et al., 1991)

$$a_y(\lambda) = a_g(\lambda) + a_d(\lambda) = a_y(\lambda_0)\exp(-S(\lambda - \lambda_0)), \tag{16}$$

with $a_y(\lambda_0)$ the absorption coefficient at a reference wavelength λ_0.

3.3 SPECTRAL MODELS OF $a_{ph}(\lambda)$

Depending on size, abundance, and pigment composition of phytoplankton, $a_{ph}(\lambda)$ varies significantly in both magnitude and spectral shape (Bricaud and Stramski, 1990; Ciotti et al., 2002; Hoepffner and Sathyendranath, 1991; Sathyendranath et al., 1987; Stuart et al., 1998). Due to this complexity, it has been found that the spectral shape of $a_{ph}(\lambda)$ cannot be modeled accurately with a simple mathematical function.

Based on field measurements, Prieur and Sathyendranath (1981) (also Fischer and Doerffer (1987) and Bukata et al. (1981)) expressed $a_{ph}(\lambda)$ as

$$a_{ph}(\lambda) = a_{ph}(440)a_{ph}^+(\lambda), \tag{17}$$

with $a_{ph}^+(\lambda)$ the $a_{ph}(440)$-normalized $a_{ph}(\lambda)$ spectrum. The values of $a_{ph}^+(\lambda)$ derived by Prieur and Sathyendranath (1981) are provided in Table 1. $a_{ph}(440)$ was further linked to $[C]$ through (Prieur and Sathyendranath, 1981)

$$a_{ph}(440) = 0.06 [C]^{0.65}. \tag{18}$$

Recently, more hyperspectral models have been proposed. From measurements of over 1166 spectra, Bricaud et al. (1995; 1998) expressed $a_{ph}(\lambda)$ as

$$a_{ph}(\lambda) = A_{ph}(\lambda)[C]^{E_{ph}(\lambda)}. \tag{19}$$

Here $A_{ph}(\lambda)$ and $E_{ph}(\lambda)$ are empirical model spectra and their values are statistically derived and provided in Table 1.

Lee (1994) took the Prieur and Sathyendranath (1981) approach, but expanded $a_{ph}^+(\lambda)$ as

$$a_{ph}^+(\lambda) = a_0(\lambda) + a_1(\lambda)\ln(a_{ph}(440)), \tag{20}$$

to account for the second-order variations of $a_{ph}(\lambda)$ shapes. $a_0(\lambda)$ and $a_1(\lambda)$ are empirical model spectra and are listed in Table 1.

For multiple wavelengths, such as the bands of MODIS, Carder et al. (1999) empirically modeled $a_{ph}(\lambda)$ as a function of $a_{ph}(675)$,

$$a_{ph}(\lambda_i) = a_0(\lambda_i)\exp\left[a_1(\lambda_i)\tanh\left(a_2(\lambda_i)\ln\left(\frac{a_{ph}(675)}{a_3(\lambda_i)}\right)\right)\right]a_{ph}(675). \qquad (21)$$

$a_{0\text{-}3}$ are model parameters for the MODIS bands and their values are presented in Table 2. With this hyperbolic tangent function, the ratio of $a_{ph}(\lambda_i)/a_{ph}(675)$ will approach asymptotic values for very high or very low values of $a_{ph}(675)$, as indicated from field measurements.

Table 1. Empirical constants for modeling hyperspectral $a_{ph}(\lambda)$.

lambda nm	P&S 81[a] $a_{ph}^{+}(\lambda)$	L 94[b] $a_0(\lambda)$	L 94[b] $a_1(\lambda)$	B 98[c] $A_{ph}(\lambda)$	B 98[c] $E_{ph}(\lambda)$	C 02[d] $\overline{a}_{ph}^{<p>}(\lambda)$	C 02[d] $\overline{a}_{ph}^{<m>}(\lambda)$
400	0.687	0.6843	0.0205	0.0240	0.6877	1.7439	1.5388
410	0.828	0.7782	0.0129	0.0287	0.6834	2.1799	1.6328
420	0.913	0.8637	0.0064	0.0328	0.6664	2.6669	1.6883
430	0.973	0.9603	0.0017	0.0359	0.6478	3.0479	1.7760
440	1.000	1.0000	0.0000	0.0378	0.6266	3.2247	1.7937
450	0.944	0.9634	0.0060	0.0350	0.5993	3.1540	1.6663
460	0.917	0.9311	0.0109	0.0328	0.5961	2.7848	1.6197
470	0.870	0.8697	0.0157	0.0309	0.5970	2.3488	1.5649
480	0.798	0.7890	0.0152	0.0280	0.5890	2.1013	1.4460
490	0.750	0.7558	0.0256	0.0254	0.6074	1.8657	1.3721
500	0.668	0.7333	0.0559	0.0210	0.6529	1.4336	1.2776
510	0.618	0.6911	0.0865	0.0162	0.7213	0.8994	1.1490
520	0.528	0.6327	0.0981	0.0126	0.7939	0.4870	1.0497
530	0.474	0.5681	0.0969	0.0103	0.8500	0.2486	0.9539
540	0.416	0.5046	0.0899	0.0085	0.9036	0.1233	0.8569
550	0.357	0.4262	0.0781	0.0070	0.9312	0.0530	0.7423
560	0.294	0.3433	0.0659	0.0057	0.9345	0.0258	0.6071
570	0.276	0.2950	0.0596	0.0050	0.9298	0.0331	0.5030
580	0.291	0.2784	0.0581	0.0051	0.8933	0.0416	0.4738
590	0.282	0.2595	0.0540	0.0054	0.8589	0.0424	0.4568
600	0.236	0.2389	0.0495	0.0052	0.8410	0.0452	0.4116
610	0.252	0.2745	0.0578	0.0055	0.8548	0.0758	0.4397
620	0.276	0.3197	0.0674	0.0061	0.8704	0.1060	0.5047
630	0.317	0.3421	0.0718	0.0066	0.8638	0.1280	0.5354
640	0.334	0.3331	0.0685	0.0071	0.8524	0.1371	0.5418
650	0.356	0.3502	0.0713	0.0078	0.8155	0.1909	0.5593
660	0.441	0.5610	0.1128	0.0108	0.8233	0.3692	0.8106
670	0.595	0.8435	0.1595	0.0174	0.8138	0.6324	1.1880
680	0.502	0.7485	0.1388	0.0161	0.8284	0.4674	1.0699
690	0.329	0.4000	0.0812	0.0069	0.9255	0.1618	0.5052
700	0.215	0.1500	0.0300	0.0025	1.0286	0.0503	0.1614

[a]Source: Prieur and Sathyendranath (1981). [b]Source: Lee (1994).
[c]Source: Bricaud et al. (1998). [d]Source: A. Ciotti (personal communication).

Table 2. Empirical parameters for modeling $a_{ph}(\lambda)$ at MODIS bands (Carder et al. 1999).

lambda	a_0	a_1	a_2	a_3
412	2.20	0.75	-0.5	0.0112
443	3.59	0.80	-0.5	0.0112
488	2.27	0.59	-0.5	0.0112
510	1.40	0.35	-0.5	0.0112
551	0.42	-0.22	-0.5	0.0112

For all these models, $a_{ph}(\lambda)$ spectrum is determined by one variable: $[C]$ or $a_{ph}(440)$ or $a_{ph}(675)$. For different $[C]$ values, the Prieur and Sathyendranath (1981) model provides the same $a_{ph}(\lambda)$ curvature, while the later models accounted for some of the curvature variations. Figure 2 presents examples of modeled $a_{ph}(\lambda)$ spectra. Differences remain amongst these modeled $a_{ph}(\lambda)$ spectra, though all models have the same $a_{ph}(440)$ values. Methods to model $a_{ph}(\lambda)$ with more than one variable have also been proposed. Ciotti et al. (2002) developed a two-parameter model for $a_{ph}(\lambda)$:

$$a_{ph}(\lambda) = [F\,\overline{a}_{ph}^{<p>}(\lambda) + (1-F)\,\overline{a}_{ph}^{<m>}(\lambda)]\langle a_{ph} \rangle, \qquad (22)$$

where F is a parameter describing the relative abundance of picoplankton among all phytoplankton. $\langle a_{ph} \rangle$ is the mean absorption computed between 400 and 700 nm, and $\overline{a}_{ph}^{<p>,<m>}(\lambda)$ are the $\langle a_{ph} \rangle$-normalized $a_{ph}(\lambda)$ for pico- and micro-plankton, respectively. The values of these normalized $a_{ph}(\lambda)$ are also provided in Table 1.

An alternative approach is to re-write Eqs.17&20 such that the $a_{ph}(\lambda)$ model of Lee (1994) can be extended to use two variables:

$$a_{ph}(\lambda) = a_0(\lambda)\,P + a_1(\lambda)\,P_1, \qquad (23)$$

with $P = a_{ph}(440)$ and P_1 a parameter for second-order variation of $a_{ph}(\lambda)$.

Hoepffner and Sathyendranath (1991) used a series of Gaussian functions to express $a_{ph}(\lambda)$ as:

$$a_{ph}(\lambda) = \sum_{i}^{n}[C_j]\,a_j^*(\lambda_{mj})\exp\left[-\frac{(\lambda - \lambda_{mj})^2}{2\sigma_j^2}\right], \qquad (24)$$

where λ_{mj} is the position of maximum absorption for the jth Gaussian band; σ_j determines the width of the peak; $a_j^*(\lambda_{mj})$ is the specific absorption coefficient for the jth absorption band at λ_{mj}, the wavelength of maximum absorption; and $[C_j]$ is the concentration of the pigment responsible for the jth absorption band. These specific absorption coefficients are believed to be independent of algal species, where package effects do not significantly affect the absorption properties of phytoplankton communities in seawater. (Hoepffner and Sathyendranath, 1991). In a similar approach, Stramski et al. (2001) modeled $a_{ph}(\lambda)$ by summarizing the contributions of 18 planktonic components.

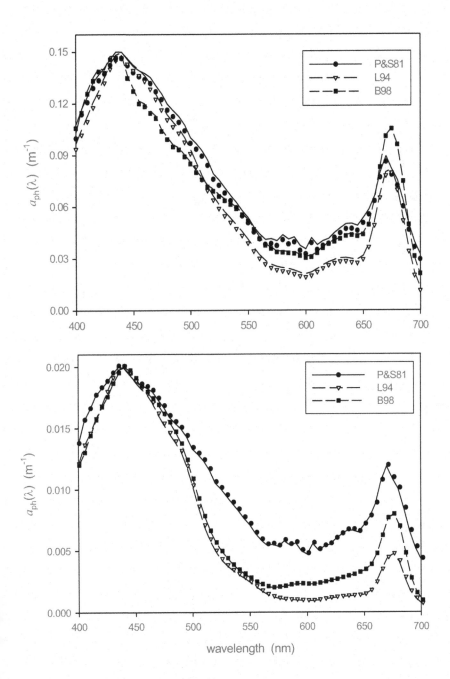

Figure 2. Examples of modeled hyperspectral $a_{ph}(\lambda)$ spectra. Prieur and Sathyendranath (1981) for Eq. (17) (P&S81); Lee (1994) for Eq. (20) (L94); Bricaud et al. (1998) for Eq. (19) (B98). Top: $a_{ph}(440\text{nm}) = 0.15$ m^{-1}. Bottom: $a_{ph}(440\text{nm}) = 0.020$ m^{-1}.

3.4 HYPERSPECTRAL MODELS OF THE BACKSCATTERING COEFFICIENT OF PARTICLES ($b_{bp}(\lambda)$)

It is generally believed that $b_{bp}(\lambda)$ can be mathematically described as (Gordon et al., 1980b; Sathyendranath et al., 1989b; Smith and Baker, 1981)

$$b_{bp}(\lambda) = b_{bp}(\lambda_0)\left(\frac{\lambda_0}{\lambda}\right)^{\eta}, \tag{25}$$

with λ_0 a reference wavelength. $b_{bp}(\lambda_0)$ varies in orders of magnitude for different waters, while η lies in a range $0 - 2.0$ for most oceanic and coastal waters (Bricaud et al., 1981; Lee et al., 1996a; Sathyendranath et al., 1989b; Smith and Baker, 1981).

Roesler and Boss (2003) recently modeled the backscattering coefficient of phytoplankton, $b_{b\text{-}ph}(\lambda)$, based on Mie scattering theory and sample measurements (Ahn et al., 1992; Bricaud and Morel, 1986; Morel and Bricaud, 1981) using the expression

$$b_{b-ph}(\lambda) = \sigma\left(c_{ph}(\lambda_0)\left(\frac{\lambda_0}{\lambda}\right)^{\psi} - a_{ph}(\lambda)\right). \tag{26}$$

Here σ is the backscattering to total scattering ratio and c_{ph} is the beam attenuation coefficient of phytoplankton. After accounting for the contributions of mineral particles, detritus, and 18 planktonic components, Stramski et al. (2001) found that, generally, more than 80% of $b_{bp}(\lambda)$ came from the contributions of mineral particles, which can be well described using Eq. (25). Phytoplankton contributes only a few percent to the total $b_{bp}(\lambda)$.

3.5 HYPERSPECTRAL MODEL OF $\rho(\lambda)$

$\rho(\lambda)$ is usually expressed as (Clark et al., 1987; Lee et al., 1999)

$$\rho(\lambda) = B \rho_x^+(550), \tag{27}$$

where B is the bottom reflectance at 550 nm and $\rho_x^+(550)$ the 550-normalized $\rho(\lambda)$ for different substrates on the bottom.

Based on the above modeling processes, if the one-variable $a_{ph}(\lambda)$ model is used, Eq. (14) becomes

$$\begin{aligned} R_{rs}(\lambda_1) &= Fun\{a_w(\lambda_1), b_{bw}(\lambda_1), \boldsymbol{P}, \boldsymbol{G}, \boldsymbol{S}, \boldsymbol{X}, \boldsymbol{\eta}, \boldsymbol{B}, \boldsymbol{H}\} \\ R_{rs}(\lambda_2) &= Fun\{a_w(\lambda_2), b_{bw}(\lambda_2), \boldsymbol{P}, \boldsymbol{G}, \boldsymbol{S}, \boldsymbol{X}, \boldsymbol{\eta}, \boldsymbol{B}, \boldsymbol{H}\} \\ &\vdots \\ R_{rs}(\lambda_n) &= Fun\{a_w(\lambda_n), b_{bw}(\lambda_n), \boldsymbol{P}, \boldsymbol{G}, \boldsymbol{S}, \boldsymbol{X}, \boldsymbol{\eta}, \boldsymbol{B}, \boldsymbol{H}\} \end{aligned} \tag{28}$$

with P for $a_{ph}(440)$ (or $[C]$ or $a_{ph}(675)$), G for $a_y(\lambda_0)$, and X for $b_{bp}(\lambda_0)$, respectively.

Now, for the series of n equations, there are only five (or seven) unknowns to be derived for optically deep (or shallow) waters. Theoretically, if the sensor has seven or more channels covering the necessary spectral range, the seven unknowns can be analytically derived from $R_{rs}(\lambda)$. Obviously, there are two effects which result from the above modeling processes: First, adding constraints among the different wavelengths makes it possible to have definite solutions to Eq. (14). Second, since no model is perfect, the added spectral models automatically introduce uncertainties in the derived values.

4. Analytical/Semi-analytical Methods of Solving Equation 28

Many methods have been developed for the retrieval of those unknowns (IOCCG, 2000). In the following, for brevity, a few analytical methods that use Eq. (28) as their basis are briefly introduced. Other methods that can be executed without Eq. (28), such as the principal component analysis (Fischer et al., 1986; Mueller, 1976; Neumann et al., 1995), spectral-curvature algorithms (Barnard et al., 1999; Campbell and Esaias, 1983; Hoge and Swift, 1986), and neural-network approaches (Lee et al., 1998c; Schiller and Doerffer, 1999) are not included here.

4.1 SPECTRAL OPTIMIZATION

Since $R_{rs}(\lambda)$ spectrum represents the cumulative effects of the five (or seven) variables, these variables could then be derived simultaneously from known $R_{rs}(\lambda)$, applying a widely used method called spectral optimization. Basically, the technique compares an $R_{rs}(\lambda)$ spectrum modeled by Eq. (28) with the measured $R_{rs}(\lambda)$ spectrum, with an index (ε) quantifying the difference between the two spectra

$$\varepsilon = \sum_j (R_{rs}(\lambda_j) - R_{rs}(\lambda_j))^2. \tag{29}$$

There are different forms in expressing ε (Lee et al., 1999; Roesler and Perry, 1995), but the fundamental basics are the same. To further reduce uncertainties and minimize numerical compensation among these variables, the spectral-shape parameter S and η are pre-determined and kept out of the spectral-optimization process, therefore there are only three (or five) unknowns left for deep (or shallow) waters. Generic computer programs have been developed to evaluate values of ε by varying the values of (P, G, X, B, H). When ε reaches a minimum, i.e. the modeled $R_{rs}(\lambda)$ best matches the measured $R_{rs}(\lambda)$ curve, the set of (P, G, X, B, H) value is then considered the derived set of environmental properties that make up the measured $R_{rs}(\lambda)$, and Eq. (28) is solved.

Earliest examples of using such an approach can be found in Fisher and Doerffer (1987) and Doerffer and Fisher (1994) for processing CZCS data off coastal areas, where the traditional simple spectral-ratio approach cannot provide accurate results. More recent examples can be found in Bukata et al. (1991a; 1991b) for analyzing data of turbid lakes, Lee et al. (1996a) and Roesler and Perry (1995) for hyperspectral data obtained from open ocean and coastal waters, Garver and Siegel (1997) for processing data from moored instruments, and Maritorena et al. (2000) for SeaWiFS data. The major difference among those processing methods rest on the way of handling $a_{ph}(\lambda)$ and values of η. Most use $a_{ph}(\lambda)$ spectral shapes from in situ measurements and apply a constant η value (Garver and Siegel, 1997; Maritorena et al., 2000; Roesler and Perry, 1995), while some use empirical $a_{ph}(\lambda)$ models with varying η values (Lee et al., 1999; Lee et al., 1996a). Figure 3 shows a comparison of $a(440)$ derived using $R_{rs}(440)/R_{rs}(555)$ ratio versus $a(440)$ derived from spectral optimization, with all values compared with those from in-water measurements (Lee, 1994). Clearly, as demonstrated by Bukata et al. (1991a), optimization provides more reliable results compared to simple ratio approach.

Spectral optimization is a very useful method for shallow-water remote sensing (Lee et al., 1999), where R_{rs} signal is a combination of contributions from the water column and the bottom. To resolve these contributions, it is necessary to retrieve properties of the water column and the bottom simultaneously. Figure 4 presents examples of spectral-optimization derived water absorption coefficient and bottom depth using

AVIRIS data collected over Tampa Bay (Florida) (Lee et al., 2001), where the water is so turbid that the traditional simple approach (Lyzenga, 1978; Philpot, 1989) cannot be applied.

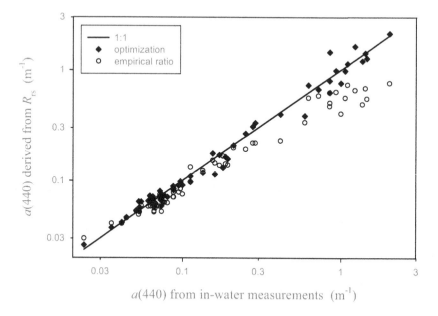

Figure 3. Spectral-ratio and spectral optimization derived $a(440 \text{ nm})$ compared with $a(440 \text{ nm})$ from in-water measurements (Lee, 1994). Reprinted with permission of the author.

4.2 LINEAR MATRIX INVERSION TO SOLVE $R_{RS}(\lambda)$ OF DEEP WATERS

Using Eq. (17) or a single Gaussian function to model $a_{ph}(\lambda)$, Hoge and Lyon (1996) rearranged Eq. (28) of optically deep waters at band λ_i as

$$P \exp\left(-\frac{(\lambda_i - \lambda_0)^2}{2g^2} \right) + G \exp(-S(\lambda_i - \lambda_0)) + X\, v(\lambda_i)\left(\frac{\lambda_0}{\lambda_i} \right)^{\eta} = h(\lambda_i). \tag{30}$$

Here, $v(\lambda_i)$ and $h(\lambda_i)$ are spectra determined by $R_{rs}(\lambda_i)$, $a_w(\lambda_i)$, and $b_{bw}(\lambda_i)$ (Hoge and Lyon, 1996). In this kind of formulation, Eq. (30) becomes a series of linear equations with three variables: P, G, and X. After fixing g to 700 nm, S as 0.018 nm^{-1} and η varied through an empirical relationship using $R_{rs}(490)/R_{rs}(555)$, values of P, G, and X are derived by linear matrix inversion using $R_{rs}(\lambda_i)$ at 412, 490, and 555 nm (Hoge et al., 2001). Since the spectral variability of the primary IOPs ($a_{ph}(\lambda)$, $a_y(\lambda)$, and $b_{bp}(\lambda)$) are quite significant, this inversion technique depends on using wavelengths that have minimal relative variation in each IOP spectral model. Over-determined linear inversions (with more than three $R_{rs}(\lambda_i)$ as inputs) can be applied to minimize the effect of errors in the inputs or models. However, the benefits gained in least-square fitting of more input data may be offset by the fact that the additional wavelengths may not co-vary spectrally as the model suggests (P. Lyon, pers. comm.).

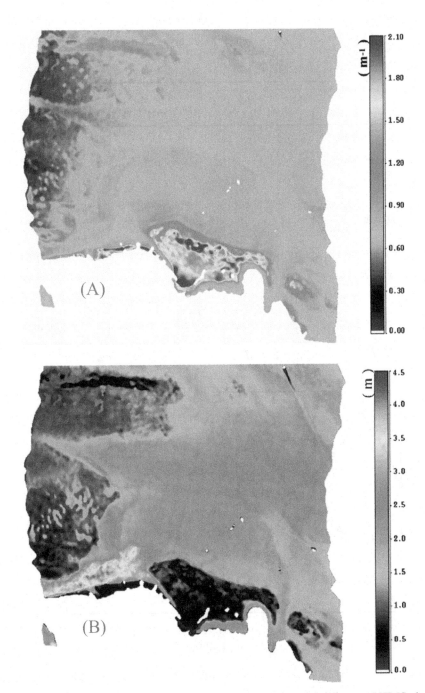

Figure 4. Properties of the water column and bottom derived from AVIRIS data by spectral optimization (Lee et al., 2001). Water absorption coefficient at 440 nm (A). Bottom depth (B). Reproduced by permission of the author and publisher (American Geophysical Union).

4.3 ITERATIVE METHOD TO SOLVE $R_{RS}(\lambda)$ OF DEEP WATERS

To derive $a_{ph}(675)$ (which can further be converted to concentrations of chlorophyll *a*) and $a_y(400)$ using data from MODIS, Carder et al. (1999) developed an iterative method to algebraically solve Eq. (28). Generally, Eq. (28) is rearranged as:

$$\frac{R_{rs}(412)}{R_{rs}(443)} = \frac{b_{bw}(412) + X(412)}{b_{bw}(443) + X(443)} \frac{a_w(443) + \{a_{ph}(675)\}_{443} + \{a_y(400)\}_{443}}{a_w(412) + \{a_{ph}(675)\}_{412} + \{a_y(400)\}_{412}}$$

$$\frac{R_{rs}(443)}{R_{rs}(551)} = \frac{b_{bw}(443) + X(443)}{b_{bw}(555) + X(551)} \frac{a_w(551) + \{a_{ph}(675)\}_{551} + \{a_y(400)\}_{551}}{a_w(443) + \{a_{ph}(675)\}_{443} + \{a_y(400)\}_{443}}$$

(31)

after $a_{ph}(\lambda_i)$ is modeled by Eq. (21), and $a_y(\lambda_i)$ is modeled by Eq. (16) with an *S* value of 0.022 nm^{-1}. Values of *X* and η are empirically derived from measured $R_{rs}(\lambda)$ (Carder et al., 1999). In this set of equations, the two unknowns ($a_{ph}(675)$ and $a_y(400)$) are derived iteratively. This method was designed for $a_{ph}(675)$ less than 0.03 m^{-1} (equivalent to [C] of ~1.5 – 2.0 mg/m^3). Empirical spectral-ratio algorithms were used for values beyond this limit.

4.4 QUASI-ANALYTICAL INVERSION OF DEEP-WATER $R_{RS}(\Lambda)$

In the above methods, a basic requirement is to have spectral models for $a_{ph}(\lambda)$ and $a_y(\lambda)$ in the derivation process. Since both spectra vary regionally and/or temporally, there is no guarantee that the modeled spectral shape will match the spectral shape under study. To minimize this uncertainty and to avoid the extensive computation time required by spectral optimization, Lee et al. (2002a) developed a quasi-analytical algorithm (QAA) to analytically invert $R_{rs}(\lambda)$ for optically deep waters.

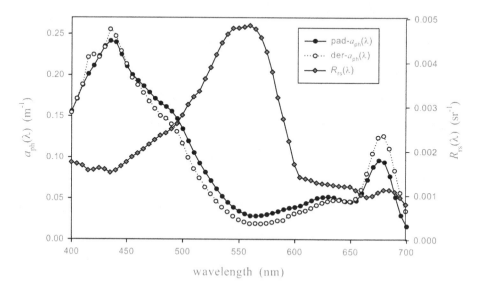

Figure 5. Example of measured versus remotely derived hyperspectral $a_{ph}(\lambda)$ specra. pad- $a_{ph}(\lambda)$ is the $a_{ph}(\lambda)$ from pad (GF/F) measurement; der- $a_{ph}(\lambda)$ is the $a_{ph}(\lambda)$ derived from $R_{rs}(\lambda)$ (Lee and Carder, 2004). Reprinted with permission of the author and publisher (Elsevier).

QAA first estimates (empirically or by neural networks) total absorption coefficient at a longer wavelength (λ_0, 555 or 640 nm, for instance) where $R_{rs}(\lambda_0)$ is well measured. At this wavelength, the total absorption coefficient is dominated by the contribution of pure water, so the errors in its estimation are limited. With $a(\lambda_0)$ known, $b_{bp}(\lambda_0)$ is analytically calculated from $r_{rs}(\lambda_0)$ by Eq. (2). $b_{bp}(\lambda)$ spectrum is then derived by applying the spectral $b_{bp}(\lambda)$ model (Eq. (25)) and estimating (empirically or by neural networks) the value of η from $R_{rs}(\lambda)$. This estimated $b_{bp}(\lambda)$ spectrum is substituted back into Eq. (2), $a(\lambda)$ spectrum is therefore calculated from measured $r_{rs}(\lambda)$. In this derivation sequence, there is no spectral model needed for $a_{ph}(\lambda)$ and $a_y(\lambda)$, which then reduces uncertainties in the derived total absorption coefficients. Further, when $a(\lambda)$ is known, it can be decomposed into $a_{ph}(\lambda)$ and $a_y(\lambda)$ after applying models regarding two ratios: $a_{ph}(410)/a_{ph}(440)$ and $a_y(410)/a_y(440)$ (Lee et al., 2002a). Figure 5 presents an example of QAA-derived $a_{ph}(\lambda)$ versus measured $a_{ph}(\lambda)$ from water samples (Lee and Carder, 2004). Note that when hyperspectral $a_{ph}(\lambda)$ is known, both major and minor pigments could be potentially derived using the approach developed by Hoepffner and Sathyendranath (1993).

5. Advantages and Drawbacks of Hyperspectral Remote Sensing

In recent years, hyperspectral remote sensing has gained much attention. An example in which hyperspectral information may be particularly useful is in distinguishing the subtle spectral signatures amongst different phytoplankton classes or species (Bidigare et al., 1989; Hoepffner and Sathyendranath, 1993; Millie et al., 1997). These hyperspectral techniques require spectral information from water samples that accurately represent the ranges of spectral variability found *in situ*. The example shown in Fig. 5, though, provides evidence to support potential application of the techniques to remotely measured data. However, further developments in sensor technologies and atmospheric correction techniques are required in order to obtain remotely sensed data of comparable quality to the data currently available from traditional *in situ* techniques. Signals from the water contribute only up to ~20% of the total signal collected at the satellite altitude, and partitioning the spectral domain too thinly will lead to reduced signal-to-noise ratios within a given spectral band. To separate the subtle spectral signatures that are desired, signal-to-noise ratio is the key.

Hyperspectral data is required for the separation of the subtle spectral signatures or minor properties. This may not be the case for those major properties, such as the unknowns in Eq. (28). After values of S and η are pre-determined, only three (or five) major unknowns left to be derived. This seems to suggest that if a sensor can produce high-quality water-color data at three (or five) spectral bands then it is sufficient for the derivation of those properties. In reality, however, due to limitations from sensor technology and atmospheric effects, there is always noise present in the measured $R_{rs}(\lambda)$. This is particularly true for wavelengths with high absorption and low backscattering where the true $R_{rs}(\lambda)$ is so low that its value can scarcely be measured remotely. To overcome the effects of measurement noise, redundant measurements at extra bands are always necessary.

Past (CZCS), current (e.g., SeaWiFS, MODIS, MERIS), and future satellite sensors all have at least four spectral bands in the visible domain. Assuming R_{rs} from all these sensors were error free, it is not clear if all these sensors would provide the same quality of retrievals for those major properties. To partially answer this question, Lee and Carder (2002) used field measured $R_{rs}(\lambda)$, from both optically deep and shallow waters, to analyze the influence of sensor configuration on retrievals.

For a series of real and "laboratory" sensors, such as the SeaWiFS, MODIS, MERIS, and sensors with spectral bands spaced every 5 nm (E5), every 10 nm (E10), and every 20 nm (E20), the major properties (such as absorption coefficient, bathymetry, etc.) of the water column and the bottom were derived from the R_{rs} spectra that are measured by these sensors. All derivations used the same spectral-optimization method (Section 4.1) and spectral IOP models, with the only difference in the number of spectral bands (and a few slightly different band positions). From the derived water column and bottom properties, there are some interesting findings (Lee and Carder, 2002): 1) For both optically deep and shallow waters, the performance of E10 is almost identical to that of E5; 2) E20 is slightly worse than E10, but better than MERIS for optically shallow waters; 3) For optically deep and clearer waters (almost all oceanic waters), SeaWiFS and MODIS operate satisfactorily in deriving the inherent optical properties of the water column; and, 4) The spectral configurations of SeaWiFS and MODIS only work for limited cases to accurately derive bathymetry of optically shallow waters (see Fig. 6).

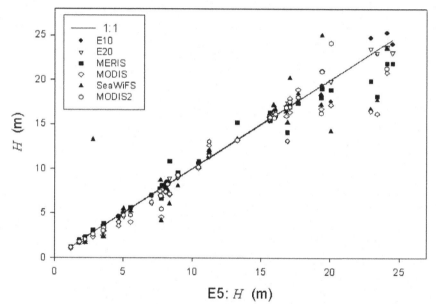

Figure 6. Comparison of retrieved bottom depth from different sensors. E5, E10, and E20 are sensors with wavelengths every 5, 10, and 20 nm respectively. MODIS2 is the MODIS configuration plus a band at 645 nm (Lee and Carder, 2002). Reprinted with permission of the author and publisher (Optical Society of America).

These results clearly indicate that ocean color data from extra bands improve the quality of retrieval in ocean color remote sensing, especially for complicated coastal or optically shallow waters. Further, it suggests that if the main objective of satellite observation is to estimate the major properties, such as total absorption coefficients or mass of suspended sediments, a sensor with 15 or so well positioned bands covering the 400 - 800 nm range is adequate (extra bands are needed for atmospheric corrections). Therefore, an ideal and cost effective sensor for such purposes may be one with enough bands but with high spatial resolution and high signal-to-noise ratios, instead of a large number of spectral bands.

6. Conclusions

During the past two decades, significant progress has been achieved in spectral remote sensing of marine environments. These achievements include better measurements of the water absorption coefficient, improvements in semi-analytical models for remote-sensing reflectance of both optically deep and optically shallow waters, and models for describing spectra of the absorption and backscattering coefficients of water-column constituents. All of these are important components needed for modeling and understanding the variation of ocean color and for retrieving semi-analytically water column, bottom properties, or both. These achievements also include improved algorithms to retrieve water absorption and backscattering coefficients from measurements of ocean color. Note that these properties can be converted to concentrations of water constituents. Because absorption and backscattering of the water medium are affected by all water constituents and water color is determined by the total absorption and backscattering properties, a change of chlorophyll concentration alone may not alter the appearance of water color (Sathyendranath et al., 1989b). Conversely, for both oceanic blue water and coastal green water, a change of ocean color may not necessarily result in changes in chlorophyll concentration (Bricaud et al., 1981; Carder et al., 1989; Siegel and Michaels, 1996). These relationships point out the intrinsic limitations of the empirical ocean color algorithms (O'Reilly et al., 1998) that use simple spectral ratios to derive chlorophyll concentration from measurements of water color. Since chlorophyll concentration derived using such a process contains large uncertainties, it is better to use the analytically or semi-analytically derived IOPs for applications in physical or biogeochemical oceanography, such as estimations of heat transfer (Lewis et al., 1990; McClain et al., 1996; Morel and Antoine, 1994; Ohlmann et al., 2000) and primary production (Behrenfeld and Falkowski, 1997; Lee et al., 1996b; Marra et al., 1992; Platt, 1986; Sathyendranath et al., 1989a; Zaneveld et al., 1993). Though progress has been made in the remote-sensing retrieval of IOPs (section 4), we still have difficulty selecting the parameters that determine the spectral shapes of particle backscattering, CDOM-detritus absorption, and pigment absorption. All of these properties influence (sometimes very significantly) the separation of water constituents. To improve our understanding of coastal environments, a better quantification of these minor parameters is desired, and the separation of marine environments into different biogeochemical provinces (Platt and Sathyendranath, 1988) may provide clues that are helpful in the selection of these minor parameters. However, the method to maximize the potential of using hyperspectral data to effectively retrieve important marine properties is still under investigation.

7. Acknowledgement

ZPL is extremely grateful to Dr. Susanne Craig for comments on an earlier draft of this manuscript.

8. References:

Ahn, Y.H., A. Bricaud, and A. Morel. 1992. Light backscattering efficiency and related properties of some phytoplanktonters, Deep-Sea Research I, 39:1835-1855.

Austin, R.W. 1974. Inherent spectral radiance signatures of the ocean surface, Ocean Color Analysis, SIO Ref. 7410.

Austin, R.W. 1980. Gulf of Mexico, ocean-colour surface-truth measurements, Boundary-Layer Meteorology, 18:269-285.

Austin, R.W., and T.J. Petzold,. 1981. The determination of the diffuse attenuation coefficient of sea water using the coastal zone color scanner, in Oceanography from Space, edited by J.F.R. Gower, Plenum Press, New York, pg. 239-256.

Barnard, A.H., J.R. Zaneveld, and W.S. Pegau. 1999. In situ determination of the remotely sensed reflectance and the absorption coefficient: closure and inversion, Applied Optics, 38:5108-5117.

Behrenfeld, M.J., and P.G. Falkowski. 1997. Photosynthetic rates derived from satellite-based chlorophyll concentration, Limnology and Oceanography, 42:1-20.

Bidigare, R.R., J.H. Morrow, and D.A. Kiefer. 1989. Derivative analysis of spectral absorption by photosynthetic pigments in the western Sargasso Sea, Journal Marine Research, 47:323-341.

Bricaud, A., M. Babin, A. Morel, and H. Claustre. 1995. Variability in the chlorophyll-specific absorption coefficients of naturnal phytoplankton: Analysis and parameterization, Journal Geophysical Research, 100:13321-13332.

Bricaud, A., and A. Morel. 1986. Light attenuation and scattering by phytoplanktonic cells: a theoretical modeling, Applied Optics, 25:571-580.

Bricaud, A., A. Morel, M. Babin, K. Allali, and H. Claustre. 1998. Variations of light absorption by suspended particles with chlorophyll a concentration in oceanic (case 1) waters: Analysis and implications for bio-optical models, Journal Geophysical Research, 103:31033-31044.

Bricaud, A., A. Morel, and L. Prieur. 1981. Absorption by dissolved organic matter of the sea (yellow substance) in the UV and visible domains, Limnology and Oceanography, 26:43-53.

Bricaud, A., and D. Stramski. 1990. Spectral absorption coefficients of living phytoplankton and nonalgal biogenous matter: A comparison between the Peru upwelling area and the Sargasso Sea, Limnology and Oceanography, 35:562-582.

Bukata, R.P., J.H. Jerome, J.E. Bruton, S.C. Jain, and H.H. Zwick. 1981. Optical water quality model of Lake Ontario. 1: Determination of the optical cross sections of organic and inorganic particulates in Lake Ontario, Applied Optics, 20:1696.

Bukata, R.P., J.H. Jerome, K.Y. Kondratyev, and D.V. Pozdnyakov. 1991a. Estimation of organic and inorganic matter in inland waters: Optical cross sections of Lakes Ontario and Ladoga, Journal Great Lakes Research, 17:461-469.

Bukata, R.P., J.H. Jerome, K.Y. Kondratyev, and D.V. Pozdnyakov. 1991b. Satellite monitoring of optically-active components of inland waters: an essential input to regional climate impact studies, Journal Great Lakes Research, 17:470-478.

Bukata, R.P., J.H. Jerome, K.Y. Kondratyev, and D.V. Pozdnyakov. 1995. Optical Properties and Remote Sensing of Inland and Coastal Waters, CRC Press, Boca Raton, FL.

Campbell, J., and W.E. Esaias. 1983. Basis for spectral curvature algorithms in remote sensing of chlorophyll, Applied Optics, 22:1084-1093.

Carder, K.L., F.R. Chen, Z.P. Lee, S.K. Hawes, and D. Kamykowski. 1999. Semianalytic Moderate-Resolution Imaging Spectrometer algorithms for chlorophyll-a and absorption with bio-optical domains based on nitrate-depletion temperatures, Journal of Geophysical Research, 104:5403-5421.

Carder, K.L., S.K. Hawes, K.A. Baker, R.C. Smith, R.G. Steward, and B.G. Mitchell. 1991. Reflectance model for quantifying chlorophyll a in the presence of productivity degradation products, Journal of Geophysical Research, 96:20599-20611.

Carder, K.L., and R.G. Steward. 1985. A remote-sensing reflectance model of a red tide dinoflagellate off West Florida, Limnology and Oceanography, 30:286-298.

Carder, K.L., R.G. Steward, G.R. Harvey, and P.B. Ortner. 1989. Marine humic and fulvic acids: their effects on remote sensing of ocean chlorophyll, Limnology and Oceanography, 34:68-81.

Carder, K.L., R.G. Steward, J.H. Paul, and G.A. Vargo. 1986. Relationships between chlorophyll and ocean color constituents as they affect remote-sensing reflectance models, Limnology and Oceanography, 31:403-413.

Ciotti, A.M., M.R. Lewis, and J.J. Cullen. 2002. Assessment of the relationships between domininant cell size in natural phytoplankton communities and spectral shape of the absorption coefficient, Limnology and Oceanography, 47:404-417.

Clark, D.K. 1981. Phytoplankton algorithm for the Nimbus-7 CZCS, in Oceanography from space, edited by J.R.F. Gower, Plenum Press, New York, pg. 227-238.

Clark, R.K., T.H. Fay, and C.L. Walker. 1987. Bathymetry calculations with Landsat 4 TM imagery under a generalized ratio assumption, Applied Optics, 26:4036-4038.

Dekker, A.G., R.J. Vos, and S.W.M. Peters. 2001. Analytical algorithms for lake water TSM estimation for retrospective analysis of TM and SPOT sensor data, International Journal Remote Sensing, 23:15-36.

Doerffer, R., and J. Fisher. 1994. Concentrations of chlorophyll, suspended matter, and gelbstoff in case II waters derived form satellite coastal zone color scanner data with inverse modeling methods, Journal Geophysical Research, 99:7475-7466.

Fischer, J., and R. Doerffer. 1987. An inverse technique for remote detection of suspended matter, phytoplankton and yellow substance from CZCS measurements, Advances. Space Research, 7:21-26.

Fischer, J., R. Doerffer, and H. Grassl. 1986. Factor analysis of multispectral radiances over coastal and open ocean water based on radiative transfer calculations, Applied Optics, 25:448-456.

Garver, S.A., and D. Siegel. 1997. Inherent optical property inversion of ocean color spectra and its biogeochemical interpretation 1. Time series from the Sargasso Sea, Journal Geophysical Research, 102:18607-18625.

Gordon, H.R. 1979. Diffuse reflectance of the ocean: the theory of its augmentation by chl a fluorescence at 685nm, Applied Optics, 18:1161-1166.

Gordon, H.R. 1994. Modeling and simulating radiative transfer in the ocean, in Ocean Optics, edited by R.W. Spinrad, K.L. Carder, and M.J. Perry, Oxford University, New York.

Gordon, H.R., O.B. Brown, and M.M. Jacobs. 1975. Computed relationship between the inherent and apparent optical properties of a flat homogeneous ocean, Applied Optics, 14:417-427.

Gordon, H.R., D.K. Clark, J.L. Mueller, and W.A. Hovis. 1980a. Phytoplankton pigments from the Nimbus-7 coastal Zone Color Scanner: Comparisons with surface measurements, Science, 210:63-66.

Gordon, H.R., R.C. Smith, and J.R.V. Zaneveld. 1980b. Introduction to ocean optics, in Ocean Optics VI, Proc. SPIE 208, pg. 1-43.

Gordon, H.R., D.K. Clark, J.W. Brown, O.B. Brown, R.H. Evans, and W.W. Broenkow. 1983. Phytoplankton pigment concentrations in the Middle Atlantic Bight: Comparison of ship determinations and CZCS estimates, Applied Optics, 22:20-36.

Gordon, H.R., and A. Morel. 1983. Remote assessment of ocean color for interpretation of satellite visible imagery: A review. Springer-Verlag, New York, 44 pp.

Gordon, H.R., O.B. Brown, R.H. Evans, J.W. Brown, R.C. Smith, K.S. Baker, and D.K. Clark. 1988. A semianalytic radiance model of ocean color, Journal Geophysical Research, 93:10,909-10,924.

Gordon, H.R., and M. Wang. 1994. Retrieval of water-leaving radiance and aerosol optical thickness over oceans with SeaWiFS: A preliminary algorithm, Applied Optics, 33:443-452.

Gregg, W.W., and K.L. Carder. 1990. A simple spectral solar irradiance model for cloudless maritime atmospheres, Limnology and Oceanography, 35:1657-1675.

Hawes, S.K., K.L. Carder, and G.R. Harvey. 1992. Quantum fluorescence efficiencies of marine humic and fulvic acids: effects on ocean color and fluorometric detection, in Ocean Optics, pg. 212-223.

Hoepffner, N., and S. Sathyendranath. 1991. Effect of pigment composition on absorption properties of phytoplankton, Marine Ecology Progress Series, 73:11-23.

Hoepffner, N., and S. Sathyendranath. 1993. Determination of the major groups of phytoplankton pigments from the absorption spectra of total particulate matter, Journal Geophysical Research, 98:22789-22803.

Hoge, F.E., and P.E. Lyon. 1996. Satellite retrieval of inherent optical properties by linear matrix inversion of oceanic radiance models: an analysis of model and radiance measurement errors, Journal Geophysical Research, 101:16631-16648.

Hoge, F.E., and R.N. Swift. 1986. Chlorophyll pigment concentration using spectral curvature algorithms: an evaluation of present and proposed satellite ocean color sensor bands, Applied Optics, 25:3677-3682.

Hoge, F.E., C.W. Wright, P.E. Lyon, R.N. Swift, and J.K. Yungel. 2001. Inherent optical properties imagery of the western North Atlantic Ocean: Horizontal spatial variability of the upper mixed layer, Journal Geophysical Research, 106:31129-31140.

Hojerslev, N.K. 2001. Analytic remote-sensing optical algorithms requiring simple and practical field parameter inputs, Applied Optics, 40:4870-4874.

IOCCG. 2000.Remote Sensing of Ocean Colour in Coastal, and Other Optically-Complex, Waters, in Reports of the International Ocean-Colour Coordinating Group, No.3, edited by S. Sathyendranath, IOCCG, Dartmouth, Canada.

Jerlov, N.G. 1976. Marine Optics, Elsevier, New York.

Jerome, J.H., R.P. Bukata, and J.R. Miller. 1996. Remote sensing reflectance and its relationship to optical properties of natural waters, International Journal of Remote Sensing, 17:3135-3155.

Kahru, M., and B.G. Mitchell. 1999. Empirical chlorophyll algorithm and preliminary SeaWiFS validation for the California Current, International Journal of Remote Sensing, 20:3423-3429.

Kirk, J.T.O. 1991. Volume scattering function, average cosines, and the underwater light field, Limnology and Oceanography, 36:455-467.

Kirk, J.T.O. 1994. Light & Photosynthesis in Aquatic Ecosystems, University Press, Cambridge.

Lee, Z.P. 1994. Visible-infrared Remote-sensing Model and Applications for Ocean Waters, Ph. D thesis, University of South Florida, St. Petersburg, Florida..

Lee, Z.P., and K.L. Carder. 2002. Effect of spectral band numbers on the retrieval of water column and bottom properties from ocean color data, Applied Optics, 41:2191-2201.

Lee, Z.P., and K.L. Carder. 2004. Absorption spectrum of phytoplankton pigments derived from hyperspectral remote-sensing reflectance, Remote Sensing Environment, 89(3):361-368.

Lee, Z.P., K.L. Carder, S.K. Hawes, R.G. Steward, T.G. Peacock, and C.O. Davis. 1994. A model for interpretation of hyperspectral remote sensing reflectance, Applied Optics, 33:5721-5732.

Lee, Z.P., K.L. Carder, R.G. Steward, T.G. Peacock, C.O. Davis, and J.L. Mueller. 1996a. Remote-sensing reflectance and inherent optical properties of oceanic waters derived from above-water measurements, in Ocean Optics XIII, edited by S.G. Ackleson, and R. Frouin, Proc. SPIE 2963, pg. 160-166.

Lee, Z.P., K.L. Carder, R.G. Steward, and M.J. Perry. 1996b. Estimating primary production at depth from remote sensing, Applied Optics, 35:463-474.

Lee, Z.P., K.L. Carder, C.D. Mobley, R.G. Steward, and J.S. Patch. 1998a. Hyperspectral remote sensing for shallow waters. 1. A semianalytical model, Applied Optics, 37:6329-6338.

Lee, Z.P., K.L. Carder, R.G. Steward, T.G. Peacock, C.O. Davis, and J.S. Patch. 1998b. An empirical algorithm for light absorption by ocean water based on color, Journal Geophysical Research, 103:27967-27978.

Lee, Z.P., M.R. Zhang, K.L. Carder, and L.O. Hall. 1998c. A neural network approach to deriving optical properties and depths of shallow waters, in Ocean Optics XIV, Kona, HI.

Lee, Z.P., K.L. Carder, C.D. Mobley, R.G. Steward, and J.S. Patch. 1999. Hyperspectral remote sensing for shallow waters: 2. Deriving bottom depths and water properties by optimization, Applied Optics, 38:3831-3843.

Lee, Z.P., K.L. Carder, R.F. Chen, and T.G. Peacock. 2001. Properties of the water column and bottom derived from AVIRIS data, Journal Geophysical Research, 106:11639-11652.

Lee, Z.P., K.L. Carder, and R. Arnone. 2002a. Deriving inherent optical properties from water color: A multi-band quasi-analytical algorithm for optically deep waters, Applied Optics, 41:5755-5772.

Lee, Z.P., K.L. Carder, and K.P. Du. 2002b. Influence of particle scattering on the model parameter of remote-sensing reflectance, in Ocean Optics XVI, Santa Fe, New Mexico.

Lewis, M.R., M. Carr, G. Feldman, W. Esaias, and C. McMclain. 1990. Influence of Penetrating solar radiation on the heat budget of the equatorial pacific ocean, Nature, 347:543-545.

Lyzenga, D.R. 1978. Passive remote-sensing techniques for mapping water depth and bottom features, Applied Optics, 17:379-383.

Lyzenga, D.R. 1981. Remote sensing of bottom reflectance and water attenuation parameters in shallow water using aircraft and Landsat data, International Journal of Remote Sensing, 2:71-82.

Maritorena, S., A. Morel, and B. Gentili. 1994. Diffuse reflectance of oceanic shallow waters: influence of water depth and bottom albedo, Limnology and Oceanography, 39:1689-1703.

Maritorena, S., D.A. Siegel, and A.R. Peterson. 2000. Optimization of a semianalytical ocean color model for global-scale applications, Applied Optics, 41:2705-2714.

Marra, J., T. Dickey, W.S. Chamberlin, C. Ho, T. Granata, D.A. Kiefer, C. Langdon, R.C. Smith, K.S. Baker, R.R. Bidigare, and M. Hamilton. 1992. Estimation of seasonal primary production from moored optical sensors in the Sargasso Sea, Deep-Sea Research, 97:7399-7412.

Marshall, B.R., and R.C. Smith. 1990. Raman scattering and in-water ocean properties, Applied Optics, 29:71-84.

McClain, C.R., K. Arrigo, K.-S. Tai, and D. Turk. 1996. Observations and simulations of physical and biological process at ocean weather station P, 1951-1980, Journal Geophysical Research, 101:3697-3713.

Millie, D.F., O.M. Schofied, G.J. Kirkpatrick, G. Johnsen, P.A. Tester, and B.T. Vinyard. 1997. Detection of harmful algal blooms using photopigments and absorption signature: A case study of the Florida red tide dinoflagellate, Gymnodinium breve, Limnology and Oceanography, 42:1240-1251.

Mitchell, B.G., and O. Holm-Hansen. 1990. Bio-optical properties of Antarctic Peninsula waters: differentiation from temperate ocean models, Deep-Sea Research, 38:1009-1028.

Mitchell, B.G., and M. Kahru. 1998. Algorithms for SeaWiFS standard products developed with the CalCOFI big-optical data set, Scripps Institute of Oceanography, La Jolla, CA 92093, USA, pg. 133-147.

Mobley, C.D. 1994. Light and Water: radiative transfer in natural waters, Academic Press, New York.

Mobley, C.D. 1995. Hydrolight 3.0 Users' Guide, SRI International, Menlo Park, Calif.

Mobley, C.D., L.K. Sundman, and E. Boss. 2002. Phase function effects on oceanic light fields, Applied Optics, 41:1035-1050.

Morel, A. 1974. Optical properties of pure water and pure sea water, in Optical aspects of oceanography, edited by N.G. Jerlov, and Nielsen, E. S., Academic, New York, pg. 1-24.

Morel, A., and D. Antoine. 1994. Heating rate within the upper ocean in relation to its bio-optical state, Journal of Physical Oceanography, 24:1652-1665.

Morel, A., D. Antoine, and B. Gentili. 2002. Bidirectional reflectance of oceanic waters: accounting for Raman emission and varying particle scattering phase function, Applied Optics, 41:6289-6306.

Morel, A., and A. Bricaud. 1981. Theoretical results concerning the optics of phytoplankton, with special reference to remote sensing applications, in Oceanography from space, edited by J.F.R. Gower, Plenum, New York.

Morel, A., and B. Gentili. 1991. Diffuse reflectance of oceanic waters: its dependence on sun angle as influenced by the molecular scattering contribution, Applied Optics, 30:4427-4438.

Morel, A., and B. Gentili. 1993. Diffuse reflectance of oceanic waters (2): Bi-directional aspects, Applied Optics, 32:6864-6879.

Morel, A., and L. Prieur. 1977. Analysis of variations in ocean color, Limnology and Oceanography, 22:709-722.

Mueller, J.L. 1976. Ocean color spectra measured off the Oregon coast: characteristic vectors, Applied Optics, 15:394-402.

Nelson, J.R., and C.Y. Robertson. 1993. Detrital spectral absorption: Laboratory studies of visible light effects on phytodetritus absorption, bacterial spectral signal, and comparison to field measurements, Journal Marine Research, 51:181-207.

Neumann, A., H. Krawczyk, and T. Walzel. 1995. A complex approach to quantitative interpretation of spectral high resolution imagery, in Third Thematic Conference on Remote Sensing for Marine and Coastal Environments, Seattle, USA.

Ohlmann, J.C., D.A. Siegel, and C.D. Mobley. 2000. Ocean radiant heating. Part I: Optical influences, Journal of Physical Oceanography, 30:1833-1848.

O'Reilly, J., S. Maritorena, B.G. Mitchell, D. Siegel, K.L. Carder, S. Garver, M. Kahru, and C. McClain. 1998. Ocean color chlorophyll algorithms for SeaWiFS, Journal Geophysical Research, 103:24937-24953.

Peacock, T.G., K.L. Carder, and C.O. Davis, and R. G. Steward. 1990. Effects of fluorescence and water Raman scattering on models of remote-sensing reflectance, in Ocean Optics X, SPIE, 1990, pg. 303-319.

Petzold, T.J. 1972. Volume scattering functions for selected natural waters, Scripps Institution Oceanography. pg. 72-78.

Philpot, W.D. 1989. Bathymetric mapping with passive multispectral imagery, Applied Optics, 28:1569-1578.

Platt, T. 1986. Primary production of ocean water column as a function of surface light intensity: algorithms for remote sensing, Deep-Sea Research, 33:149-163.

Platt, T., and S. Sathyendranath. 1988. Oceanic primary production: estimation by remote sensing at local and regional scales, Science, 241:1613-1620.

Pope, R., and E. Fry. 1997. Absorption spectrum (380 - 700 nm) of pure waters: II. Integrating cavity measurements, Applied Optics, 36:8710-8723.

Preisendorfer, R.W. 1976. Hydrologic optics vol. 1: introduction, National Technical Information Service, Springfield.

Prieur, L., and S. Sathyendranath. 1981. An optical classification of coastal and oceanic waters based on the specific spectral absorption curves of phytoplankton pigments, dissolved organic matter, and other particulate materials, Limnology and Oceanography, 26:671-689.

Roesler, C.S., and E. Boss. 2003. Spectral beam attenuation coefficient retrieved from ocean color inversion, Geophysical Research Letters, 30(9):1468.

Roesler, C.S., and M.J. Perry. 1995. In situ phytoplankton absorption, fluorescence emission, and particulate backscattering spectra determined from reflectance, Journal Geophysical Research, 100:13279-13294.

Roesler, C.S., M.J. Perry, and K.L. Carder. 1989. Modeling in situ phytoplankton absorption from total absorption spectra in productive inland marine waters, Limnology and Oceanography, 34:1510-1523.

Sathyendranath, S., G. Cota, V. Stuart, M. Maass, and T. Platt. 2001. Remote sensing of phytoplankton pigments: a comparison of empirical and theoretical approaches, International Journal of Remote Sensing, 22:249-273.

Sathyendranath, S., L. Lazzara, and L. Prieur. 1987. Variations in the spectral values of specific absorption of phytoplankton, Limnology and Oceanography, 32:403-415.

Sathyendranath, S., T. Platt, C.M. Caverhill, R.E. Warnock, and M.R. Lewis. 1989a. Remote sensing of oceanic primary production: computations using a spectral model, Deep-Sea Research, 36:431-453.

Sathyendranath, S., L. Prieur, and A. Morel. 1989b. A three-component model of ocean colour and its application to remote sensing of phytoplankton pigments in coastal waters, International Journal of Remote Sensing, 10:1373-1394.

Sathyendranath, S., and T. Platt. 1997. Analytic model of ocean color, Applied Optics, 36:2620-2629.

Sathyendranath, S., and T. Platt. 1998. Ocean color model incorporating transspectral processes, Applied Optics, 37:2216-2227.

Schiller, H., and R. Doerffer. 1999. Neural network for emulation of an inverse model -- operational derivation of Case II water properties from MERIS data, International Journal of Remote Sensing, 20:1735-1746.

Siegel, D., and A.F. Michaels. 1996. Quantification of non-algal light attenuation in the Sargasso Sea: Implications for biogeochemistry and remote sensing, Deep-Sea Research, 43(2-3):321-345.

Smith, R.C., and K.S. Baker. 1981. Optical properties of the clearest natural waters, Applied Optics, 20:177-184.

Smyth, T.J., S.B. Groom, D.G. Cummings, and C.A. Llewellyn. 2002. Comparison of SeaWiFS bio-optical chlorophyll-a algorithms within the OMEXII programme, International Journal of Remote Sensing, 23:2321-2326.

Stavn, R.H., and A.D. Weidemann. 1990. Raman scattering effects at the shorter visible wavelengths in clear ocean waters, in Ocean Optics X, SPIE, pg. 94-100.

Stramski, D., A. Bricaud, and A. Morel. 2001. Modeling the inherent optical properties of the ocean based on the detailed composition of the planktonic community, Applied Optics, 40:2929-2945.

Stuart, V., S. Sathyendranath, T. Platt, H. Maass, and B. Irwin. 1998. Pigments and species composition of natural phytoplankton populations: effect on the absorption spectra, Journal Plankton Research, 20:187-217.

Stumpf, R.P., and J.R. Pennock. 1991. Remote estimation of the diffuse attenuation coefficient in a moderately turbid estuary, Remote Sensing Evironment, 38:183-191.

Voss, K.J., C.D. Mobley, L.K. Sundman, J.E. Ivey, and C.H. Mazel. 2003. The spectral upwelling radiance distribution in optically shallow waters, Limnology and Oceanography, 48:364-373.

Yentsch, C.S., and D.A. Phinney. 1985. Spectral fluorescence: an ataxonomic tool for studying the structure of phytoplankton populations, Journal of Plankton Research, 7:617-632.

Yentsch, C.S., and C.M. Yentsch. 1979. Fluorescence spectral signatures: the characterization of phytoplankton populations by the use of excitation and emission spectra, Journal Marine Research, 37:471-483.

Zaneveld, J.R.V. 1982. Remote sensed reflectance and its dependence on vertical structure: a theoretical derivation, Applied Optics, 21:4146-4150.

Zaneveld, J.R.V. 1995. A theoretical derivation of the dependence of the remotely sensed reflectance of the ocean on the inherent optical properties, Journal Geophysical Research, 100:13135-13142.

Zaneveld, J.R.V., J.C. Kitchen, and J.L. Mueller. 1993.Vertical structure of productivity and its vertical integration as derived from remotely sensed observations, Limnology and Oceanography, 38:1384-1393.

Chapter 9

COMPUTATIONAL INTELLIGENCE AND ITS APPLICATION IN REMOTE SENSING

HABTOM RESSOM, [2]RICHARD L. MILLER, [3]PADMA NATARAJAN, AND [3]WAYNE H. SLADE

[1]Lombardi Comprehensive Cancer Center, Biostatistics Shared Resource, Department of Oncology, Georgetown University Medical Center, Washington, DC, 20057 USA
[2]National Aeronautics and Space Administration, Earth Science Applications Directorate, Stennis Space Center, MS, 39529 USA
[3]Department of Electrical and Computer Engineering, University of Maine, Orono, ME, 04473 USA

1. Introduction

Remote sensing observations provide a new global perspective of the Earth environment. Measurements from airborne and space borne sensor systems help scientists gain a better understanding of the complex interactions between the Earth's atmosphere, oceans, ice regions and land surfaces, as well as human-induced change due to population growth and human activities. These remote sensing measurements are widely used in geographical, meteorological, and environmental studies.

Technological advancements have resulted in an increase in the number of observation platforms and sensor capabilities (e.g., spectral and spatial resolution). This trend will continue and soon will produce an unprecedented volume of data. Information extracted from these datasets will support national research agendas and national applications that will exert an ever-increasing requirement for shorter processing times and greater data and algorithm accuracies. Hence, advanced mathematical techniques are needed to effectively analyze data generated from the rapidly growing remote sensing technology.

For most geophysical retrieval algorithms, adding additional information to improve the measurement of *in situ* properties is not a simple task because of the nonlinear nature of the problem as well as computational difficulties. Moreover, most current mathematical techniques generally require a high level of scientific knowledge of the physical system to accurately analyze remotely sensed data. In contrast, computational intelligence (CI) techniques such as artificial neural networks, genetic algorithms, and fuzzy logic systems, provide the capability to better examine complex data without requiring detailed knowledge about the underlying physical system. For example, CI techniques have been used to accurately estimate bio-optical parameters in complex coastal aquatic environments from remotely sensed data by employing special features such as the ability to *learn* from data, adaptive behavior, handling of non-linear systems, flexibility towards the choice of inputs, and resilience against noise.

For example, in satellite remote sensing of ocean color, most algorithms are based on regression (or empirical) models that use power and/or cubic polynomials to relate ratios of remotely sensed reflectance to bio-optical parameters such as chlorophyll

R.L. Miller et al. (eds.), Remote Sensing of Coastal Aquatic Environments, 205–227.

concentration. Though these algorithms have been successful in open ocean water, there has been limited success in optically complex coastal waters due in-part to the high variability of bio-optical relationships in these environments and limited spectral characteristics of ocean color satellites. As coastal regions are disproportionately important in terms of direct human impact and in their role in large-scale biogeochemical processes, there is a critical need for accurate remote sensing algorithms for the coastal margin. To date, only a few regional models exist for coastal waters, mainly due to a lack of representative data and a lack of sufficient analytical techniques. In contrast, CI-based techniques have great potential in developing ocean color algorithms that can effectively estimate many important parameters in complex coastal waters. The fact that CI techniques require little knowledge about the underlying relationships also make them ideal candidates for building global models.

In the following sections we present a detailed overview of CI definitions, common paradigms, and model development procedures. We then present the application of CI paradigms in estimating coastal bio-optical parameters such as chlorophyll a (chl *a*), color dissolved organic matter (CDOM), suspended sediments, and phytoplankton primary production. It must be noted that CI is an expansive discipline that embodies a vast array of terminology and techniques that may be new and complex to many marine scientists. We provide here only a concise description of the more salient features of CI and its application to remote sensing. The reader is directed to Haykin (1999), Mitchell (1998), Yen and Langari (1999), and Engelbrecht (2002) for a more thorough treatise on the subject.

2. Background on Computational Intelligence

Computational intelligence (CI) is the study of adaptive mechanisms to enable or facilitate intelligent behavior in complex and changing environments. These mechanisms include mathematical models that exhibit an ability to learn or adapt to new situations, to generalize, abstract, discover and associate. The three paradigms of CI discussed in this chapter are artificial neural networks (NN), genetic algorithms (GA), and fuzzy systems (FS). Each of these CI paradigms has its origins in biological systems. NN model biological neural systems, FS originated from studies of how organisms interact with their environment, and GA model natural evolution (Englebrecht, 2002). NN and GA deal primarily with numeric data. They can be used to extract knowledge from data and to solve optimization problems. FS accommodate numeric and semantic information and can be used to transform human expert knowledge into a mathematical description. Hybrid systems use a combination of these biologically and linguistically inspired computational models to take advantage of their collective features.

2.1 NEURAL NETWORKS

Neural networks (NN) generally consist of a number of interconnected processing elements known as neurons. The way neurons are interconnected or how the inter-neuron connections are arranged determine the architecture of a neural network. The way the strengths of the connections (known as weights or synaptic weights) are adjusted, or trained to achieve a desired overall behavior of the network, is governed by the learning algorithm used. NN are classified according to their architecture and learning algorithms.

The most popular architecture is a *feedforward neural network*, where the neurons are grouped into layers. All connections are *feedforward*; that is, they allow information transfer only from an earlier layer to the next consecutive layers. Neurons within a layer are not connected, and neurons in non adjacent layers are not connected. Input signals are presented to the network via an "input layer." The nodes in the input layers do not process input signals but pass them to one or more "hidden layers" where the actual processing is done via a system of weighted "connections". The hidden layers then link to an "output layer" which provides the outputs of the network.

Figure 1 depicts a feedforward neural network that has four layers: an input layer, two hidden layers, and an output layer. As shown in the figure, the inputs to the network are $x_1, x_2, \ldots, x_n$, and the network outputs are $y_1, y_2, \ldots, y_m$. Figure 2 illustrates the details of a single neuron, where net_1, known as activation level, is the sum of the weighted inputs to the neuron; and $f(.)$ represents an activation function, which transforms the activation level of a neuron into an output signal. Typically, an activation function could be a threshold function, a sigmoid function (an S-shaped, symmetric function that is continuous and differentiable), a Gaussian function, or a linear function. For example, the output of the neuron in Fig. 2 (i.e., the output of the first neuron in the first hidden layer of the network in Fig. 1) can be written as:

$$o_1 = f(net_1) = f\left(\sum_{j=1}^{n} w_{1j} x_j\right), \tag{1}$$

where the synaptic weights $w_{11}, w_{12}, \ldots, w_{1n}$ define the strength of connection between the neuron and its inputs. Such synaptic weights exist between all pairs of neurons in each successive layer of the network. They are adapted during learning to yield the desired outputs at the output layer of the network.

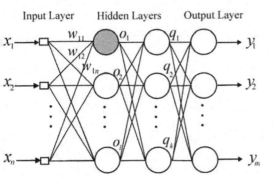

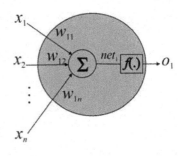

Figure 1. Multilayer perceptron network.   **Figure 2.** Details of a neuron.

A multilayer feedforward network, whose neurons in the hidden layers have sigmoidal activation functions, is known as a multilayer perceptron (MLP) network. MLP networks are capable of learning complex input-output mapping. That is, given a set of inputs and desired outputs, an adequately chosen MLP network can emulate the mechanism that produces the data through learning.

In a *supervised learning* paradigm, the network uses training examples (specifically, the desired outputs for a given set of inputs) to determine how well it has learned, and to guide adjustments of the synaptic weights in order to reduce its overall error. An example of a supervised learning rule is the back-propagation algorithm (Rumelhart and McClelland, 1986), which is developed to train MLP networks based on the principle of steepest gradient method. The training of a neural network is complete when a pre-specified stopping criterion is fulfilled. A typical stopping criterion is the performance of the network on a validation dataset, which is a portion of the training examples that was not used for updating the weights.

In contrast to supervised learning, *unsupervised learning* discovers patterns or features in the input data with no help from a teacher, essentially performing a clustering of the input space. Typical unsupervised learning rules include Hebbian learning, principal component learning, and Kohonen learning. These rules are applied in training self-organizing neural network architectures such as self-organizing maps (SOM) and learning vector quantizers. While neural networks developed for both the supervised and unsupervised learning paradigms have performed very well in their respective application fields, improvements have been developed by combining the two paradigms. A typical example is a radial basis function (RBF) network discussed below.

RBF networks are special feedforward networks that have a single hidden layer. The activation functions of the neurons in the hidden layer are radial basis functions, while the neurons in the output layer have simple linear activation functions. Radial basis functions are a set of predominantly nonlinear functions such as Gaussian functions that are built up into one function. Each Gaussian function responds only to a small region of the input space where the Gaussian is centered. Hence, while an MLP network uses hyperplanes defined by weighted sums as arguments to sigmoidal functions, an RBF network uses hyperellipsoids to partition the input space in which measurements or observations are made. Thus, RBF networks find the input to output map using local approximators. The key to a successful implementation of RBF networks is to find suitable centers for the Gaussian functions. Theoretically, the centers and width of the Gaussian functions can be determined with supervised learning, for example, steepest gradient method. They can also be determined through unsupervised learning by clustering the training data points. Once the centers and width of the Gaussian functions are obtained, the weights for the connection between the hidden layer and the output layer are easily and quickly determined using methods such as linear least squares, as the output neurons are simple linear combiners.

If a neural network is to perform a time series prediction or build a model of a dynamical system, it is important to establish a form of expansion so that the network contains some type of memory element. This can be achieved by applying time-delayed inputs to feedforward networks. Alternatively, a *recurrent neural network* can be built, in which the outputs of some neurons are fed back to the same neurons or to other neurons in the same or proceeding layers. Thus, signals can flow in both forward and backward directions. Recurrent neural networks have a dynamic memory - their outputs at a given instant reflect the current model input as well as previous inputs and outputs.

2.2 GENETIC ALGORITHMS

In contrast to gradient search methods, which sometimes lead the solution to local optima, genetic algorithms (GA) are global optimization algorithms that originated from mechanics of natural genetics and selection (Holland, 1975; Goldberg, 1989). They provide a method of problem solving that is based on genetic evolution. Based on probabilistic decisions, they exploit historic information to guide the search for better solutions in the problem space. Using a direct analogy of natural evolution, they work with a population of chromosomes, where each chromosome encodes a possible solution to the problem as a string of bits (0100101010), list of real values (0.2, 6.5, 4.1, 1.3), string of characters (B2, A3, C1, A1), or some other representation.

The algorithm starts by randomly creating a population of chromosomes. The chromosomes in the population are evaluated by a fitness function indicating how good encoded solutions appear with respect to the problem under consideration. The algorithm takes the chromosomes with higher relative fitness (chosen as parents) and then creates a new generation of chromosomes using genetic operators such as crossover or mutation. In crossover, new offspring are created from two parents by swapping a portion of their strings. In mutation, offspring are identical to their parents, but have random changes in portions of their strings. The algorithm repeats the above steps until a predefined number of generations or fitness value is reached.

Genetic algorithms therefore begin with a random process and arrive at an optimized solution. They are thus well suited for those tasks that seek global optimization. They are highly effective in situations where many inputs interact to produce a large number of possible outputs or solutions. They are a robust search method requiring little information to search effectively in a large or poorly understood search space.

2.3 FUZZY SYSTEMS

Fuzzy logic is a superset of conventional two-valued (Boolean) logic that has been extended to handle the concept of partial truth. Thus, in fuzzy logic, the truth-value of a statement is defined in the continuous interval between 0 (completely false) and 1 (completely true). It enables designers to simulate human thinking by quantifying concepts such as hot, cold, far, near, soon, high and low. For example, the interpretation of "high" temperature in describing weather condition is somewhat arbitrary. How can one define "high" temperature? Using a cutoff temperature (say, 25 °C) is not sufficient; 24 °C would not be considered "high", while 15 °C is not as "high" as 24 °C. A more accurate definition of "high" temperature may be to consider 25 °C and above completely "high" (truth-value of 1), 15 °C and below completely "not high" (truth-value of 0), and the range between 15 °C and 25 °C to be partially "high" in a linear relationship (see Fig. 3).

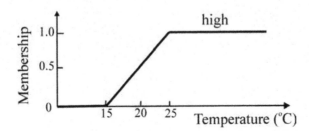

Figure 3. Membership function for high temperature.

Thus, in contrast to a traditional set theory that requires elements to be either part of a set or not, fuzzy logic allows an element to belong to a set to a certain degree of certainty. A membership function is used to associate a degree of membership of each of the elements of the domain to a fuzzy set. The degree of membership to a fuzzy set indicates the certainty that the element belongs to that set. For example, Fig. 3 shows the membership function for the fuzzy set temperature "high".

Besides membership functions, a fuzzy system consists of a set of fuzzy rules. A fuzzy rule has two components, an *if* part (also referred to as premise) and a *then* part (also referred to as conclusion). Such rules can be used to represent knowledge and association, which are inexact and imprecise in nature, expressed in qualitative values that a human can easily understand. For example, one might say, "if temperature is high and humidity low, then the weather is medium hot."

Figure 4 depicts a fuzzy system that has four principal units: fuzzification, knowledge base, decision-making (inference), and defuzzification. The fuzzy system accepts a set of inputs (x_1, x_2, ... x_n) as its information about the outside world (also referred to as crisp data). The fuzzification unit converts these inputs into fuzzy sets based up on fuzzy values, such as "high", "medium", and "low." In this unit, the membership functions defined on the input variables are applied to their actual values, to determine the degree of truth for each rule premise. The knowledge base contains member functions defining fuzzy values and a set of fuzzy rules. The inference unit executes these fuzzy rules. The truth-value for the premise of each rule is computed, and applied to the conclusion part of each rule. When all the rules are executed, a fuzzy region will be created for each output variable (y_1, y_2,..., y_m). This results in one fuzzy subset to be assigned to each output variable for each rule. With the process of defuzzification, a discrete value of each output (crisp data) as a solution will be generated.

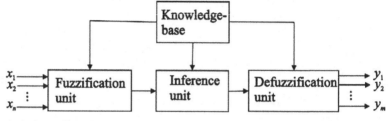

Figure 4. A fuzzy-based model of a system.

Thus, a fuzzy system can provide us with a quickly implemented nonlinear model of a complex process, described by a set of humanly understandable statements that express typical relationships between various input levels and desired outputs. Zadeh (1994) observed that as a system becomes more complex, the need to describe it with precision becomes less important. He also observed that precision in complex systems is often unnecessary and that it is possible to describe a system's behavior and control laws linguistically. Hence, in modeling complex systems using fuzzy logic, the underlying mechanics are represented linguistically rather than mathematically. In ordinary system identification, for example, parameters are the coefficients in a functional system model. In fuzzy model, the parameters are those in the membership of the fuzzy sets. These fuzzy terms, e.g. temperature "high", define general categories, but not rigid, fixed collections.

2.4 HYBRID SYSTEMS

Hybrid systems use a combination of CI paradigms to take advantage of their collective features. For example, a neuro-fuzzy system (NFS) combines attributes of neural networks with those of fuzzy systems. Neural networks and fuzzy logic have some common features such as distributed representation of knowledge, model-free estimation, ability to handle data with uncertainty and imprecision, etc. Fuzzy logic has tolerance for imprecision of data, while neural networks have tolerance for noisy data (Medsker, 1995). A neural network's learning capability provides a good way to adjust expert's knowledge. For example, it can be used to automatically generate fuzzy rules and membership functions to meet certain specifications. This reduces the design time and cost. On the other hand, the fuzzy logic approach can be applied to enhance the generalization capability of a neural network by providing more reliable output when extrapolation is needed beyond the limits of the training data (Lin, 1994).

The advantage of a NFS is that new fuzzy rules can be learned from data as opposed to the fuzzy system where the fuzzy rules are selected intuitively. The learning procedure operates on local information, and causes only local modifications in the underlying fuzzy system. Also, a NFS offers a great possibility of interpreting the learned relationship linguistically. Hence, it offers better transparency to the user than NN, alleviating the "black box" problem. A number of researchers investigated different architectures of NFS (Lin, 1994; Medsker, 1995). These architectures have been applied in many applications such as process control.

Genetic algorithms can be used to optimize network architecture, tune weights of a neural network, and select an optimal subset of input variables. For example, in input variable selection, a genetic algorithm searches for a subset of input variables that produce the highest accuracy by starting with a small subset of inputs and adding input variables according to network performance. They can also be used to design optimal fuzzy systems, particularly when a fuzzy system involves discontinuous membership functions (e.g. trapezoidal membership function). Such membership functions cannot be trained using the standard gradient descent method, as they are not differentiable.

3. Model Development using Computational Intelligence

Besides selecting a suitable CI paradigm, developing an intelligent model involves four steps: data preparation, model structure selection, learning, and model evaluation. As illustrated in Fig. 5, these steps are repeated until the last step results in a satisfactory performance.

3.1 DATA PREPARATION

Many times the "raw" data are not the best data to use for modeling a CI paradigm. Hence, in using CI paradigms to solve real-world problems, it is important to transform raw data into a form acceptable to the paradigm. The first step is to decide on what the inputs and outputs are. Inputs that are not relevant for modeling should be excluded. The next step is to process the data in order to handle missing data, remove outliers, and to normalize and scale the data into acceptable range. Furthermore, depending on the CI paradigm selected, transformation of the data may be necessary. For example, to build a model using NN non-numeric data must be transformed to numeric data. Neural network training can be made more efficient if certain preprocessing steps are performed on the network inputs and targets.

Handling missing values. In training a NN, a value is required for each input parameter. Although self-organizing networks do not suffer under these problems, in supervised neural networks, missing values are a problem. It is common that real-world datasets have missing values. Several options are available to handle missing values such as removing the entire pattern if there is a missing value or replace each missing value with an average value.

Screening outliers. An outlier is a data pattern that deviates substantially from the data distribution. Outliers can have severe effects on accuracy. The problem can be addressed by

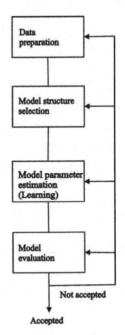

Figure 5. Modeling steps.

removing outliers through statistical techniques or by using a robust objective function that is not influenced by outliers.

Data standardization: In training NN, performance can be improved if inputs are scaled to the active domain of the activation functions. Min-max scaling preserves the relationships among the original data. Mean centering is more appropriate when the data contain no biases. Variance scaling is appropriate when training data are measured with different units. Z-score normalization is a combination of mean centering and variance scaling, and can be very useful when there are outliers present in the data.

Input variable selection: Modeling real-life data with a large number of input variables is a complex and time consuming task. This is particularly true for neural networks where irrelevant inputs can significantly increase learning complexity. The selection of input variables is a major challenge in intelligent modeling. Problems can occur due to improper selection of inputs such as increase in computational complexity and memory requirements as the input dimensionality grows. Also, the learning process becomes more difficult in the presence of irrelevant inputs, causing mis-convergence and poor model accuracy. Hence, input variable selection is aimed at choosing the most salient inputs that are most predictive of a given outcome. The objective of variable selection is two-fold: improving the prediction performance of the modles and providing a better understanding of the underlying concepts that generated the data.

Dimensionality reduction. Dimensionality reduction is essential in exploratory data analysis, where the purpose is to map data onto a low-dimensional space for improved visualization. It also reduces the complexity of a problem and makes it *easier* to build a CI paradigm. Dimensionality reduction can be accomplished through feature selection, where an optimum subset of features derived from the input variables is selected. Thus, feature selection methods keep only useful features and discard others. Note that feature selection is distinct from variable selection, because the former constructs new features out of the original variables. One well-known linear transformation used to reduce model dimensionality is principal component analysis (PCA). PCA reduces input dimensionality by transforming the input variables to a new set of variables (features), which captures most of the information in the original data. The new variables are computed as a linear combination of the original variables.

3.2 MODEL STRUCTURE SELECTION

We define model structure selection to include the choice of a neural network architecture, fuzzy rules, membership functions, fuzzy operators, genetic operators, and coding scheme. The selection of neural network architecture includes choosing activation functions, appropriate number of layers, number of neurons in each layer, and the interconnection of the neurons and the layers. It is also known that too many neurons degrade the effectiveness of the model, as the number of connection weights in the model may cause overfitting and loss of a model's generalization. Too few hidden neurons may not capture the full complexity of the data. Many theories have been suggested for finding the optimal number of neurons in the hidden layer (Moody, 1992; Amari, 1995; and Maass, 1995). However, most users employ trial-and-error methods, in which the training of NN starts with a small number of hidden neurons, and additional neurons are gradually added until some performance goal is satisfied. Genetic algorithms (GA) have also been used to determine the optimal number of hidden neurons by "evolutionary" trial–and-error.

Key selections in implementing genetic algorithms include: (i) the fitness function, (ii) the genetic representation or the chromosome coding scheme, and (iii) the genetic operators such as mutation and crossover.

In fuzzy logic-based modeling, model structure selection includes the choice of (i) shapes of the membership functions (trapezoidal, Gaussian, etc.) of fuzzy sets; (ii) a qualitative set of rules that can model the underlying process; and (iii) fuzzy operators to handle rule conjunctions such as "and" in the following fuzzy rule: "if temperature is high and humidity low, then the weather is medium hot." Fuzzy operators include intersection, union, and complement. The fuzzy intersection operation is mathematically equivalent to the "and", the union to "or" and complement to "not" operation, respectively. The structure of the fuzzy rules could cause an exponential growth in the number of rules as the number of inputs increases, resulting in unwieldy rule bases. Hence, it is important to select an optimal set of fuzzy rules.

3.3 LEARNING

The predominant feature of CI paradigms is that they learn from data. We define learning as the process of finding the free parameters of a CI model (e.g. determining the weights of a neural network). Depending on the environment in which the neural network operates, two types of learning paradigms are commonly used, supervised and unsupervised. Supervised learning uses an external teacher to produce a desired output; the model learns from training examples. Conversely, unsupervised learning does not involve an external teacher; the network discovers collective properties of the inputs by self-organization. Additionally, two types of learning styles have been applied in neural network training. The first approach is referred to as pattern learning, where the weights of the network are adapted immediately after each pattern is fed in. The other approach, known as batch learning, takes the entire training dataset and updates the network parameters after the entire batch of data has been processed.

An example of a supervised learning paradigm is the back-propagation algorithm (Rumelhart and McClelland, 1986), which is commonly used to train MLP networks. The back-propagation algorithm consists of two passes, forward and backward. During the forward pass, the input signals are applied to the network and their effect propagates through the different layers and generates the network outputs at the output layer. Note that the synaptic weights of the network are kept fixed during the forward pass. During the backward pass, the synaptic weights are adjusted in accordance with an error correction rule. The network outputs are subtracted from the desired outputs to produce error signals. The error signals are propagated backward through the network against the direction of the synaptic connections to determine how the synaptic weights should be adjusted in order to decrease a pre-defined error function. Typical error function is the sum of square of the error signals between the network outputs and the desired outputs. Thus, during the backward pass, the back-propagation algorithm alters the synaptic weights in the direction of the steepest gradient of the error function to reduce the error function. The forward and backward passes are repeated until a pre-specified stopping criterion is achieved or the error function is reduced significantly.

A number of methods are proposed to improve the performance of the steepest gradient method described above, which is based on the first derivatives of the error function with respect to the synaptic weights. Newton's method is used to speed-up training by employing second derivatives of the error function with respect to the synaptic weights. The Gauss-Newton method is designed to approach second order training speed without calculating the second derivates. The Levenberg-Marquardt learning algorithm speeds up the learning process as well as produces enhanced learning performance by combining the standard gradient technique with the Gauss-Newton method.

An example of an unsupervised learning paradigm is the Kohonen learning rule (Kohonen, 2001), which is used for training a self-organizing map (SOM). The design of SOM starts with defining a geometric configuration for the partitions in a one- or two-dimensional grid. Then, random weight vectors are assigned to each partition. During training, an input pattern (input vector) is picked randomly. The weight vector closest to the profile is identified. The identified weight vector and its neighbors are adjusted to look similar to the input vector. This process is repeated until the weight vectors converge. During operation, SOM maps input patterns to the relevant partitions based on the reference vectors to which they are most similar.

3.4 MODEL EVALUATION

After learning is completed, a CI paradigm is evaluated for its performance through testing. The purpose of this testing is to prove the adequacy or to detect the inadequacy of the created model. The latter could arise from an inappropriate selection of network topology, too few or too many neurons, or from insufficient training or overtraining. Incorrect input node assignments, noisy data, errors in the program code, or several other effects may also cause a poor fit. The aim of model evaluation is to insure that the model fit is acceptable, that the model satisfies the desired accuracy requirements, and that it serves as a general model. A general model is one whose input-output relationships (derived from the training dataset) apply equally well to new sets of data (previously unseen test data) from the same problem not included in the training set. The main goal of intelligent modeling is thus the generalization to new data of the relationships learned on the training set.

4. Learning Tasks

CI paradigms learn from data to accomplish a task. The major *learning tasks* include pattern recognition, pattern association, function approximation, estimation, filtering, smoothing, and prediction.

4.1 PATTERN RECOGNITION

A pattern recognition task involves feature extraction and clustering or classification. Clustering involves the comparison of input patterns to determine common properties inherent in the data (with no information other than the observed values). It is a useful exploratory technique for analysis of large volume and high-dimensional data when there is no *a priori* information about existing common properties. Depending on how they cluster data, we can distinguish clustering algorithms into hierarchical and partitioning (non-hierarchical clustering) methods. Hierarchical clustering organizes the input patterns in a hierarchical tree structure, which allows detecting higher order relationships between clusters of patterns. The non-hierarchical or partitioning methods begin from a pre-defined number of clusters. Iterative reallocation of cluster members minimizes the overall within-cluster dispersion. An example of a non-hierarchical method is the SOM that has been widely used to the partition large volume and high-dimensional datasets.

Clustering algorithms, such as SOM, assign each data point to only one of the clusters. Thus, they are referred to as hard clustering algorithms. The clusters formed by these algorithms may not reflect the description of real data, where boundaries between subgroups might be fuzzy. A fuzzy clustering algorithm assigns a certain degree of closeness or similarity to each object in a cluster. Bezdek (1981) developed a fuzzy clustering algorithm known as fuzzy c-means, which is a fuzzy extension of the classic K-means clustering method. In fuzzy c-means clustering, each data point belongs to a cluster to a degree that is specified by a membership function. The algorithm iteratively updates the cluster centers and the membership function's values for each data point. This iteration is based on minimizing an objective or a distance function that represents the distance from any given data point to a cluster center weighted by that data point's membership function value.

Clustering techniques can be used to group pixels of remotely sensed images. In particular, when *a priori* labels on the pixels are not available, clustering in the spectral domain enables us to group pixels based on the observed reflectance in various bands of a multispectral scanner. Spatial information about the pixels is also utilized for clustering. For example, neighboring pixels in a two-dimensional image are more likely to belong to the same cluster than pixels that are farther apart. SOM have been very popular in such applications. They arrange feature vectors according to their internal similarity, creating a continuous topological map of the input space. In this topological map, the vectors that are similar in the input space are grouped together, or clustered, while those vectors that are different are kept far apart.

Classification involves the automated grouping of all, or selected, features into pre-specified categories. Despite the considerable developments made recently, the accuracy with which thematic maps may be derived from remotely sensed data is often still judged to be too low for operational use (Wilkinson, 1996). Several studies have described the limitations of the conventional statistical image classification techniques. These limitations include an assumption that the data are normally distributed, a requirement of a large number of training samples, and the limitation of incorporating low-level ancillary data, etc. (Tso and Mather, 2001).

CI paradigms offer an alternative to statistical methods in the classification of remotely sensed images. Performing appropriate feature extraction helps in achieving accurate classification. For example, SOM can be used for feature extraction followed by a supervised network, such as MLP and RBF, for classification, thereby accomplishing a pattern recognition task.

4.2 PATTERN ASSOCIATION

Pattern association takes two forms, autoassociation or hetroassociation. In the former case, the network is trained to recognize the input signals (input patterns) presented to it. In other words, the desired outputs are equal to the input pattern. In hetroassociation, the input patterns are paired with another arbitrary set of patterns. A typical example is estimating a set of variables from other but correlated variables. Consider a set of difficult to measure or infrequently measurable variables (primary variables, e.g. chl *a*) and a set of variables that can be measured frequently (secondary variables, e.g. remotely sensed reflectance measurements). If the secondary and primary variables are related, then a CI paradigm can be trained from historical data to estimate the primary variables from the secondary variables. The resulting CI paradigm will serve as a virtual sensor that provides an estimate of the primary variables at the measurement frequency of the secondary variables. For example, an adequately trained neural network can be used to estimate chl *a* at the measurement frequency of remotely sensed reflectance. Neural networks accomplish such task without requiring knowledge about the underlying relationship between chl *a* and reflectance measurements, providing better accuracy than traditional regression analysis.

4.3 FUNCTION APPROXIMATION AND OTHER LEARNING TASKS

Consider a nonlinear input mapping described by the functional relationship $d=G(x)$, where x is the input vector and d is the output vector. The vector-valued function G is assumed to be unknown. Given a set of examples, a CI paradigm can be trained to approximate the unknown function G such that the new function $\hat{G}$, describing the input-output mapping realized by the CI model, is sufficiently close enough to G. An example of a function approximation problem is building an inverse model from an existing forward model (radiance model) that relates inherent optical properties (IOP) and apparent optical properties (AOP). IOP are fundamental properties of the water volume such as spectral absorption coefficient and spectral scattering coefficient. AOP are properties that depend on the distribution and geometry of incident light such as reflectance or radiance distributions. The forward model is used to generate a representative dataset of IOP and AOP that can be used as training examples. Using the training examples, a CI paradigm builds a model that approximates an inverse function, mapping AOP to IOP. The resulting model can be used to estimate optically active constituents such as chl *a*, color dissolved organic matter, and suspended sediment concentrations by relating them empirically to IOP.

Other learning tasks of CI paradigms include filtering, smoothing, and time-series prediction. In filtering tasks, CI paradigms are used to extract information about a prescribed quantity of interest from a set of noisy data. Filtering uses data measured up to and including time n to extract information about a quantity of interest at discrete time n. Smoothing uses data measured after time n, as well as that measured up to time n. In time-series prediction, the task is to forecast the quantity of interest at some time $n + n_0$ in the future by using data measured up to and including time n.

5. Software Tools for Building CI-Based Models

A number of software tools are available for CI-based model building including NeuroSolutions, NeuralWorks, Stuttgart Neural Network Simulator (SNNS), SPLICER, and *fuzzy*TECH. These and other tools provide graphical interfaces allowing the user to easily upload and preprocess the data, select model structure and parameters, and evaluate the models. While all of these tools offer ample capabilities to develop models based on specific CI paradigms, MATLAB (The MathWorks Inc.) offers an integrated programming environment for design and implementation of various CI paradigms, and has been commonly used in CI research and development. It offers powerful numeric computation and visualization tools for the Unix, Macintosh, and Windows platforms. It consists of application-specific functions grouped into toolboxes, each extending the core MATLAB language. The toolbox functions can be easily viewed and modified.

The MATLAB Neural Network Toolbox offers comprehensive support for the design, training, and simulation of many proven neural network paradigms (e.g. MLP, RBF, SOM, etc.) and is completely customizable. The MATLAB Fuzzy Logic Toolbox features graphical interfaces for guiding the user through the steps of fuzzy inference system design and visualization. In addition, it implements fuzzy c-means clustering and neuro-fuzzy systems. The MATLAB Genetic Algorithm and Direct Search Toolbox features graphical user interfaces and command-line functions for solving problems using genetic algorithms. It provides numerous options for creation, fitness scaling, selection, crossover, and mutation.

6. Ocean Color Remote Sensing

Ocean color remote sensing, a focus of research worldwide, has received tremendous impetus due to greatly improved satellite-borne optical sensors. Accurate estimates of bio-optical parameters of the ocean, such as phytoplankton pigment (namely chl *a*), colored dissolved organic matter (CDOM), and suspended particulate concentrations are critical for various studies. In addition to providing insight into global ocean-atmosphere exchange (especially for carbon), these variables provide vital information on phytoplankton abundance, biological productivity, and coastal zone water quality and dynamics. Furthermore, long-term observations of phytoplankton can provide useful information on ocean circulation, marine pollution and nutrient dynamics. Phytoplankton biomass is usually expressed in terms of the concentration of chl *a*, the primary optical constituent of phytoplankton. Therefore, chl *a* concentration is a good indicator of the amount of phytoplankton present in the ocean. Chlorophyll *a* can be used to effectively monitor phytoplankton abundance and the resulting effect on water quality.

6.1 CONVENTIONAL OCEAN COLOR ANALYSIS

Numerous bio-optical algorithms have been proposed in the literature for estimating chl *a* from normalized water leaving radiance or remote sensing reflectance. The simplest are band-ratio algorithms (Gordon and Morel, 1983 for CZCS; O'Reilly et al., 1998 and O'Reilly et al., 2000 for Sea-viewing Wide Field-of-view Sensor (SeaWiFS) data). Most of these algorithms are empirically derived and generally restricted to Case 1 waters. O'Reilly et al. (2000) proposed an updated two-band (OC2v4) and four-band (OC4v4) algorithms. They suggested that the OC4v4 is expected to perform well when applied to satellite-derived water leaving radiances both in oligotrophic and eutrophic conditions. Subsequently, NASA has adopted the OC4v4 algorithm for global SeaWiFS processing. Other band ratio algorithms and mathematical models for chl *a* retrieval in Case 1 and coastal waters include those proposed by Arenz et al. (1996), Siegel et al. (1994), George (1997), Sathyendranath et al. (1989), and Wernarnd et al. (1997). Due to the complexity and seasonal variability in river plume and coastal waters, the band-ratio schemes often perform poorly (Hu et al., 2002).

The second category of inversion algorithm is semi-analytical (Roesler and Perry, 1995; Garver and Seigel, 1997; Carder et al., 1999; Doerffer and Fischer, 1994, Lee et al., 1999; Gordon et al., 1988; Hoge et al., 1995) because analytical and empirical formulae are used to obtain the spectral absorption and backscattering coefficients of various water constituents. For example, the Carder et al. (1999) algorithm uses three bands at 412, 443, and 555 nm to separate CDOM from chl *a* based on the strong absorption at 412 nm by CDOM. This algorithm is currently applied operationally as a primary option to process data from the Moderate Resolution Imaging Spectroradiometer (MODIS) (Salomonson et al., 1989; Esaias et al., 1998).

Several of the semi-analytic inversion algorithms (Doerffer and Fischer, 1994; Roesler and Perry, 1995; Lee et al., 1999) employ optimization techniques, where parameters in the algorithm (for example, chl *a*, CDOM, and suspended sediment concentration) are adjusted until a prescribed error criterion between modeled and measured remote sensing spectra is minimized. The limitation of this type of approach is the computational complexity, which presents an obstacle for efficient operational processing of satellite imagery. The Roesler and Perry (1995) model includes fluorescence emission and considers both total absorption and total backscatter to be a linear combination of non-dimensional basis vectors and scalar magnitudes for each optically active substance (OAS). The magnitudes are determined using Levenberg-Marquardt optimization, and the algorithm has been demonstrated on *in situ* bio-optical measurements.

Colored dissolved organic matter (CDOM), also known as yellow substance or Gelbstoff, is an important component in governing light propagation in coastal and open ocean waters (Bricaud et al., 1981; Morel, 1988; Siegel and Michaels, 1996). CDOM comprises a vast array of molecules of varying size, which absorb strongly in the UV and blue (Schwarz et al., 2002; Bricaud et al., 1981) portions of the electromagnetic spectrum. Overlaps of pigment absorption spectra with CDOM absorption complicate the use of chlorophyll *a* retrieval algorithms that are based on limited spectral bands (Carder et al., 1991; O'Reilly et al., 1998). Even at low CDOM concentrations, the chl *a* absorption peak of phytoplankton corresponds to significant CDOM absorption near 443 nm. While initially seen as a disturbance to chl *a* satellite algorithms, CDOM is an important bio-optical parameter, representing a fraction of the large dissolved organic carbon (DOC) reservoir (Hedges et al. 2000), which affects productivity by modifying availability of both nutrients (McCarthy et al. 1997, Arrigo and Brown, 1996) and light (Kirk 1994, Neale et al. 1998).

Mapping suspended sediment concentrations and patterns in coastal and inland water environments is important to monitor coastal water quality. Suspended solids in water produce visible changes in the surface of the waters and in the reflected solar radiation. Such changes in the spectral signal from surface water can be captured by satellite sensors (Ritchie and Cooper, 1988). Automated field methods to monitor suspended sediment concentration often require extensive maintenance due bio-fouling and the effects of prolonged deployments. Mapping suspended sediment concentrations using satellite images has been described in previous studies (Topliss et al., 1990; Bhargava and Mariam, 1992; Khan et al., 1992; Froidefond et al., 1993; Tassan, 1993). Although most operational mapping tools and procedures for suspended sediment can be costly and time intensive, Miller et al. (Chapter 11) describe an effective and inexpensive approach to mapping suspended particulates using MODIS 250 m data.

6.2 OCEAN COLOR ALGORITHMS USING CI-BASED TECHNIQUES

Empirical and semi-analytical algorithms have inherent limitations because of the non-linear nature of the regressions involved – employing non-linear regression methods requires *a priori* knowledge of the nature of the non-linear behavior, which is generally not available. Another major drawback with these models is that they have been designed for specific regions and times since they were derived from data collected from specific waters at specific times.

In contrast, computational intelligence-based techniques, such as neural networks, have the capability to address several of these challenges, and can be used to accurately and efficiently estimate bio-optical parameters in coastal waters from remotely sensed data. Several neural network models have recently been proposed for bio-optical models including Keiner and Yan (1998), Keiner and Brown (1999), Buckton et al. (1999), Schiller and Doerffer (1999), Baruah et al. (2000); Kishino et al. (2001); Musavi et al. (2001), and Slade et al. (2004).

Given the success of neural networks in modeling non-linear systems, they are a logical candidate for inversion models of ocean color data to determine CDOM and suspended sediment concentration, as well as phytoplankton abundance, in complex coastal water environments. We will now provide an overview of several studies that have applied neural networks and other computational intelligence techniques to ocean color analysis. Then an example using a neural network for the estimation of chl *a* concentration from multispectral remote sensing reflectance (Rrs) measurements is given.

Recently, Schiller and Doerffer (1999) used a neural network based model to derive the concentrations of phytoplankton pigment, suspended matter and CDOM, from "Rayleigh corrected" top of atmosphere reflectances over turbid coastal waters. The inverse model proposed in their paper is based on a neural network of a two-flow radiative transfer model (Doerffer and Fischer, 1994), using the 16 spectral bands of the Medium Resolution Imaging Spectroradiometer (MERIS) flown aboard the European Space Agency Envisat satellite. The training dataset was simulated from the two-flow radiative transfer model by varying the parameters (phytoplankton pigment, suspended matter, CDOM absorption at 420 nm, and atmospheric visibility) across a wide range of values intended to represent both clear waters as well turbid coastal and estuarine waters. Variations in training data parameters was accomplished systematically by dividing the range of each parameter across its log(min) to log(max) range with 11 equidistant steps, leading to a training dataset of 12^4 patterns. A test dataset was developed by simulating a dataset of 5000 patterns with parameter vectors varied using a random uniform distribution. The neural network presented was an MLP network with 16 inputs for the MERIS radiance bands, 2 hidden layers (45 and 12 neurons), and 4 outputs corresponding to the parameters to be inverted. Schiller and Doerffer (1994) note that this model was designed for the North Sea, but that, if the optical parameters of other regions are known, then a new dataset could be simulated, and the neural network can be adaptively trained. Further information on the operational MERIS Case 2 bio-optical algorithm can be found in the MERIS ATBD (Doerffer and Schiller, 1998).

Similarly, Buckton et al. (1999) used neural network techniques for inversion of Case 2 MERIS radiances. This inversion model is based on simulated data as well, using empirical ocean and atmospheric models to relate chl *a* concentration, sediment load, CDOM absorption, atmospheric optical thickness, and solar/satellite geometry to top of atmosphere radiance. The modeled radiances are then modified by a model of the noise characteristics of the MERIS instrument. In order to reduce the dimensionality of the MERIS radiance observations and increase the computational efficiency of the algorithm, singular value decomposition is employed, compressing the radiance variation into a ten dimensional space. The reduced dimension radiances combined with solar and satellite geometry information form the input to an MLP network with 2 hidden layers (13 and 5 neurons) and 3 outputs for the constituent estimates: chl *a*, sediment concentration, and CDOM absorption.

Keiner and Yan (1998) used a neural network model to estimate surface chl *a* and sediment concentrations in Delaware Bay from Landsat Thematic Mapper (TM) radiances. Their model is based on a Case 2 dataset of *in situ* bio-optical measurements matched with geographically and temporally co-incident Landsat TM data; only 15 patterns were available for model development. A multiple regression technique, a neural network with three inputs corresponding to the Landsat TM bands, a single output, and only two hidden neurons was used. Separate networks are trained for estimation of chl *a* and suspended sediment. With so few data, particular care must be taken to avoid over fitting the training data. Keiner and Yan's (1998) neural network model achieved a high level of accuracy despite the limited TM spectral bands.

Keiner and Brown (1999) subsequently published results of using neural networks to estimate oceanic chl *a* concentrations. They designed their model around the SeaWiFS Bio-optical Algorithm Mini-workshop (SeaBAM) dataset (O'Reilly et al., 1998). SeaBAM was developed as a benchmark and intercomparison dataset for the competing

SeaWiFS chl *a* algorithms, and is mainly representative of Case 1 optical conditions. They used an MLP network with one hidden layer with ten neurons. The neural network uses the five visible SeaWiFS reflectance bands as input, and provides an estimation of chl *a* concentration at the output.

Cipollini et al. (2001) presented an inversion algorithm based on RBF networks. Their model was trained according to Orr et al. (2000) by first determining a large set of basis vectors in the input space by means of a regression tree, and subsequently using a forward selection procedure to iteratively select a good set of basis vectors. Finally, the output weights of the network were determined using a pseudoinverse technique. Training data were simulated based on a hyperspectral three constituent bio-optical model, similar to other studies, which was then used to derive reflectances for the 8 MERIS visible channels. Effort was made to simulate Case 1 and 2 data, as well as a noise component (spectrally independent) comprised of instrument noise and atmospheric correction uncertainty. The results of Cipollini et al. (2001) indicate significantly better performance in constituent retrievals compared to band ratio and multilinear techniques.

Other computational intelligence techniques such as fuzzy logic and genetic algorithms have also been successfully used. For example, fuzzy logic based clustering techniques were used for selecting and blending ocean color algorithms (Moore et al., 2001). Moore et al. (2001) used the fuzzy c-means clustering scheme to cluster *in situ* reflectance spectra. Separate sub-models were created for each cluster separately using only the *in situ* spectra within the membership of a specific cluster. Subsequently, measured spectra can be applied to each sub-model, membership weights determined for the clusters, and the blended model output is determined as the weighted recombination of the sub-model outputs.

Corsini et al. (2002) used a fuzzy model to retrieve the concentration of optically active constituents in Case 2 waters from MERIS data. The retrieved optically active constituents were chl *a* plus phaeophytin *a* mass to volume concentration, the scattering coefficient of non-chlorophyllous particles at 550 nm, and the absorption coefficient of the yellow substance at 440 nm. A dataset was generated using MERIS channel reflectance using the model. This dataset consisted of 5000 pairs of simulated subsurface spectral reflectance and the optically active constituents. From these pairs, 500 pairs were used to train a fuzzy model. The remaining 4500 pairs were used to test model performance. A fuzzy c-means clustering algorithm was used to partition the data into 12 groups. The partitioned data were used to identify a compact initial set of rules for the fuzzy model. The rules were then tuned via genetic algorithms to minimize the error between the predicted and desired outputs. The resulting fuzzy model demonstrated good predictive capabilities.

Genetic algorithms have been used to retrieve water optical properties for optically deep waters (Zhan et al., 2003) by constrained optimization of the parameters of a bio-optical model designed to minimize the error between measured and modeled reflectance spectra. The effectiveness of GA was demonstrated on *in situ* bio-optical data from several regions, including Monterey Bay, the North Atlantic, the Gulf of Mexico, the Arabian Sea, and the Bering Sea. These regions span a wide variation of constituent parameters and concentrations. MODIS spectral wavelengths were used and the GA inversion was compared to an optimization model inversion using the sequential quadratic programming method. A key to the success of GA as an optimization method for bio-optical inversions is that they more often determine a globally optimum solution without *a priori* knowledge.

We too have developed a neural network-based model to estimate chl *a* concentration. A dataset was constructed using *in situ* measurements of Rrs and chl *a* (fluorometric and HPLC) from various cruises of the SeaBAM, SeaWiFS Bio-Optical Archive and Storage System (SeaBASS), and Sensor Intercomparison and Merger for Biological and Interdisciplinary Oceanic Studies (SIMBIOS) datasets. Data from the SeaBAM dataset are detailed in (O'Reilly et al., 1998). Our dataset contains a 539 point subset of the original 919 SeaBAM data points that contained Rrs values at the SeaWiFS wavelengths of 410, 443, 490, 510, and 555 nm. This SeaBAM subset is augmented with 565 data points from the SIMBIOS program. SIMBIOS data were collected from 1996-2002, and tend to represent more optically complex coastal stations compared to the earlier SeaBAM dataset. The final dataset represents near-surface chl *a* values from 0.015 to 32.79 µg/ml, with an average concentration of 0.964 µg/ml over 1104 data points.

The dataset was divided into training, validation, and testing datasets consisting of 60%, 20%, and 20% of the original data, respectively. Several different network topologies were tested, and an MLP network with two hidden layers (six neurons per layer) trained using the Levenberg-Marquardt algorithm was found to perform well on the dataset. All data were normalized to zero mean and unity standard deviation before input to the neural network. Target chl *a* data were log-transformed to compensate for the wide range of values. The network (NN5) was evaluated using the testing data; testing data were not used for network training. The performance of the network on *in situ* data was evaluated by comparing model results to the SeaWiFS program OC4V4 results. Coefficient of determination (r^2) and root mean square error (RMSE) were used as metrics for comparison of the neural network and OC4 model (Table 1). Scatter plots of model estimates vs. *in situ* target values (for the testing dataset) are shown for the OC4 and neural network models in Figs. 6 and 7, respectively.

	Training Data (n=662)		Validation Data (n=221)		Testing Data (n=221)	
	r^2	RMSE	r^2	RMSE	r^2	RMSE
OC4V4	0.65	0.48	0.68	0.45	0.56	0.50
NN5	0.92	0.16	0.84	0.24	0.87	0.20

Table 1. Performance characteristics of the neural network and empirical OC4 models on the training, validation, and testing datasets.

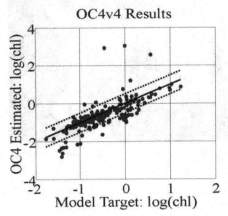

Figure 6. Scatter plot of OC4v4 empirical algorithm on testing dataset

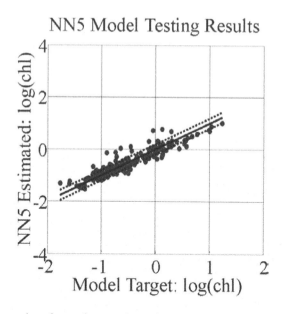

Figure 7. Scatter plot of neural network model on testing dataset.

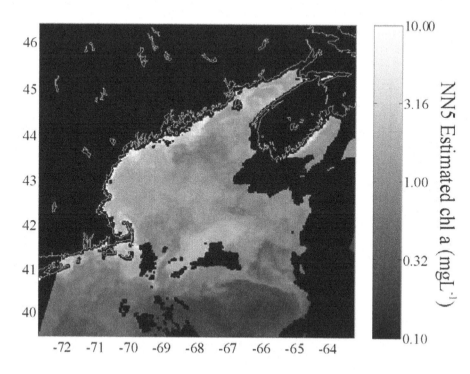

Figure 8. Neural network estimated chl *a* map for the Gulf of Maine. (Solid black regions indicate land or cloud-masked pixels).

There is a clear reduction in estimation uncertainty over several decades of target chl a measurements. The NN5 chl a estimation algorithm outperforms the operational OC4 algorithm, with a 60% and 53% improvement in RMSE and r^2, respectively. Figure 8 shows a SeaWiFS image of the Gulf of Maine that was processed using the NN5 chl a estimation algorithm. While there is no way to fully quantify the model uncertainty in the image, patterns of chl a are consistent with what is generally known about phytoplankton dynamics in the region.

While there are many choices for network architecture, in this case an uncomplicated MLP network outperforms the SeaWiFS operational OC4V4 algorithm. The resilience of neural network models to noise in training data is also an advantage when developing neural networks based on *in situ* data, where instrumentation and data processing uncertainty are present. In addition, construction of a neural network model is no more difficult or time-consuming than the development of a general empirical model. The advantages of neural network based-models over empirical algorithms in retrieving biogeochemical parameters make them excellent candidates for the next generation of ocean color algorithms.

7. Estimation of Phytoplankton Primary Production from Remotely Sensed Data

Measurement of ocean primary production has been an important objective of the SeaWiFS and MODIS programs. Direct *in situ* measurements of primary productivity are difficult and time consuming. Several empirical models have been developed to estimate primary productivity based on predictive variables, such as phytoplankton biomass and light intensity, which are more readily available than primary productivity.

For example, MODIS provides two estimates or indices of primary production using empirical annual algorithms (Esaias et al., 1998). The first index was developed by Behrenfeld and Falkowski (1997) and the second one by Howard and Yoder (1997). The model of Behrenfeld and Falkowski (1997) employs a light dependent depth resolved empirical model for carbon fixation, namely, the Vertical Generalized Production Model (VGPM). This model partitions environmental factors affecting primary production into those that influence the relative vertical distribution of primary production and those that control the optimal assimilation efficiency of the productivity profile. Modeling primary production with the VGPM requires data for five input variables including a photo-adaptive parameter (optimal rate of daily carbon fixation within a water column). As this parameter cannot be directly measured, it must be coupled to other parameters that can be detected remotely, such as sea surface temperature. Though VGPM and several other empirical models have provided useful results, the demand for improving the prediction accuracy has led to the development of neural network-based models. In this framework, neural networks have been found to offer a better solution for estimating primary productivity (Scardi, 1996; Scardi, 2001; Ressom et al., 2001). The advantage of neural networks is that they provide the flexibility to include additional parameters for predicting primary production, even though no *a priori* knowledge about the specific effect of these parameters is available.

Scardi (1996) developed a neural network based phytoplankton production model for the Mediterranean Sea. The model was constructed to use three predictive variables that are available from remote sensing: sea surface temperature, surface irradiance, and surface chlorophyll concentration. The model was based on an MLP network with 3 inputs, 4 hidden neurons and one output. The network was trained using a small dataset extracted from the Ocean Primary Productivity Working Group (OPPWG) database. This dataset included 97 primary production measurements obtained during three spring cruises in the western Mediterranean. Scardi (1996) compared his results to a simple linear model and the VGPM model. Both the VGPM and the linear model tend to underestimate primary production and their error distributions are less symmetrical than the neural network model. The mean square error (MSE) between the log-transformed network outputs and measured values was 0.059, which was considerably less than the error of the VGPM (0.130) and the linear model (0.239). This example demonstrates that a major advantage of NN is their capability to incorporate information that is difficult to handle using conventional models, and their ability to effectively include variables that tend to co-vary non-linearly with the output variable. Scardi (2001) demonstrated that co-predictors could play an important role in improving existing models by exploiting non-linear relationships between primary production and other variables that are not directly involved in the photosynthetic process. To overcome the problem of limited data a metamodeling approach was proposed that merges real and modeled information to extend the predictive ability of a neural network model.

Ressom et al. (2001) employed MLP networks to estimate phytoplankton primary production using a dataset of 300 samples extracted from the OPPWG database. An MLP network was constructed with 4 inputs, 7 neurons in the first hidden layer, 3 neurons in the second hidden layer, and 1 neuron in the output layer. The inputs to the network were chlorophyll concentration, surface irradiance, sea surface temperature, and day length. The output is integrated daily carbon fixation (i.e., the measured integrated primary production). Half of the data were used for training and the remaining half for validation testing resulting neural network model outperformed VGPM and a linear regression model with over 50% improvement in the log-transformed RMSE or coefficient of determination.

8. Summary

This chapter introduces to the fundamentals of computational intelligence and its applications in remote sensing of coastal aquatic environments. It gives insight into the use of popular computational intelligence paradigms in estimating coastal bio-optical parameters and phytoplankton primary production. Several advantages give neural networks an edge over other empirical models as candidates for the next generation of ocean color algorithms. Particular advantages are: (i) resilience to noise; (ii) ease of construction without requiring detailed knowledge of the underlying relationships between remotely sensed data and the desired biogeochemical/bio-optical parameters; (iii) the flexibility to incorporate difficult to manage variables; (iv) efficiency in dealing with non-linearity; and, (v) robustness with respect to redundant inputs. Neural networks can also efficiently cope with large amounts of data and lend themselves to process complex datasets, where different information sources are combined.

As large volume and high dimensional data are being generated by the rapidly expanding remote sensing technology, the number of reported applications of computational intelligence based techniques is steadily increasing. Some potential applications of computational intelligence in remote sensing of aquatic coastal environments could include monitoring dissolved oxygen, harmful algal blooms, coral reefs, and submerged aquatic vegetation. With the increasing demand, however, comes the need for further improvements that can make CI based implementation in remote sensing more efficient. Key improvements include: (i) enhanced computational power to handle the high dimensionality and large volume data; (ii) higher temporal, spectral, and spatial resolution and access to near real time data; (iii) increased availability of *in situ* bio-optical measurements coincident with remotely sensed data (note that like other empirical models, CI-based models are only as good as the dataset to which they are applied; hence, the quality of the data collected is very important); and, (iv) advances in computational intelligence techniques to enhance their speed and make them more accessible to the user.

9. References

Amari, S. 1995. Learning and statistical inference. In: The Handbook of Brain Theory and Neural Networks. Arbib, M.A. (Ed.). MIT Press, Cambridge, MA, 522-526.

Arenz R.F. Jr., W.M. Lewis Jr., J.F. Saunders III. 1996. Determination of chlorophyll and dissolved organic carbon from reflectance data for Colorado reservoirs. International Journal of Remote Sensing, 17(8): 1547-1566.

Arrigo, K.P., and C.W. Brown. 1996. Impact of chromophoric dissolved organic matter on UV inhibition of primary productivity in the sea. Marine Ecology Progress Series, 140:207-216.

Baruah, P.J., K. Oki, and H. Nishimura. 2000. A neural network model for estimating surface chlorophyll and sediment content at the Lake Kasumi Gaura of Japan, Proceedings, Asian Conference on Remote Sensing, 418-424.

Behrenfeld, M.J. and P.G. Falkowski. 1997. Photosynthetic rates derived from satellite based chlorophyll concentration, Limnology and Oceanography, 42(1):1-20.

Bhargava, D.S., and D.W. Mariam. 1992. Cumulative effects of salinity and sediment concentration of reflectance measurements. International Journal of Remote Sensing, 13(11):2151-2159.

Bricaud, A., A. Morel and L. Prieur. 1981. Absorption by dissolved organic matter of the sea (yellow substance) in the UV and visible domain. Limnology and Oceanography, 26:43-53.

Buckton, D., E. O'Monogan, and S. Danaher. 1999. The use of neural networks for the estimation of oceanic constituents based on the MERIS instrument. International Journal of Remote Sensing, 20(9):1841-1851.

Bezdek, J.C. 1981. Pattern Recognition with Fuzzy Objective Function Algorithms. Plenum, New York.

Carder, K.L., F.R. Chen, Z.P. Lee, S. Hawes, and D. Kamykowski. 1999. Semianalytic MODIS algorithms for chlorophyll-a and absorption with bio-optical domains based on nitrate-depletion temperatures. Journal of Geophysical Research, 104(C3):5403-5421.

Carder, K.L., S.K. Hawes, K.A. Baker, R.C. Smith, R.G. Steward, and B.G. Mitchell. 1991. Reflectance model for quantifying chlorophyll a in the presence of productivity degradation products. Journal of Geophysical Research – Oceans, 96(C11):20599-20611.

Cipollini, P., G. Corsini, M. Diani, and R. Grasso. 2001. Retrieval of sea water optically active parameters from hyperspectral data by means of generalized radial basis function neural networks. IEEE Transactions on Geoscience and Remote Sensing, 39(7):1508-1524.

Corsini, G., M. Diani, R. Grasso, B. Lazzerini, F. Marcelloni, M. Cococcioni. 2002. A fuzzy model for the retrieval of the sea water optically active constituents concentration from MERIS data. IEEE International Geoscience and Remote Sensing Symposium, vol. 1, 98-100.

Doerffer, R. and J. Fischer. 1994. Concentrations of chlorophyll suspended matter and gelbstoff in case II waters derived from satellite coastal zone color scanner data with inverse modeling methods. Journal of Geophysical Research, 99(C4):7457-7466.

Doerffer, R. and H. Schiller. 1998. Algorithm Theoretical Basis Document (ATBD 2.12): Pigment index, sediment and gelbstoff retrieval from directional water leaving radiance reflectances using inverse modelling technique. ESA Doc. No. PO-TN-MEL-GS-0005, 12-1 - 12-60.

Engelbrecht, A.P. 2002. Computational Intelligence: An Introduction. John Wiley and Sons, Inc., England.

Esaias, W.E., M.R. Abbott, I. Barton, O.B. Brown, J.W. Campbell, K.L. Carder, D.K. Clark, R.H. Evans, F.E. Hoge, H.R. Gordon, W.M. Balch, R. Letelier, and P.J. Minnett. 1998. An Overview of MODIS Capabilities for Ocean Science Observations. IEEE Transactions on Geoscience and Remote Sensing, 36(4):1250-1265.

Froidefond, J.M., P. Castaing, J.M. Jouanneau, R. Prudhomme and A. Dinet. 1993. Method for the quantification of suspended sediments from AVHRR NOAA-11 satellite data. International Journal of Remote Sensing, 14(5):885-894.

Garver, S.A. and D.A. Siegel. 1997. Inherent optical property inversion of ocean color spectra and its biogeochemical interpretation. 1. Time series from the Sargasso Sea. Journal of Geophysical. Research, 102 (C8): 18607-18625.

George D.G. 1997. The airborne remote sensing of phytoplankton chlorophyll in the lakes and tarns of the English Lake District. International Journal of Remote Sensing, 18(9):1961-1975.

Goldberg, D.E. 1989. Genetic Algorithms in Search, Optimization, and Machine Learning. Addison-Wesley, Reading, Mass.

Gordon, H.R., and A.Y. Morel. 1983. Remote assessment of ocean color for interpretation of satellite visible imagery: a review. In: Lecture Notes on Coastal and Estuarine Studies, vol. 4, M. Bowman (ed.). Springer-Verlag, New York. 1-114.

Gordon, H.R., O.B. Brown, R.H. Evans, J.W. Brown, R.C. Smith, K.S. Baker, D.K. Clark. 1988. A semianalytic radiance model of ocean color. Journal of Geophysical Research, 93:10909-10924.

Haykin, S. 1999. Neural Networks: A Comprehensive Foundation (2nd Edition), Prentice Hall, Upper Saddle River, NJ.

Hedges J.I., G. Eglinton, P.G. Hatcher et al. 2000. The molecularly-uncharacterised component of nonliving organic matter in natural environments. Organic Geochemistry, 31:945-958.

Hoge, F.E., R.N. Swift, and J.K. Yungel. 1995. Oceanic radiance model development and validation: application of airborne active-passive ocean color spectral measurements. Applied Optics, 34:3468-3476.

Holland, J.H. 1975. Adaptation in Natural and Artificial Systems. University of Michigan Press, Ann Arbor.

Howard, K.L. and J.A. Yoder. 1997. Contribution of the subtropical oceans to global primary productivity. In: Proceedings of COSPAR Colloquium, Space Remote Sensing of Subtropical Oceans (C.T. Liu, ed), COSPAR Colloquia Series vol. 8, Pergamon. pg 157-168.

Hu, C., A.L. Odriozola, J.P. Akl, F.E. Muller-Karger, R. Varela, Y. Astor, P. Swarzenski, and J.M. Froidefond. 2002. Remote sensing algorithms for river plumes: A comparison, ASLO2002, Victoria, British Columbia, Canada, 10-14.

Keiner L.E. and C.W. Brown. 1999. Estimating oceanic chlorophyll concentrations with neural networks, International Journal of Remote Sensing, 20(1):189-194.

Keiner, L.E. and X. Yan. 1998. A neural network model for estimating sea surface chlorophyll and sediments from Thematic Mapper imagery. Remote Sensing of Environment, 66:153-165.

Khan, M.A., Y.H. Fadlallah, and K.G. Al-Hinai. 1992. Thematic mapping of subtidal coastal habitats in the western Arabian Gulf using Landsat TM data - Abu Ali Bay, Saudi Arabia. International Journal of Remote Sensing, 13(4):605-614.

Kirk, J.T.O. 1994. Light and Photosynthesis in Aquatic Ecosystems, 2nd ed., Cambridge University Press: Cambridge.

Kishino, M., A. Tanaka, T. Oishi, R. Doerffer, H. Schiller. 2001. Temporal and spatial variability of chlorophyll a, suspended solids, and yellow substance in the Yellow Sea and East China Sea using ocean color sensor, Proc. SPIE, vol. 4154, pg 179-187.

Kohonen, T. 2001. Self-Organizing Maps. 3rd Edition. Springer-Verlag, Berlin, Heidelberg, New York.

Lee, Z., K.L. Carder, C.D. Mobley, R.G. Steward, and J.S. Patch. 1999. Hyperspectral remote sensing for shallow waters: Deriving bottom depths and water properties by optimization. Applied Optics, 38:3831-3843.

Lin, C.T. 1994. Neural fuzzy control systems structure and parameter learning, World Scientific Co. Ltd.

Maass, W. 1995. Vapnik-Chervonenkis dimension of neural networks. In: The Handbook of Brain Theory and Neural Networks.Arbib, M.A. (Ed.). MIT Press, Cambridge, MA, 522-526.

McCarthy M., T. Pratum, J. Hedges, R. Benner. 1997. Chemical composition of dissolved nitrogen in the ocean. Nature, 390:150-153.

Medsker, L.R. 1995. Hybrid intelligent systems. Kluwer Academic Publishers, Boston, MA, USA.

Mitchell, M.. 1998. An Introduction to Genetic Algorithms (Complex Adaptive Systems), MIT Press, Cambridge, MA.

Moody, J. 1992. The effective number of parameters: An analysis of generalization and regularization in nonlinear learning systems. In: Advances in Neural Information Processing Systems. Moody, J., S. J. Hanson, and R.P. Lippmann (Eds.). Morgan Kaufmann, San Mateo, CA, 847-854.

Moore, T.S., J.W. Campbell, and H. Feng. 2001. A fuzzy logic classification scheme for selecting and blending satellite ocean color algorithms. IEEE Transactions on Geoscience and Remote Sensing, 39(8): 1764-1776.

Morel, A. 1988. Optical modeling of the upper ocean in relation to its biogenous matter content (case 1 waters). Journal Geophysical Research, 93:10749-10768.

Musavi, M.T., R.L. Miller, H. Ressom, and P. Natarajan. 2001. Neural network-based estimation of chlorophyll-a concentration in coastal waters. In Proceedings of SPIE, 4488, pg 176-183.

Neale P.J., J.J. Cullen, and R.F. Davis. 1998. Inhibition of marine photosynthesis by ultraviolet radiation: Variable sensitivity of phytoplankton in the Weddell-Scotia Sea during the austral spring, Limnology and Oceanography, 43(3):433-448.

O'Reilly, J.E., S. Maritorena, D.A. Siegel, M.C. O'Brien, D. Toole, B.G. Mitchell, M. Kahru, F.P. Chavez, P. Strutton, G.F. Cota, S.B. Hooker, C.R. McClain, K.L. Carder, F. Müller-Karger, L. Harding, A. Magnuson, D. Phinney, G.F. Moore, J. Aiken, K.R. Arrigo, R. Letelier, M. Culver. 2000. Ocean color chlorophyll a algorithms for SeaWiFS, OC2, OC4: version 4. NASA-TM-2000-206892, 11:9-23.

O'Reilly, J.E., S. Maritorena, B.G. Mitchell, D.A. Siegel, K.L. Carder, S.A. Garver, M. Kahru, and C. McClain. 1998. Ocean color chlorophyll algorithms for SeaWiFS. Journal of Geophysical Research, 103 (C11):24,937-24,953.

Orr, M., J. Hallam, K. Takezawa, A. Murray, S. Nimomiya, M. Oide, and T. Leonard. 2000. Combining regression trees and radial basis function networks. International Journal of Neural Systems, 10(6):453-465.

Ressom, H., M.T. Musavi, P. Natarajan. 2001. Neural network-based estimation of phytoplankton primary production, Proceedings of SPIE, vol. 4488, pg 213-220.

Ritchie, J.C., and C.M. Cooper. 1988. Comparison of measured suspended sediment concentrations with suspended sediment concentrations estimated from Landsat MSS data, International Journal of Remote Sensing, 9(3):379-387.

Roesler, C.S., and M.J. Perry. 1995. In situ phytoplankton absorption, fluorescence emission, and particulate backscattering spectra determined from reflectance. Journal of Geophysical Research, 100(C7):13279 – 13294.

Rumelhart, D.E. and J.L. McClelland (Eds.). 1986. Parallel distributed processing: Explorations in the microstructure of cognition, vol. 1: Foundations, MIT press.

Salomonson, V.V., W.L. Barnes, P.W. Maymon, H.E. Montgomery, and H. Ostrow. 1989. MODIS: advanced facility instrument for studies of the Earth as a system. IEEE Transactions of Geoscience and Remote Sensing, 27: 145-152.

Sathyendranath S., L. Priuer, A. Morel. 1989. A three-component model of ocean colour and its application to remote sensing of phytoplankton pigments in coastal waters. International Journal of Remote Sensing 10 (8):1373-1394.

Scardi, M. 1996. Artificial neural networks as empirical models of phytoplankton production, Marine Ecology Progress Series, 139:289-299.

Scardi, M. 2001. Advances in neural network modeling of phytoplankton primary production. Ecological Modelling, 146(1-3):33-45.

Schiller, H. and R. Doerffer. 1999. Neural network for emulation of an inverse model – operational derivation of Case II water properties from MERIS data. International Journal of Remote Sensing, 20(9):1735-1746.

Schwarz, J.N., P. Kowalczuk, S. Kaczmarek, G. F. Cota, B. G. Mitchell, M. Kahru, F. P. Chavez, A. Cunningham, D. McKee, P. Gege, M. Kishino, D. A. Phiney, R. Raine. 2002. Two models for absorption by colored dissolved organic matter (CDOM). Oceanologia, 44(2):209-241.

Siegel H., M. Gerth, M. Beckert. 1994. The variation of optical properties in the Baltic sea and algorithms for the application of remote sensing. SPIE vol. 2258: 894-905.

Siegel, D.A., A.F. Michaels. 1996. Quantification of non-algal attenuation in the Sargasso Sea: implications for biogeochemistry and remote sensing. Deep-Sea Research II, 43:321-345.

Slade, W.H., R.L. Miller, H. Ressom, and P. Natarajan. 2004. Neural network retrieval of phytoplankton abundance from remotely-sensed ocean radiance. In Proceedings of 2nd IASTED International Conference on Neural Networks and Computational Intelligence, Grindelwald, Switzerland.

Tassan, S. 1993. An improved in-water algorithm for the determination of chlorophyll and suspended sediment concentration from Thematic Mapper data in coastal waters. International Journal of Remote Sensing, 14 (6):1221-1229.

Topliss, B.J., C.L. Almos, and P.R. Hill. 1990. Algorithms for remote sensing of high concentration inorganic suspended sediment. International Journal of Remote Sensing, 11(6):947-966.

Tso, B. and P.M. Mather. 2001. Classification Methods for Remotely Sensed Data, Published by Taylor and Francis, London.

Wernarnd, M.R., S.J. Shimwell, and J.C. Munck. 1997. A simple method of full spectrum reconstruction by a five band approach for ocean colour applications. International Journal of Remote Sensing, 18(9):1977-1986.

Wilkinson, G.G. 1996. A review of current issues in the integration of GIS and remote sensing data. International Journal of Geographical Information Systems, 10(1): 85-101.

Yen, J. and R. Langari. 1999. Fuzzy Logic: Intelligence, Control, and Information, Prentice Hall, Upper Saddle River, NJ.

Zadeh, L.A. 1994. Fuzzy Logic, Neural Networks and Soft Computing. Communications of the ACM, 37(3): 77-84.

Zhan, H., Z. Lee, P. Shi, C. Chen, and K.L. Carder. 2003. Retrieval of water optical properties for optically deep waters using genetic algorithms. IEEE Transactions of Geoscience and Remote Sensing, 41(5): 1123-1128.

CHAPTER 10

MODELING AND DATA ASSIMILATION

[1]JOHN R. MOISAN, [2]ARTHUR J. MILLER, [2]EMANUELE DI LORENZO AND [3]JOHN WILKIN

[1]*NASA, Goddard Space Flight Center, Wallops Flight Facility, Wallops Island, VA, 23337 USA*
[2]*Scripps Institution of Oceanography, University of California, San Diego, La Jolla, CA, 92093 USA*
[3]*Rutgers University, Douglas Campus, Institute of Marine and Coastal Sciences, 71 Dudley Road, New Brunswick, NJ, 08901USA*

1. Introduction

Coastal areas are by far the most complex and dynamic of all ocean regions. They are important zones for the accumulation and transformation of nutrients and sediments derived from terrestrial and atmospheric sources. These areas are also crucial fish nursery and foraging grounds and are home to the majority of ocean fish stocks that compose our fisheries. Approximately 90% of the total marine fish catch is derived from continental shelf regions, an area comprising less than 8% of the total ocean area. The proximity of the coastal ocean to terrestrial and fluvial influences complicates the underlying coastal ocean dynamics often associated with coastal regions, such as tides, coastal trapped waves, shoaling internal waves, upwelling, etc. Mankind has heavily influenced coastal regions by modifying freshwater influx patterns, altering nutrient and sediment fluxes from both fluvial and atmospheric sources, and overexploiting fisheries resources. One goal in understanding the dynamics of coastal regions is to use this knowledge to improve coastal management practices to reduce the impact of anthropogenic influences. However, gaining an understanding of the mechanisms important to answering a host of questions related to coastal ocean regions requires the coordinated use of a wide variety of data sets, remote and *in situ*, and numerical models.

Remote sensing products are becoming increasingly available for coastal ocean applications (King et al., 2003). These products are typically derived from passive reflectance measurements (e.g. sea surface temperature from the Advanced Very High Resolution Radiometer, AVHRR; phytoplankton chlorophyll and primary production from the Sea-viewing Wide Field-of-view Sensor, SeaWiFS; and, Moderate Resolution Imaging Spectroradiometer, MODIS) or active microwave radar reflectance measurements (e.g. QuickSCAT for ocean surface wind velocities, and TOPEX and Jason-1 for ocean surface altimetry measurements). The capabilities and resolution of each of these satellite sensors and data products varies between coastal and open ocean regions. For passive remotely sensed data, the high-resolution Local Area Coverage (LAC) data are available for coastal regions, while open ocean region studies have access primarily to coarser resolution Global Area Coverage (GAC) data. Passive reflectance data sets from coastal regions are more resolved, but Case 2 waters associated with coastal zones makes deriving products, like surface chlorophyll, more difficult and prone to increased uncertainties due to higher concentrations of coastal-

R.L. Miller et al. (eds.), Remote Sensing of Coastal Aquatic Environments, 229–257.
© 2005 *Springer.*

derived, optically-active scalars such as Colored Dissolved Organic Material (CDOM) and suspended sediments (Carder et al., 1999). Radar scatterometer (wind velocity) and altimeter (sea level topography) data sets involve separate issues related to data quality within coastal areas. Radar scatterometer estimates of wind, while well resolved in open ocean regions, are unable to resolve the small-scale wind field structures located near the coast, where wind measurements are required by circulation models to resolve processes such as coastal upwelling. Sea level topographic measurements are even more problematic because of the low spatial and temporal resolution of the data coupled with the noise issues due to tidal signals in coastal regions.

Because of the short time and space scales and complex dynamics associated with coastal environments, gaining further understanding of these regions remains difficult. As a result, methods are now being developed to merge satellite observations with models in an effort to understand and eventually predict the observed variability in these regions. However, many of the available coastal ocean satellite data sets have yet to be

Table 1. List of presently available satellite-derived coastal ocean variables.

Measurement	Example Satellite Sensors	Used in Numerical Modeling	References on use of satellite data in modeling studies
Sea Surface Temperature (SST)	AVHRR, MODIS	Yes	Anderson et al., 2000; Fox et al., 2001; Di Lorenzo et al., 2004; Wilkin et al., 2004
chlorophyll *a*	SeaWiFS, MODIS	Yes	Prunet et al., 1996a,b; Semovski et al., 1995; Semovski and Wozniak, 1995; Di Lorenzo et al., 2004
Primary Productivity	SeaWiFS, MODIS	No	N/A
Chlorophyll Fluorescence	MODIS	No	N/A
Total Suspended Matter	MODIS	No	N/A
Organic Matter	MODIS	No	N/A
Coccolith Concentration	MODIS	No	N/A
Rainfall	TRMM	No	N/A
Photosynthetically Available Radiation	MODIS	Yes	Spitz et al., 2001
Suspended Solids	MODIS	No	N/A
Colored Dissolved Organic Matter (CDOM)	MODIS	Yes	Bissett et al., 1999a,b, Bissett et al., 2004
Wind Velocities	QuickSCAT	Yes	Fox et al., 2001
Sea Surface Topography	TOPEX, Jason	Yes	Fox et al., 2001; Di Lorenzo et al., 2004

used in support of coastal modeling efforts (Table 1). The steps required to develop this capability are complex. The matching of observations to model variables, discerning between observational and modeling errors, properly constraining the model solutions with realistic forcing and boundary conditions and a host of other issues remain unresolved.

The path towards making use of satellite observations for coastal ocean studies is marked with a host of model and algorithm applications, ranging from the radiative transfer models that are used to interpret the satellite observations to fully three dimensional (3D) coupled numerical circulation bio-optical models. A wide array of numerical modeling activities presently makes use of satellite data to address specific coastal ocean-related questions. These models range in complexity from simple algorithms to complex systems of time and space dependent coupled partial differential equations.

Because coastal regions possess small space scale and short time varying processes and features, models play a crucial role in helping us understand the interplay between the various processes that contribute to the final observed dynamic fields. The scales of the processes that contribute to the evolution of the observed features are poorly resolved by *in situ* observations and in cloudy regions (such as coastal upwelling centers) even by satellite sensors. Numerical models are required to integrate the observations into a dynamic modeling framework in order to allow us to test hypotheses on coastal dynamics.

The status of modeling ocean processes has progressed rapidly in the last decade due to the increase in computer technologies; improved methods in computational fluid dynamics; improved knowledge in ocean circulation and biogeochemical dynamics; and, a large increase in the availability of remotely sensed data for model forcing and validation (Shchepetkin and McWilliams, 2003; Moore et al., 2004). Contemporary modeling efforts now use satellite data for a variety of purposes, ranging from model forcing fields to independent data sets for model validation (Robinson, 1996; Di Lorenzo et al., 2004; Wilkin et al., 2004). In coastal regions, where modeling efforts require high resolution data sets due to short time and space scales of coastal ocean processes, use of satellite data sets is crucial. In this chapter, the variety of ways that satellite observations are used to support modeling activities will be presented through an overview of present and anticipated future applications.

2. Diagnostic/Analytical Models

2.1 OVERVIEW OF DIAGNOSTIC MODEL DEVELOPMENT METHODOLOGIES

Satellite imagery has historically been used in coastal applications as a qualitative tool to characterize the spatial structure of coastal ocean features (Bernstein et al. 1977; Abbott and Chelton, 1991). Additional efforts to use these data have focused on characterizing seasonal and interannual variability (Thomas and Strub 1989, 1990; Strub et al., 1990). By far, the dominant approach for using satellite and field data in a quantitative sense is to develop algorithms or models that use observed relationships (empirical algorithms) that require satellite or *in situ* data as input variables to estimate scalars or processes that cannot be measured from space. A crucial application of models using satellite observations involves using Radiative Transfer Models (Zaneveld et al., Chapter 1) to estimate water-leaving radiance values near the ocean surface. These estimates are used—as shown below—to obtain estimates of optically active ocean scalars (chlorophyll *a*, colored dissolved organic material, etc.). There are a growing number of diagnostic models/algorithms presently available for use in the ocean remote sensing community (Table 2). Two applications of primary importance to coastal ocean ecosystem research are presented below.

Table 2. Diagnostic Models/Algorithms for Ocean Remote Sensing Applications on MODIS.

Estimated Scalar	Method	References
Chlorophyll *a* pigment	Empirical and Semi-Analytical Models	O'Reilly et al., 1998; 2000
Total Suspended Matter	Empirical Model	Gordon and Clark, 1980
Diffuse Attenuation Coefficient at 490nm	Empirical Model	Gordon and Clark, 1980
Chlorophyll Fluorescence	Analytical Model	Abbott et al., 1982; Abbott and Letelier, 1998
Colored Dissolved Organic Matter (CDOM)	Empirical, Semi-analytical Models	O'Reilly et al., 2000
Absorption Coefficients	Empirical, Semi-analytical Models	O'Reilly et al., 2000
Coccolith Concentration	Semi-analytical Model	Gordon et al., 1988
Primary Production	Empirical and Analytical Models	Iverson et al., 2000; Behrenfeld and Falkowski, 1997a; Howard and Yoder, 1997
Phycoerythrin	Semi-Analytical Model	Gordon et al., 1988

2.1.1 *Case 1: ocean chlorophyll a estimates*

There are three distinct types of models to estimate ocean chlorophyll *a* using satellite reflectance data: empirical, semi-analytical and analytical. Of these, only the first two have been widely implemented. Empirical models use *in situ* observations of ocean chlorophyll *a* to develop a relationship between the apparent optical property (AOP) of spectral remote-sensing reflectance $R_{rs}(\lambda)$ or normalized water-leaving radiance $L_{wn}(\lambda)$ and chlorophyll *a* concentrations. A number of these models are presented and compared in O'Reilly (1998; 2000). For instance, the OC4 model (version 4),

$$\text{Chl } a = 10^{\left(a_0 + a_1 R + a_2 R^2 + a_3 R^3\right)} + a_4,$$

where $R = \log\left(\max\left(\dfrac{R_{rs}(443)}{R_{rs}(555)}; \dfrac{R_{rs}(490)}{R_{rs}(555)}; \dfrac{R_{rs}(510)}{R_{rs}(555)}\right)\right)$, and $\quad$ (1)

$$a_0 = 0.366; a_1 = -3.067; a_2 = 1.930; a_3 = 0.649; a_4 = -1.532$$

is a five parameter model that uses a 4th order polynomial to utilize the maximum band ratio, R, of three different waveband ratios of the spectral remote-sensing reflectance $R_{rs}(\lambda)$. The maximum function causes the model to switch to alternate band ratios when the other band ratios become lower. This band-ratio switching is how many of the CZCS pigment algorithms operate (O'Reilly et al., 2000). The majority of empirical models fit radiance band ratios (converted to either logarithmic or natural log scales) to *in situ* chlorophyll *a* data using a variety of functions such as power, hyperbolic, and cubic polynomials. These color ratio algorithms work best in Case 1 waters and do poorly in coastal Case 2 waters, where the increased number of optically-active constituents add to the complexity of the ocean color problem.

Semi-analytical models, the second model type, use relationships that relate Inherent Optical Properties (IOPs), typically backscattering $b_b(\lambda)$ and absorption $a(\lambda)$ coefficients, to $R_{rs}(\lambda)$ (Garver and Seigel, 1997; Carder et al., 1999) or $L_{wn}(\lambda)$. Semi-analytical models can be implemented as either forward or inverse applications. Forward model applications use observations of IOPs, either measured directly or estimated from relationships of IOPs and in-water distributions of optically-active compounds, such as chlorophyll *a* or CDOM, to calculate $L_{wn}(\lambda)$ or $R_{rs}(\lambda)$. Inverse model applications use observations of $L_{wn}(\lambda)$ or $R_{rs}(\lambda)$ to calculate chlorophyll *a* or CDOM concentrations (Hoge et al., 1999, 2001; Garver and Siegel, 1997; Siegel et al., 2003).

The majority of ocean color models have been developed for application to Case 1 waters, with few exceptions (Doerffer and Fischer, 1994; Carder et al., 1999). A more detailed presentation of the issues involved in applying these models to Case 2 waters found in coastal regions is presented in Muller-Karger et al. (Chapter 5).

Several new computational techniques are now being used to refine and further develop the techniques for using satellite observations to estimate chlorophyll *a*. Chapter 9 of this book presents an overview of the various computational methods now under employ or development. The historical methodologies used to develop ocean color models are based upon either empirical formulations or subjectively defined relationships, such as waveband ratios. Artificial neural network techniques are now being used to retrieve chlorophyll *a* from $R_{rs}(\lambda)$ (Gross et al., 2000; Zhang et al., 2003) and to support the merger of ocean color data from multiple satellite missions (Kwiatkowska and Fargion, 2003). Only a few applications (Tanaka et al., 2000) have applied this technique to Case 2 waters. A recent global application of the semianalytical inverse ocean color model of Garver and Siegel (1997) uses a data assimilation technique called "simulated annealing" to optimize the IOP model parameters (Maritorena et al., 2002). The multiple satellite merger effort being supported by the National Aeronautic and Space Administration's (NASA) Sensor Intercomparison and Merger for Biological and Interdisciplinary Studies (SIMBIOS) project is using spectral data assimilation and simulated annealing techniques to develop a merged satellite data product. The utility of this new application is that it not only provides estimates but also provides the related uncertainties for a number of ocean color products.

2.1.2 *Case 2: satellite-based models for phytoplankton primary production*

Similar to models used to estimate chlorophyll *a* from spectral remote-sensing reflectance, $R_{rs}(\lambda)$, primary production models also fall into three distinct categories, empirical, semi-analytical and analytical. Empirical models that use chlorophyll *a* estimates to predict primary production were developed prior to the capability to measure chlorophyll *a* using satellites. Balch et al. (1989a) presents a history of the development of these initial model efforts (Table 3) that began with Ryther and Yentch (1957), and Balch and Byrne (1994) define the various problems encountered in estimating primary production from space.

Analytical models for calculating the depth-integrated, daily primary production, Π, [mg C m^{-2}] that resolve spectral, temporal and depth variability, termed WRMs for Wavelength Resolving Models (Falkowski et al., 1998), attempt to incorporate all of the

Table 3. Models for phytoplankton primary production estimates.

Reference	Type[1]	Product	Required Data Input
Ryther and Yentch, 1957	E	photosynthetic rate at light saturation	chlorophyll a
Talling, 1957a, b	E	depth-integrated primary production	irradiance, I_K
Lorenzen, 1970	E	depth-integrated primary production	surface chlorophyll a
Smith and Baker, 1978	E	primary production	chlorophyll a
Smith et al., 1982	E	primary production	chlorophyll a
Brown et al., 1985	E	mean euphotic zone production	chlorophyll a
Eppley et al., 1985	E	depth-integrated primary production	chlorophyll a, temperature, day length
Platt (1986)	E	primary production	surface light intensity, chlorophyll a
Balch et al., 1989a	S-A	surface and depth-integrated primary production	pigments and temperature
Balch et al., 1989b	S-A	surface and depth-integrated primary production	pigments, temperature, and light
Behrenfeld and Falkowski, 1997a,b	S-A	depth-integrated primary production	temperature and chlorophyll

[1] E: Empirical; S-A: Semi-analytical; A: Analytical

physical, bio-optical and physiological processes that are involved in regulating net primary production. The explicit analytical model for

$$\Pi = \int_0^{z_{eu}} \int_{0hrs}^{24hrs} \int_{350nm}^{700nm} 12\,\Phi(\lambda,t,z)\,E_0(\lambda,t,z)\,a_{ph}^*(\lambda,t,z)\,d\lambda\,Chl(t,z)\,dt\,dz$$

$$-\int_0^{z_{eu}} \int_{0hrs}^{24hrs} R(t,z)\,dt\,dz, \tag{2}$$

where $\phi(\lambda,t,z)$, is the quantum yield for photosynthesis for available radiance [mol C (mol quanta)$^{-1}$], $E_o(\lambda,t,z)$ is the available incident spectral solar radiance [mol quanta m^{-2} s^{-1} nm^{-1}], $a_{ph}^*(\lambda,t,z)$ is the chlorophyll-specific absorption [m^2 (mg chla)$^{-1}$], $Chl(t,z)$ is the *in situ* concentration of chlorophyll a [mg chla m^{-3}], and $R(t,z)$ is the loss term for respiration [mg C m^{-3} s^{-1}] due to losses of fixed carbon from photosynthetic respiratory processes and nighttime respiration. The factor of 12 is a simple conversion term [12 mg C (mol C)$^{-1}$].

These models are termed semi-empirical because, as with the ocean color models, they require empirical relationships to provide model closure, in this case relationships that link the variables $\Phi(\lambda,t,z)$ and $a_{ph}^*(\lambda,t,z)$ to environmental conditions, such as light or temperature, that can be estimated by satellite observations or simulated/predicted using advanced numerical models (Moisan, 1993; Bissett et al., 1999b). For instance, Moisan and Mitchell (1999) developed a modified version of the WRM, similar to that developed earlier by Kiefer and Mitchell (1983), such that the quantum yield for growth and the chlorophyll absorption relationships were quantified using temperature and light dependent empirical relationships. Using these relationships, it is possible to obtain estimates of daily primary production using satellite-based measurements of sea surface temperature, chlorophyll and photosynthetically available radiance (PAR).

By integrating the WRM equation over the visible spectrum of available radiance the equation is modified into a Wavelength-Integrated Model (WIM). The actual integration of equation 13.2 must be carried out using the integration by parts technique, creating a more complex model equation. In practice this is not done (Falkowski et al., 1998). Instead, a daily primary production models is developed that is devoid of wavelength-dependent terms such that

$$\Pi = \int_0^{z_{eu}} \int_{0hrs}^{24hrs} \left(\varphi(t,z) \, PAR(t,z) \, Chl(t,z) - R(t,z) \right) dt \, dz \ , \tag{3}$$

where $\varphi(t,z)$ is the chlorophyll a specific quantum yield of photosynthesis for absorbed PAR [mg C (mol quanta mg chla m^{-3})$^{-1}$] similar to the product of $\Phi(\lambda,t,z)$ and $a^*_{ph}(\lambda,t,z)$ and $PAR(t,z)$ is the Photosynthetically Available Radiance [mol quanta m^{-2} s^{-1}].

Further reductions can be made to these primary production models by creating a Time-Integrated Model (TIM), such that

$$\Pi = \int_0^{z_{eu}} \left(P^b(z) \, \overline{PAR(z)} \, \overline{Chl(z)} \right) dz \ , \tag{4}$$

where $P^b(z)$ is the daily-integrated chlorophyll-normalized photosynthetic rate at depth that incorporates the respiration and quantum yield terms [mg C (mol quanta mg chla m^{-3})$^{-1}$], $\overline{PAR(z)}$ and $\overline{Chl(z)}$ are the depth-varying, time-averaged photosynthetically available radiance and chlorophyll a, respectively. This model can be further simplified by integrating over the depth interval between the surface and the depth of the euphotic zone. The resulting Depth-Integrated Models (DIMs) become simplified to the level that the equations fall into the category of the numerous other empirical-based models from earlier efforts (See Table 3). A round-robin comparison of the ability of a number of primary production models based upon surface chlorophyll, temperature, and irradiance is presented by Campbell et al. (2002).

The majority of the ocean phytoplankton primary production models, if not all, are based upon tedious, primarily subjective, efforts to develop relationships that utilize satellite-derived observations of key environmental variables such as surface chlorophyll, temperature, and radiance. As with the development of ocean color algorithms, new applied math and computational techniques are becoming available that offer new avenues for creating more sophisticated—though likely more complex—phytoplankton primary production models. Some of these applications include neural networks, genetic algorithms, genetic programming, and fuzzy logic (See Chapter 9), and the host of available data assimilation techniques. With the establishment of high quality primary productivity data sets (Balch et al., 1992), more accurate and sophisticated productivity models are under development.

3. Deterministic Models

Beyond the realm of diagnostic models or algorithms that are used with satellite data sets to estimate ocean variables such as chlorophyll a are the more sophisticated dynamic models that have been developed to characterize the time evolution of ocean variables. These models are comprised of systems of coupled ordinary or partial differential equations and are solved through the use of numerical integration techniques and high performance computers. Several reviews (Franks, 1995; Hofmann and Lascara, 1998) present overviews on the variety of interdisciplinary, coupled circulation biological models that have been used for marine ecosystem research. These modeling efforts range from simple time-resolved (spatially-homogeneous) box models, to depth-resolved, one-dimensional (1D) models, to fully integrated coupled 3D circulation biogeochemical models.

A wealth of biogeochemical models have been developed for open ocean regions such as Ocean Weather Station Papa (Fasham, 1995; Antoine and Morel, 1995; McClain et al., 1996; Signorini et al., 2001), Burmuda-Atlantic Time-series Station (BATS) region (Fasham et al., 1990; Doney et al., 1996; Hurtt and Armstrong, 1996; Spitz et al., 1998; Hurtt and Armstrong, 1999; Bissett et al. 1999a,b; Spitz et al., 2001), Equatorial Pacific (Christian et al, 2002a,b), and the Northeast North Atlantic (Fasham et al., 1999). The majority of these efforts were in support of the recent scientific program called the Joint Global Ocean Flux Study (JGOFS) that focused on developing an understanding of the processes controlling carbon and nitrogen fluxes in the open ocean in order to close the carbon and nitrogen budgets in these regions.

3.1 BOX MODELS

Box models continue to play an important role in addressing specific coastal ocean science questions, especially those related to carbon and nutrient fluxes (Gordon et al., 1996), and they support the development of more complex (1D and 3D) models by providing a simple numerical environment to test new model formulations (Moisan et al., 2002). Also included in the box model category are the bulk mixed-layer model applications of Fasham et al. (1990) that have been successfully applied to several data assimilation studies in open ocean regions (Spitz et al., 1998, 2001). Box models have historically been used in coastal regions as tools to study ecosystem and trophic level interactions (Hofmann and Lascara, 1998, Olivieri and Chavez, 2000), biogeochemical cycling between deep and surface waters of the ocean (Broecker and Peng, 1982) and volume, salt and heat conservation in coastal embayments and larger inland seas (Pickard and Emery, 1990). The latter effort uses simple mass balance equations for salt and water to derive information on turnover time scales or residence times for coastal estuaries and bays. These residence time scales are an important parameter for estimating the fraction of carbon and nutrient fluxes from fluvial sources that reach the coastal ocean (Nixon et al., 1996). Because present coastal ocean models do not resolve these small estuaries and bays, and forcing fields for coastal inputs of nutrients and carbon are limited to regions not affected by tidal influence, estimates of these turnover time scales are important for linking the fluvial nutrient and carbon fluxes to the coastal models at the appropriate level.

3.2 ONE-DIMENSIONAL (VERTICAL) BIOGEOCHEMICAL MODELS

The primary forcing conditions that control the time evolution of biogeochemical processes in many regions of the ocean occur through vertical processes such as advection and diffusion of nutrients, vertical attenuation of solar radiation and sinking of particles. Circulation and diffusion processes are three-dimensional in nature, but it has been demonstrated (Gill and Niiler, 1973; Moisan and Niiler, 1998) that, over large enough spatial scales, the seasonal variability of physical features such as temperature, salinity, nutrients, etc. is forced primarily through vertical processes. Many open ocean modeling efforts have made use of this quality to develop 1D, vertical models for studying open ocean biogeochemical processes (McGillicuddy et al., 1995; Doney et al., 1996; Bissett et al., 1999a,b). It is worthwhile to note here that the work of Bissett et al. (1999a,b) is the first to simulate both apparent and inherent optical properties.

While 1D models are ideally suited for open ocean applications, they have also been applied to several coastal regions to investigate processes such as dissolved organic matter cycling (Anderson and le B. Williams, 1998), benthic denitrification and nitrogen cycling (Balzer et al., 1998), and plankton ecosystems for upwelling regions (Moloney, et al., 1991; Moisan and Hofmann, 1996). The application of 1D models to coastal biogeochemical studies (Moisan and Hofmann, 1996; Soetaert et al., 2001) requires additional physical constraints in order to resolve those processes, such as upwelling, that are not commonly encountered in open ocean applications, with the exception of open ocean upwelling areas such as the Equatorial Pacific (Friedrichs, 2001).

Vertical 1D biogeochemical models are composed of systems of coupled partial differential equations that govern the time and space distribution of the non-conservative scalars, such as nutrients, phytoplankton, detritus, dissolved organics, etc. The general form of this equation is written as:

$$\frac{\partial B}{\partial t} = \frac{\partial}{\partial z} K_z \frac{\partial B}{\partial z} - \frac{\partial(Bw)}{\partial z} - w_{sink} \frac{\partial B}{\partial z} - \tau_{nudge}(B - B_{clim}) + S_B, \qquad (5)$$

where B is a non-conservative quantity (one of the variables in the biogeochemical model), K_z is the depth-dependent, vertical eddy kinematic diffusivity, w is the depth-dependent vertical (upwelling/downwelling) velocity of the fluid, and w_{sink} is the vertical sinking rate of the biogeochemical components. An additional term τ_{nudge} is also used at times to specify the time scale over which the biogeochemical components are nudged back to the background climatologies of the individual model components, B_{clim}. The net local source and sink terms S_B can be prescribed to simulate processes such as nitrogen fixation, denitrification, or loss of material to higher trophic levels. One additional item to note is that the vertical advection term in coastal applications should be written in the flux divergence form so that the effects due to strong vertical divergence/convergence in the vertical velocity field ($\partial w / \partial z$) are appropriately accounted, and mass and volume are conserved.

Several applications of these 1D models focus on coastal ocean regions. For instance, Moisan and Hofmann (1996) use a food web model coupled to a multi-nutrient (nitrate, ammonia, silicate) biogeochemical model to compare the cycling of nitrogen in both an onshore and offshore region of the California Current System. Model results from 1D models are commonly used to assist in parameterization of 3D coastal models (Moisan et al., 1996; Vichi et al., 2003).

Efforts at modeling nitrogen budgets in the coastal regions are now coupling benthic and pelagic biogeochemical models. An overview of the approaches used in pelagic and benthic biogeochemical model coupling is presented by Soetaert et al. (2000). Soetaert et al. (2001) developed a 1D model to study biogeochemical processes as part of the Ocean Margin Exchange in the Northern Gulf of Biscay (OMEX). In this study, the model resolves the vertical structure of water column (NO_3, NH_4, O_2, Phytoplankton C and N, Detrital C, Detrital N, Zooplankton C, and Suspended matter) and sediment (NO_3, NH_4 and O_2) constituents. The addition of a vertically resolved sediment model, while providing important estimates on rate of nutrient regeneration and denitrification rates, has significant drawbacks due to the higher spatial resolution (mm vs m) required to resolve vertical sediment processes. For instance, in the Soetaert et al. (2001) modeling effort, the time step for the fully coupled sediment-pelagic model was 10 times slower than for the pelagic model alone. However, resolving these processes is especially critical in coastal modeling because of the uncertainty in the total amount of denitrification that occurs in coastal regions. Inclusion of sediment processes gives models a capacity to predict the total source of inorganic nitrogen through sediment remineralization as well as to predict the amount of organic nitrogen lost through denitrification processes (Seitzinger and Giblin, 1996, Balzer et al., 1998).

Recently, Moisan et al. (2004) used a fully coupled biogeochemical model with nitrogen, carbon, and oxygen pathways (Fig. 1) to study carbon and nutrient pathways in the California Current System as a tool to configure a fully 3D coupled circulation biogeochemical model (Stolzenbach et al., 2004). One of the important components of this effort was the development of a particle coagulation model. In addition, oxygen profile data were used to help parameterize the two competing processes of particle sinking and remineralization processes—a fast sinking, rapidly remineralizing material is similar to a slow sinking, slowly remineralizing material, in that both release nutrients back into the water column at a similar rate. As in the work of Oguz et al. (2000) and other studies, oxygen profiles play an important role in optimizing the values of the parameters that influence the vertical flux of carbon and nitrogen.

In the Moisan et al. (2004) model (Fig. 2), the microbial loop dynamics from Spitz et al. (2001) with bacteria, dissolved organic carbon and dissolved organic nitrogen and an inorganic carbon cycling model with alkalinity and dissolved inorganic carbon (following Ocean Carbon-cycle Model Intercomparison Project [OCMIP] guidelines, Orr, 1999; Lewis and Wallace, 1998) were added to the pelagic model from Stolzenbach et al. (2004). The Stolzenbach et al. model (2002) contained multiple nutrients (NH_4, NO_3), phytoplankton, a dynamic phytoplankton carbon to chlorophyll ratio (θ), zooplankton, small and large detrital nitrogen. In its present state, the model is able to fully resolve pelagic carbon and nitrogen processes, but a total of 15 model components are required. As more aspects of the coastal carbon are investigated the number of components is expected to grow significantly. For instance, these models continue to exclude phosphate cycle dynamics, yet the debate continues as to whether specific coastal regions are phosphate versus nitrate limited (Hecky and Kilman, 1984).

While it remains unlikely that 1D models will provide the ocean carbon cycling community with the final estimates that are required for closing the global carbon and nutrient budgets, 1D models will continue to play an important role in developing new model configurations and in parameterization of the more complex, 3D models.

3.3 THREE-DIMENSIONAL COUPLED CIRCULATION/BIOGEOCHEMICAL MODELS

Applications of three-dimensional (3D) ocean circulation models have typically been carried out under limited spatial domains, but the number of applications of the models has risen to encompass a significant fraction of the global continental margin domain (Fig. 3). These 3D models are presently being configured to resolve circulation processes along many continental margin regions of the world ocean (Table 4). These coupled models generally employ circulation models that use standard primitive equations to solve the time varying structure of the circulation fields (Haidvogel and

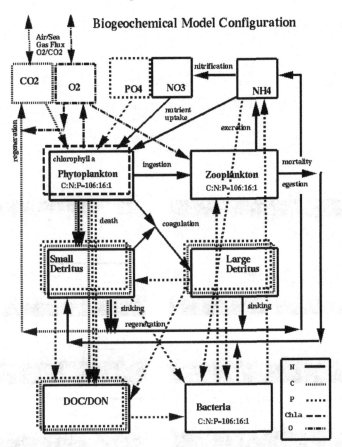

Figure 1. Present biogeochemical model configuration for U.S. West Coast Carbon Cycle modeling program. Not shown are the boxes for Oxygen, TIC, and Alkalinity. Carbon, Nitrogen, Phosphate, and Oxygen pathways are presently being resolved. All of the model pathways follow stoichiometric balances while maintaining variable C:N:P ratios for different model variables.

Beckmann, 1999). The mode of integration primarily occurs using finite difference techniques, although other efforts have employed finite-element techniques (Lynch et al., 1996). In recent years, additional refinements to grid structures have incorporated grid nesting schemes that allow for enhanced resolution of the circulation processes in regions along the coast (Weingartner et al., 2002, Marchesiello et al., 2001 and 2002). Additionally, other coastal circulation models are using model outputs from basin or global scale model domains as open boundary conditions (Mantua et al., 2002).

The role of the physical circulation model in the coupled modeling system is to provide the biogeochemical model with information on the effects of vertical and horizontal advection and diffusion. In the 3D coupled models, the biogeochemical model is composed of a system of coupled partial differential equations that govern the time and space distribution of the non-conservative scalars, such as nutrients, phytoplankton, detritus, dissolved organics, etc. The general form of the equations is

$$\frac{\partial B}{\partial t} = \nabla \bullet K \nabla B - (\vec{v} + \vec{v}_{sink}) \bullet \nabla B - \tau_{nudge} (B - B_{clim}) + S_B, \tag{6}$$

where K is the eddy kinematic diffusivity, B is a non-conservative quantity (one of the seven variables in the biogeochemical model), $\vec{v}$ is the 3D velocity of the fluid, and $\vec{v}_{sink}$ is the vertical sinking rate of the biogeochemical components. The velocity, $\vec{v}$, and kinematic diffusivity, K, fields are obtained from the 3D circulation model. The term τ_{nudge} is an optional nudging term that describes the rate that the biogeochemical components are nudged back to the background climatologies of the individual model

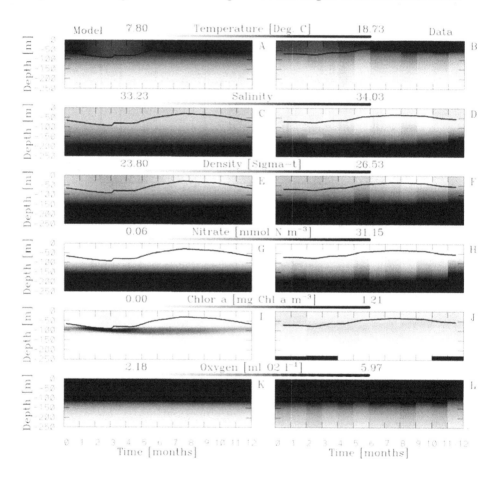

Figure 2. A comparison of the results (left panels) from a vertical (1D) coupled physical biogeochemical model to actual data (right panels) from an offshore region of the California Cooperative Fisheries Investigation (CalCOFI) domain (Moisan et al., 2004). For much of the world oceans, including coastal regions, the vertical structure of many of the biogeochemical features, such as nutriclines or chlorophyll maximums, are maintained by vertically dependent processes, such as light attenuation and vertical nutrient diffusion. The thick black line indicates the mixed layer depth.

components, B_{clim}, and S_B are the net local source and sink terms that describe the interlinking biogeochemical pathways between each of the model variables. The number of components used in the coupled 3D models has ranged from as low as 4 (Gregg and Walsch, 1992) to as high as 66 (Bissett et al., 2004).

3.3.1 *Biogeochemical processes*

There are a host of processes that must be resolved in any 3D coupled model that is focused on simulating coastal ocean dynamics. At present, most of the coupled circulation biogeochemical applications address specific questions that relate only in part to the full coastal biogeochemical system. In order to correctly simulate the full biogeochemical system, even at a minimum or gross level of resolution, a number of critical biogeochemical model components must be included. The key components include: (a) pelagic ecosystem processes; (b) microbial loop processes; (c) multiple nutrients (NH_4, NO_3, SiO_4, PO_4, Fe); (d) detrital (non-Redfield) dynamics; (e) dissolved organic material (non-Redfield) dynamics; (f) marsh and submerged aquatic vegetation (SAV) processes; (g) benthic/sediment layer processes; and, (h) inorganic carbon dynamics that specifically follow Ocean Carbon-cycle Model Intercomparison Project (OCMIP, Orr, 1999; Lewis and Wallace, 1998) conventions/guidelines. With respect to coastal systems, it is important that the processes of nitrification and denitrification be specifically included. On the global scale, denitrification in coastal regions plays an important role in balancing the rate of ocean di-nitrogen fixation (Galloway et al., 1996). Also, processes in coastal regions play an important role in determining the amount of P that ultimately makes it way into the ocean interior. As mentioned in the 1D modeling section, the debate on the relative roles that P and N dynamics play in global ocean productivity on long time scales continues, and developing an appropriate coastal model to address this is crucial for correct carbon budget estimates on long as well as short time scales. One note of caution, creating models capable of simulating non-Redfield ratio dynamics—i.e. C:N:P ratios vary for DOM and POM depending of relative

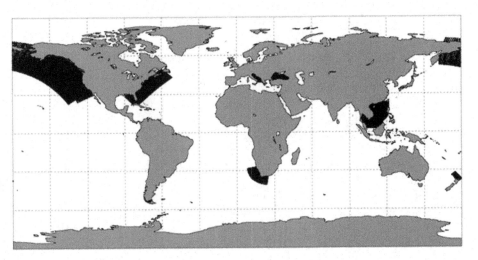

Figure 3. A global chart depicting a number of currently implemented coastal, coupled 3D circulation/biogeochemical models. See Table 4 for a complete listing of the modeling efforts represented within this figure.

Table 4. Coastal Three-Dimensional Coupled Circulation/Ecosystem Models.

Coastal Region	Physical Model	Ecosystem Model	Application	References
Alaska/U.S. West Coast	ROMS	NPZD	Fish Population Dynamics	Weingartner et al., 2002
Alaskan Gyre	ROMS	NPZD	Zooplankton Dynamics	Weingartner et al., 2002
U.S. West Coast	ROMS	Moisan et al., 2004	Biogeochemistry Carbon Cycle	Stolzenbach et al., 2004
U.S. East Coast	ROMS	Moisan et al., 2004	Biogeochemistry Carbon Cycle	Moisan et al., 2004
LEO-15 New Jersey	ROMS	ECOSIM	Bio-optics, Biogeochemistry	Bissett et al., 2004
California Bight	ROMS	Moisan et al., 2004	Ecosystem, Data Assimilation	Miller et al., 2000
South Africa	ROMS	Individual Based Model, NPZ	Small Pelagic Fish Recruitement	Mullon et al., 2002; Penven et al., 2001
South China Sea	Semtner	Fasham et al., 1990	Ecosystem, Nutrient Flux	Liu et al., 2002
Black Sea	POM	Complex, see: Oguz et al., 2002	Ecology, Biogeochemistry	Oguz et al., 1995 and 2002
Adriatic Sea	POM	Complex, see: Vichi et al., 2003	Nitrate Dynamics, Biogeochemistry	Vichi, 2002

recycling/production rates—rapidly increases the level of complexity or number of model components with even a simple 8 component model (Fig. 1). Care must be given to include only those processes that are crucial for accurate simulations in order to reduce the computational requirements and analysis complexity.

3.3.2 *Forcing and boundary conditions*

A wide range of forcing and boundary conditions are required to carry out reasonable and appropriate simulations of these 3D coupled models. The most difficult issue arising from trying to implement these models is open boundary conditions along the lateral open boundaries. Some recent efforts to develop improved open boundary conditions have been used by Marchesiello et al. (2001) in U.S. West Coast simulations. Other efforts have used larger basin scale circulation models to provide the necessary boundary conditions to force their coastal model using a 1-way nested configuration—information only passes from the large-scale model to the smaller scale coastal model.

Air-sea boundary conditions, besides requiring the general suite of heat and salt fluxes, also require estimates of airshed wet/dry deposition for specific model constituents such as NH_4 or POC/PON, and detailed parameterization for air-sea gas transfer. Because of the real lack of adequate data on all of the wet/dry deposition fluxes (Prospero et al., 1996), these have either been ignored or poorly estimated. Air-sea flux of gases is traditionally carried out using the "gradient method." This method contains significant errors (factor of 2; Takahashi et al., 1997) due to wind speed parameterized estimates of the piston pumping velocity. In coastal regions, additional factors, such as increased surfactant levels, may contribute to the total error.

Forcing or boundary conditions at the coast should account for river inflows and consider groundwater (especially in karst plane regions) inflow into the continental shelf regions. Given that the populations of humans in the coastal regions continues to increase and land use practices continue to change and add stress to these regions, a number of factors should be considered when addressing this flux category. In a number of countries, freshwater flux recordings have been taken over the past tens of years and can be used to develop simple forcing conditions such as local climatologies. The situation changes when it comes to assessing the amount of fluvial DOM, POM, or nutrient load that would ultimately reach the ocean margins. This issue is not a simple one to address and will require a significant level of effort to resolve. However, it is a crucial step in developing models that will be of practical use for land management issues.

4. Data Assimilation Efforts using Deterministic Models

4.1 DATA ASSIMILATION FOR 1D BIOGEOCHEMICAL MODELS

The term Data Assimilation encompasses a wide range of applications. For instance, some aspects of data assimilation are simple data insertion or melding techniques where model solutions are nudged back into data compliance or blended with observations using techniques that guarantee dynamical consistency during periods when data are available (Robinson, 1996; Lozano et al., 1996; Robinson et al., 1996; Anderson et al., 2000; Anderson and Robinson, 2001). Other methods seek to minimize the errors between model solutions and observations (satellite or otherwise) by carrying out parameter or initial condition optimization. Efforts to use data assimilation to develop biogeochemical models have primarily focused on using data assimilation techniques for model parameter estimation (Table 5).

A large comparative data assimilation effort by Vallino (2000) tested the ability of various data assimilation methods to assimilate mesocosm experiment data into a marine ecosystem model. The results of this effort demonstrated a number of ongoing concerns facing implementation of data assimilation into ecosystem model development. Of the 12 data assimilation methods tested in this effort, no two solutions to the parameter set were similar. To date, there is no data assimilation method that has been accepted as being more appropriate than any other method. Indeed, no one method is available that can guarantee the parameter set solutions correspond to the parameter set that provides the global minimum error.

Table 5. Parameter optimization methods used in ecosystem models.

DA Method	References
Adjoint	Lawson et al. 1995, 1996; Matear and Holloway, 1995; Prunet et al., 1996a,b; Spitz et al., 1998, 2001; McGillicuddy et al., 1998
Markov chain Monte Carlo	Harmon and Challenor, 1997
Simulated Annealing	Matear, 1995, Hurtt and Armstrong, 1996, 1999; Vallino, 2000
Conjugate Direction/Gradient	Fasham and Evans, 1995; Vallino, 2000
Variational Assimilation	Prunet et al., 1996a,b
Simplex Algorithm	Vallino, 2000
Genetic Algorithm	Vallino, 2000

There are several important considerations that need to be addressed when assimilating data into models. A workshop to investigate the issues facing data assimilation of biological data in 3D coupled models identified a number of factors that must be dealt with for proper data assimilation applications (Robinson and Lermusiaux, 2000). With regard to the use of satellite data in data assimilation applications, there are three specific concerns that need to be addressed.

First, it is important to determine how the satellite estimated or *in situ* measured data equate to the model variables being simulated. This is especially true for ecosystem models where the definitions of model variables may be dramatically different from the working definition of the data being collected. For instance, models that attempt to simulate bacterial dynamics often represent bacteria in terms of the amount of nitrogen per unit volume [mmol N m^{-3}] whereas *in situ* data sets measure bacteria in terms of cell per unit volume (Spitz et al., 2001). Conversions between bacterial cell counts are not straightforward because at present there is no universally accepted value for converting bacterial cell counts into nitrogen or carbon biomass (Carlson et al., 1996). In addition, a number of studies have shown that up to 25% of the enumerated bacteria may in fact be prochlorophytes rather than the assumed heterotrophic bacteria (Sieracki et al., 1995; Carlson et al., 1996); and, it is now widely recognized that the metabolic state, dead vs. living vs. senescent (LeBaron et al., 2001), of individual bacteria cells varies widely with the majority of the cells being either senescent or dead.

With regard to the use of satellite data sets, the primary concern is in matching satellite estimates of chlorophyll *a* to estimates of phytoplankton nitrogen—the limiting nutrient and primary currency used within ocean biogeochemical models, e.g. Fasham et al., 1990. Early attempts of using nitrogen-based models to simulate *in situ* chlorophyll *a* simply assume constant C:Chl *a* [mmol C mg Chl *a*$^{-1}$] or N:Chl *a* [mmol N mg Chl *a*$^{-1}$] ratios for carrying out the conversions to chlorophyll *a* concentrations (Hofmann and Ambler, 1988). In fact, the majority of ocean biogeochemical models in use today continue to use constant C:Chl *a* ratios (McClain et al., 1996; Signorini et al., 2001), though a number of modeling efforts have begun to include dynamic chlorophyll *a* pools (Doney et al., 1996; Spitz et al., 1998; Hurtt and Armstrong, 1999; Bissett et al., 1999a,b; Spitz et al., 2001). The wide variability of the phytoplankton chlorophyll *a* to carbon ratio (Geider, 1987) makes it an important parameter to resolve in order to carry out comparisons between model solutions and observational data sets. Only recently have modeling efforts included dynamic chlorophyll *a* to carbon ratios (Spitz et al., 1998,2001) for both ecosystem modeling and data assimilation application.

The second issue concerns the matching of model solutions to satellite observations. There are primarily two methods presently being used to achieve this. One method involves matching model-simulated variables, e.g. chlorophyll *a*, to satellite-derived values. Smith (1981) presents an integral approach to compare *in situ* profiles of chlorophyll *a*, $Chla(z)$ [mg Chl a m^{-3}], with satellite-derived estimates of ocean chlorophyll *a*, $Chla(z)_{sat}$ such that

$$Chla_{sat} = \int_0^{Z_{90}} Chla(z) e^{-2\int_0^z K(z)dz} dz \Big/ \int_0^{Z_{90}} e^{-2\int_0^z K(z)dz} dz , \qquad (7)$$

where Z_{90} is the depth [m] above which 90% of the upwelling light field is generated, and $K(z)$ is the diffuse attenuation coefficients [m^{-1}]. An alternative, more sophisticated (but presently unimplemented) method is to directly simulate the IOPs, such as backscattering $b_b(\lambda)$ and absorption $a(\lambda)$ coefficients (e.g., Bissett et al., 2004), and use a forward optical model (e.g., Garver and Seigel, 1997; Carder et al., 1999) to predict the spectral remote-sensed normalized water-leaving radiance $L_{wn}(\lambda)$. The modeled quantities of $L_{wn}(\lambda)$ are then available for directly comparing against satellite remotely-sensed $L_{wn}(\lambda)$ values.

The third issue is related to differentiating between model and satellite errors. Both model solutions and satellite data have significant sources of error. Even direct comparison of model solutions and satellite data against *in situ* observations remains complex (Fig. 4). This is because the errors associated with model solutions and satellite calibrations are highly variable in space and time. This is especially true for satellite data in coastal regions, where the presence of CDOM and suspended sediments create Case 2 water conditions that ocean color models/algorithms perform poorly within, and where coastal ocean models are unable to resolve many of the smaller scale features and additional biogeochemical coastal processes. Any comparisons between model performance and satellite or *in situ* observations must be carried out using valid statistical techniques, and all comparisons should avoid the extreme temptation to subjectively note that the model solutions look good when compared to satellite or *in situ* observations.

4.2 ASSIMILATION OF SATELLITE DATA INTO COASTAL OCEAN MODELS

In principal, assimilating satellite data into coastal ocean models is no different than any other data type. Huge volumes of satellite data, however, normally require smoothing or sub-sampling to reduce the total number of observations for easier manipulation and for removing redundancies among closely spaced data when the model grid is coarser than the satellite data grid.

Additional considerations arise when carrying out spatial comparisons between data and model grids of multiple resolutions, as often occurs. A statistical approach is required to allow the model errors to be separated into locational versus quantity errors, the latter being more related to data assimilation cost function determination (Pontius, 2002).

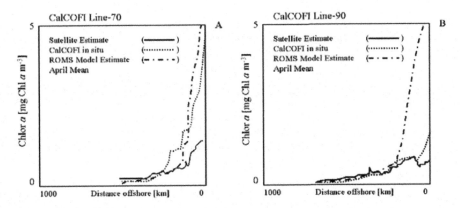

Figure 4. The April mean chlorophyll *a* estimates obtained from a coupled 3D circulation biogeochemical model is compared to SeaWiFS satellite estimates and *in situ* observations collected along two California Cooperative Fisheries Investigation (CalCOFI) lines. In the coastal regions, the 3D model estimates along the CalCOFI Line-70 (A) agree better with the *in situ* observations than do the SeaWiFS satellite estimates. The reverse is observed along the CalCOFI Line-90 (B).

Depending on the application of the assimilated solution, different techniques may be used to constrain the models (Bennett 1992; Wunsch 1996; Lermusiaux and Robinson 1999). These techniques are broadly grouped into "strong constraints" (where no artificial forcing terms are imposed on the dynamics) and "weak constraints" (where artificial forcing is allowed) formalisms. Strong constraints are more appealing when one wishes to diagnose the dynamics of an evolving flow according to the dynamics allowed by the model. They are also more useful when making forecasts, since the artificial forcing is not predictable. Weak constraints are more appealing when one wishes to make the most accurate maps of the fields. They are also useful in initializing forecasts.

Among the weak constraint techniques are 'direct insertion', where model values are simply changed to observed values at some time step, 'nudging', using a relaxation term to force the model over some time interval towards observed values, and 'blending', 'optimal interpolation' or 'Kalman filter', where model states and observed fields are melded together in suboptimal or optimal ways, sequentially at each timestep that has new data. One can also formulate the 'representer method' (Bennett, 2002) in a weak constraints formalism in which the artificial forcing field is determined for the entire state at each time step. More sophisticated techniques for predicting the error fields of the model, even non-linearly, have also been developed, such as the error-subspace estimation (ESSE) technique of Lermusiaux and Robinson (1999).

Among the strong constraints techniques are '4D variational assimilation' (4DVAR) in which a cost function (generalized data mismatch) is minimized based on assessing its curvature and directing the correction to the state vector downgradient until convergence is reached. This can be accomplished with the adjoint of the forward ocean model, which evaluates the gradient for the entire state vector, or with Green's functions, which evaluates the gradient for only a portion of the state vector variance, or with the Kalman smoother. The representer technique can also be formulated in a strong constraints framework.

4.2.1 *Case 1: CalCOFI and the Southern California Bight*

The Southern California Bight (SCB) encompasses part of the southern California Current System (CCS) and is an especially data-rich region because the California Cooperative Oceanic Fisheries Investigations (CalCOFI) Program has been collecting physical-biological data there for over 50 years (e.g. Chelton et al. 1982; Roemmich 1992; Hickey 1993; Roemmich and McGowan 1995; Hickey 1998; McGowan et al., 2003). The non-synoptic time and space resolution (roughly 1 month and 70 km, respectively, for a typical cruise track) of this in situ sampling, however, is inadequate to properly resolve the vigorous mesoscale circulation features in the region. And it is precisely these features that control the dominant biological and physical changes via localized upwelling cells, meandering fronts and filaments, and thermocline eddies (Strub and James, 2000; Swenson and Niiler, 1996).

Remotely sensed sea level height, SST, and ocean color provide a more detailed view of the region, but are limited in that they sample only surface features, have limited resolution, and do not extend right up to the coast. Combining the in situ and satellite data with data assimilation techniques in ocean models of this region is a perfect marriage of data and technique.

The CalCOFI field and satellite data has been used in strong constraints data assimilation strategies (a Green's function inverse method) to examine the short-term evolution of mesoscale features, e.g., within the time span of a single CalCOFI cruise (Miller et al., 2000; Di Lorenzo et al., 2004). With the advent of many new techniques for obtaining quasi-synoptic, high space and time resolution observations (e.g., CODAR estimates of surface currents, drifter platforms for salinity, atmospheric pressure and biological measurements, subsurface glider measurements of the upper ocean, etc.), it is of great importance now to develop procedures that can make use of this combined suite of observations to synthesize a complete picture of coupled physical-biological activity. An adjoint model 4DVAR approach with ROMS (Moore et al., 2004) is now being tested with CalCOFI and satellite data sets in the SCB. While adjoint data assimilation techniques have been widely used in 1D biogeochemical model applications, they have yet to be applied to 4D coupled models. This is primarily due to the large computational costs associated with carrying out numerous model runs.

Several practical issues of assimilating satellite data into the flow fields of the SCB were addressed in an identical twin experiment in which a model run was used to create synthetic data that are sampled and treated like observations (Fig. 5c). The relative importance of using hydrographic data and TOPEX altimetry was assessed in these fits using a Green's function inverse method (Miller and Cornuelle, 1999). The ocean model was first run from an initial condition derived from an objective analysis of the synthetic hydrographic data. This forward run was then sampled at the data points to determine an initial model-data mismatch with the synthetic hydrographic data. The linear inverse method was then used to correct the initial conditions and the model was re-run from this new starting point (Fig. 5a,b). Table 6 shows the linearly predicted error variance reduction and the actual variance reduction when running the fully non-linear model.

The correction using only hydrography as a constraint on the flow reduced the hydrographic error variance (model-data mismatch) by roughly 60%. The correction using both hydrography and TOPEX altimetry improved the mismatch with sea level only in the areas where no hydrographic data was available. It failed to improve the mismatch in the region with the synthetic hydrographic data, showing the importance of subsurface information on constraining the total flow field.

Smaller-scale structures that are not sampled by either sampling scheme could not be accounted for by the inverse solution. The inverse solution was only able to reconstruct the well-sampled larger-scale features of the eddy field, not the smaller scale eddies. These larger scale features were, however, dynamically important because they produced realistic large-scale flow fields which have predictive skill at leads of several months.

Table 6. Error variance reduction in identical twin data assimilation experiments

Expected reduction variance		True non-linear variance reduction	
CASE T, S (Fig. 1a)			
Total	65%	Total	61%
Salinity	58%	Salinity	45%
Temperature	61%	Temperature	56%
CASE T, S, SSH (Fig. 2a)			
Total	89%	Total	71%
Salinity	50%	Salinity	35%
Temperature	58%	Temperature	50%

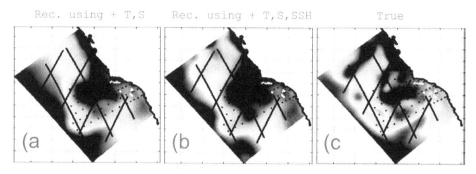

Figure 5. Sea level maps from an identical twin data assimilation experiment corresponding to a CalCOFI cruise. Three-week CalCOFI cruise sampling of hydrography is marked by dots. 10-day repeat cycle of TOPEX altimetry sampling is marked by solid lines. (a) Average sea level over the there week period for the model run from initial conditions corrected by the inverse technique assimilating only hydrography data. (b) As in (a) but assimilating both hydrography and altimetric sea level. (c) The "true" sea level pattern from the base run. Note the small-scale structures in (c) that are unable to be measured by CalCOFI or TOPEX.

4.2.2 *Case 2: New Jersey Long-term Ecosystem Observatory (LEO)*

Using circulation models to carry out ocean forecasts has been the goal of the U.S. Navy's Operational Forecast effort for some time. A recent review of the state of this effort is presented in a special issue of Oceanography magazine (Oceanography, Vol. 15(1), 2002). Research into developing methodologies for incorporation of real time data sets into these models is ongoing. As the Integrated Ocean Observing System (Ocean.US, 2002) continues to develop and *in situ* data sets become more readily available—especially in real-time, the need for and utility of coastal ocean forecasting capabilities will increase.

A coastal forecasting effort that uses satellite data and *in situ* coastal ocean observations from the Long-term Ecosystem Observatory (LEO) on the New Jersey coast has recently demonstrated the capability of a regional coastal ocean model to assimilate data and generate model forecasts in support of real-time adaptive sampling strategies (Wilkin et al., 2004). In this effort, the Regional Ocean Modeling System (ROMS) model, a 3D ocean circulation model, was used in conjunction with the U.S. Navy's Coupled Ocean Atmosphere Mesoscale Prediction System (COAMPS) to provide high-resolution model solutions of the circulation field, and ocean temperature and salinity. The ocean circulation model region included the New York Bight and New Jersey shelf with an average horizontal grid resolution of 1 km and a higher resolved (300 m) grid region of 30 km by 30 km in the vicinity of the LEO observational area. Details on the configuration, forcing, and boundary conditions are presented by Wilkin et al. (2004).

The model was used with real time data from the annual Coastal Predictive Skill Experiments that occurred between 1998 and 2001. The Coastal Predictive Skill Experiments were a series of modeling and field experiments that incorporated the model forecasts into the decision-making processes for scheduling/developing the ship-based field survey for subsequent field campaigns. In addition to the shipboard data sets that provided temperature, salinity and density profiles from CTD deployments, high resolution and long-range radar (CODAR) systems provided surface current estimates, and satellites provided surface maps of temperature that were available for assimilation into the model. Additional observations of temperature from thermistors placed onto moorings, and horizontal current profiles from several bottom-mounted Acoustic Doppler Current Profilers (ADCPs) were used as independent observations to validate the model predictions.

Two data assimilation techniques, nudging and simple sub-optimal intermittent melding, were tested for generation of the 3-day forecasts that were made available to the oceanographic field survey team. Data nudging methods simply push the model solution towards the observations over some given time scale (e.g. equation 13.5). Data melding methods use weighted sums of the objectively mapped observations and forecasts to re-initialize the model at given periods in time. More recent data assimilation developments to ROMS now allow for use of adjoint and tangent linear data assimilation methods (Moore et al., 2004).

Because CODAR data sets only provide information on surface current, assimilation of these observations can introduce significant vertical shear into the horizontal velocity fields, and can severely hamper forecast skill. In order to reduce this impact, Wilkin et al. (2004) used a modified projection scheme based on the correlations between the CODAR surface data and the ADCP current profiles. This vertical extrapolation method provides a statistically based approach to extend the CODAR data sets throughout the water column and reduce the introduction of unwanted vertical sheer. While Wilkin et al. (2004) also assimilated satellite derived SST, the effect of assimilating surface scalar values such as SST may not be as critical as it is for assimilation of momentum data. In fact, SST observations have long been used in the modeling community for alternative air-sea boundary conditions when heat flux estimates are not available.

The use of real time observations to support ocean forecast efforts is a recent development for oceanographic research and field support applications. Of the two data assimilation methods tested by Wilkin et al. (2004) the assimilation by data melding provided model solutions with greater forecast skill. The introduction of increasingly sophisticated 3D data assimilation techniques (Moore et al., 2004), improved methods to adequately assimilate surface ocean observations, and increases in coastal ocean observations will support improvement of data assimilation applications in coastal regions, e.g. increased forecast skill and duration times.

5. Future Directions

As computational capabilities grow so too will the ability to carry out more effective modeling and data assimilation studies. However, a number of critical areas need to be further developed for coastal ocean modeling and data assimilation efforts to evolve to a level that can support operational and forecast applications.

Because of the complexity of coastal regions, modeling studies are crucial tools for the synthesis of available knowledge (ie, data), for developing and testing new theories and challenging old paradigms, and for providing decision making tools to coastal managers. A coastal ocean observation program is currently being developed (Ocean.US, 2002) that call for the parallel development of coastal modeling efforts. The use of Observation System Simulation Experiments (OSSEs) needs to be encouraged for support of any observation development program effort. For instance, 3D coastal models have already been used to demonstrate the unlikelihood—due to prohibitively high costs—of using ocean moorings to estimate cross-shelf fluxes of carbon from coastal regions (Fig. 6).

There is no doubt that for biogeochemical modeling and data assimilation efforts, data limitation will remain a reality. Even with successful implementation of an ocean observing system, the ocean will remain under-sampled. Because of this, care must be taken in development of any observational system so that it is used to provide the most benefit within the limited available funds. Modeling studies can provide insight in how such an observation system might be developed. There has been some discussion about the use of Observational Systems Simulation Experiments (OSSEs) and their role in validating data assimilation techniques under specific present or proposed observing systems. An additional role of these OSSEs should be to link them with data assimilation "twin experiments" to design an optimal observation system that uses a cost-function based upon both the error estimates and the actual costs (in dollars) for the specific observing system. The goal with such an exercise would be to minimize both error from the modeling effort and financial costs of the observation effort.

Assimilation of ocean color satellite observations into coastal ocean biogeochemical models is still in its infancy. This is partly due to the lack of adequately developed coastal ocean color products. Coastal regions contain a diverse suite of biogeochemical

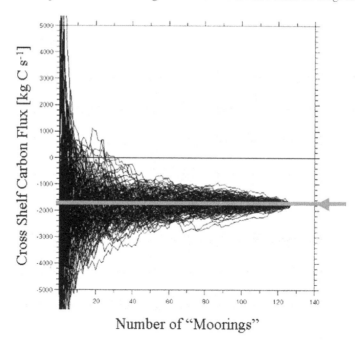

Number of "Moorings"

Figure 6. A composite plot of a suite of cross-shelf carbon flux estimates made by randomly choosing an increasing number of grid points "simulated moorings" from a 3D coupled circulation/biogeochemical model (Stolzenbach et al., 2004) to calculate/estimate the flux. As the number of "moorings" increases, the estimate approaches the model solution (thick gray bar with arrow). Of interest to observational oceanographers is the large range and sign changes observed at low (~5)—yet typical—numbers of "moorings."

processes, and the bio-optical signature of the allochthonous materials (typically CDOM and sediments) varies in space and time. Because of this, regional algorithms should be developed that use the available higher-resolution (LAC) data sets and Case 2 ocean color algorithms that have been optimized or developed for specific coastal regions. Doing so would provide an improved ocean color data set that modeling studies could then use for validation.

All of the model/algorithm applications presented within this chapter have been developed primarily using subjectively defined or chosen algorithms or equations. There are a number of new applied math techniques that will allow for objective development of model equations or algorithms. Several of these (e.g., neural networks, genetic algorithms, fuzzy logic) have been presented in Ressom et al. (Chapter 9). As these non-subjective techniques become available, new ocean color algorithms and model equations will be developed that are based relationships that have been optimization using data sets. These new techniques will allow for the optimization of both parameters and model equations, thereby removing the subjectivity that now exists in developing model equations.

While open source code development has become popularly supported within the computer software community, it continues to be less accepted within the science community. There is no doubt that open source code development provides an optimized path for creating software applications (e.g. models). It is of benefit to both the science funding agencies and associated scientists to adopt this manner of code/model development. What will be interesting to observe is the path that is taken to develop and support such an effort.

6. Acknowledgements

This research was supported by the National Aeronautics and Space Agency (NASA) under project number NAGW-3128. AJM and EDL were support by ONR (N00014-99-1-0045), NASA (NAG5-9788) and NOAA (NA17RJ1231 through ECPC).

7. References

Abbott, M. R. and D. B. Chelton. 1991. Advances in passive remote sensing of the ocean, *Review of Geophysics*, Supplement:571-589.

Abbott, M. R., P. J. Richerson, and T. M. Powell. 1982. In situ response of phytoplankton fluorescence to rapid variations in light, Limnology and Oceanography, 27:218-225.

Abbott, M. R. and R. M. Letelier. 1998. Decorrelation scales of chlorophyll as observed from bio-optical drifters in the California Current, Deep-Sea Research, 45:1639-1668.

Anderson, L. A., A. R. Robinson, and C. J. Lozano. 2000. Physical and biological modeling in the Gulf Stream region: 1. Data assimilation methodology, Deep-Sea Research I, 47:1787-1827.

Anderson, L. A. and A. R. Robinson. 2001. Physical and biological modeling in the Gulf Stream region Part II. Physical and biological processes, Deep-Sea Research I, 48:1139-1168.

Anderson, T. R. and P. J. le B. Williams. 1998. Modelling the seasonal cycle of dissolved organic carbon at Station E1 in the English Channel, Estuarine, Coastal and Shelf Science, 46:93-109.

Antoine, D. and A. Morel. 1995. Modelling the seasonal course of the upper ocean pCO_2 (2) Validation of the model and sensitivity studies, Tellus, 47B:103-121.

Balch, W. M., R. W. Eppley, and M. R. Abbott. 1989a. Remote sensing of primary production—I. A comparison of empirical and semi-analytical algorithms, Deep-Sea Research, 36:281-295.

Balch, W. M., R. W. Eppley, and M. R. Abbott. 1989b. Remote sensing of primary production—II. A semi-analytic algorithm based on pigments, temperature and light, Deep-Sea Research, 36:1201-1217.

Balch, W., R. Evans, J. Brown, G. Feldman, C. McClain, and W. Esaias. 1992. The remote sensing of ocean primary productivity: Use of new data compilation to test satellite algorithms, Journal of Geophysical Research, 97:2279-2293.

Balch, W. M. and C. F. Byrne. 1994. Factors affecting the estimate of primary production from space, Journal of Geophysical Research, 99:7555-7570.

Balzer, W., W. Helder, E. Epping, L. Lohse, and S. Otto. 1998. Benthic denitrification and nitrogen cycling at the slope and rise of the N. W. European Continental Margin (Goban Spur), Progress in Oceanography, 42:111-126.

Behrenfeld, M. J. and P. G. Falkowski. 1997a. A consumer's guide to phytoplankton primary productivity models, Limnology and Oceanography, 42:1479-1491.

Behrenfeld, M. J. and P. G. Falkowski. 1997b. Photosynthetic rates derived from satellite-based chlorophyll concentration, Limnology and Oceanography, 42:1-20.

Bennett, A. F., 1992: Inverse Methods in Physical Oceanography. Monographs on Mechanics and Applied Mathematics. Cambridge University Press, 346 pp., reprinted 1999, 2003.

Bennett, A. F., 2002. Inverse Modeling of the Ocean and Atmosphere. Cambridge University Press, 234 pp.

Bernstein, R. L., L. Breaker, and R. Whritner. 1997. California current eddy formation: Ship, air, and satellite results, Science, 195:353-359.

Bissett, W. P., J. J. Walsh, D. A. Dieterle, and K. L. Carder. 1999a. Carbon cycling in the upper waters of the Sargasso Sea: I. Numerical simulation of differential carbon and nitrogen fluxes, Deep-Sea Research I, 46:205-269.

Bissett, W. P., K. L. Carder, J. J. Walsh, and D. A. Dieterle. 1999b. Carbon cycling in the upper waters of the Sargasso Sea: II. Numerical simulation of differential carbon and nitrogen fluxes, Deep-Sea Research I, 46:271-317.

Bissett, W. P., H. Arango, R. Arnone, R. Chant, S. Glenn, C. Mobley, M. A. Moline, O. M. Schofield, R. Steward, and J. Wilkin. 2004. The prediction of hyperspectral inherent optical properties in the New Jersey Bight, In Prep.

Broecker, W. S. and T.-H. Peng. 1982. Tracers in the Sea, Lamont-Doherty Geological Observatory, Columbia University, 690 pp.

Brown, O. B., R. H. Evans, J. W. Brown, H. R. Gordon, R. C. Smith, and K. S. Baker. 1985. Phytoplankton blooming off the U.S. East Coast: A satellite description, Science, 229:163-167.

Carlson, C. A., H. W. Ducklow, and T. D. Sleeter. 1996. Stocks and dynamics of bacterioplankton in the northwestern Sargasso Sea, Deep-Sea Research II, 43:491-515.

Campbell, J., D. Antoine, R. Armstrong, K. Arrigo, W. Balch, R. Barber, M. Behrenfeld, R. Bidigare, J. Bishop, M.-E. Carr, W. Esaias, P. Falkowski, N. Hoepffner, R. Iverson, D. Kiefer, S. Lorenz, J. Marra, A. Morel, J. Ryan, V. Vedernikov, K. Waters, C. Yentch, and J. Yoder. 2002. Comparison of algorithms for estimating ocean primary production from surface chlorophyll, temperature, and irradiance, Global Biogeochemical Cycles, 16(3), 10.1029/2001GB001444.

Carder, K. L., F. R. Chen, Z. P. Lee, S. K. Hawes, and D. Kamykowski. 1999. Semianalytic moderate-resolution algorithms for chlorophyll a and absorption with bio-optical domains based on nitrate-depletion temperatures, Journal of Geophysical Research, 104:5403-5421.

Chelton, D. B., P. A. Bernal, and J. A. McGowan. 1982. Large-scale interannual physical and biological interaction in the California Current, Journal of Marine Research, 40:1095-1125.

Christian, J. R., M. A. Verschell, R. Murtugudde, A. J. Busalacchi, and C. R. McClain. 2002a. Biogeochemical modelling of the tropical Pacific Ocean. I: Seasonal and interannual variability, Deep-Sea Research II, 49:509-543.

Christian, J. R., M. A. Verschell, R. Murtugudde, A. J. Busalacchi, and C. R. McClain. 2002b. Biogeochemical modelling of the tropical Pacific Ocean. II: Iron biogeochemistry, Deep-Sea Research II, 49:545-565.

Di Lorenzo, E., A. J. Miller, D. J. Neilson, B. D. Cornuelle, and J. R. Moisan. 2004. Modeling observed California Current mesoscale eddies and the ecosystem response, International Journal of Remote Sensing, 25:1307-1312.

Doerffer, R. and J. Fischer. 1994. Concentrations of chlorophyll, suspended matter, and gelbstoff in case II waters derived from satellite coastal zone color scanner data with inverse modeling methods, Journal of Geophysical Research, 99:7457-7466.

Doney, S. C., D. M. Glover, and R. J. Najjar. 1996. A new coupled, one-dimensional biological-physical model for the upper ocean: Applications to the JGOFS Bermuda Atlantic Time-series Station (BATS) site, Deep-Sea Research II, 43:591-624.

Eppley, R. W., E. Stewart, M. R. Abbott, and U. Heyman. 1985. Estimating primary production from satellite chlorophyll. Introduction to regional differences and statistics for the Southern California Bight, Journal of Plankton Research, 7:57-70.

Falkowski, P. G., M. J. Behrenfeld, W. E. Esaias, W. Balch, J. W. Campbell, R. L. Iverson, D. A. Kiefer, A. Morel, and J. A. Yoder. 1998. Satellite primary productivity data and algorithm development: A science plan for Mission to Planet Earth, SeaWiFS Tech. Rep. Ser., Vol. 42, S. F. Hooker (ed.). NASA/TM-1998-1045566, pp. 36.

Fasham, M. J. R., H. W. Ducklow, and S. W. McKelvie. 1990. A nitrogen-based model of plankton dynamics in the ocean mixed layer, Journal of Marine Research, 48:591-639.

Fasham, M. J. R. 1995. Variations in the seasonal cycle of biological production in subarctic ocean: A model sensitivity analysis, Deep-Sea Research I, 42:1111-1149.

Fasham, M. J. R. and G. T. Evans. 1995. The use of optimization techniques to model marine ecosystem dynamics at the JGOFS station at 47° N 20° W, Philosophical Transaction of the Royal Society of London B, 348:203-209.

Fasham, M. J. R., P. W. Boyd, and G. Savidge. 1999. Modeling the relative contributions of autotrophs and heterotrophs to carbon flow at a Lagrangian JGOFS station in the Northeast Atlantic: The importance of DOC, Limnology and Oceanography, 44:80-94.

Franks, P. J. S. 1995. Coupled physical-biological models in oceanography, Reviews in Geophysics, Supplement:1177-1187.

Friedrichs, M. A. M. 2001. A data assimilative marine ecosystem model of the central equatorial Pacific: Numerical twin experiments. Journal of Marine Research, 59:859-894.

Fox, D. N., W. J. Teague, C. N. Barron, M. R. Carnes, and C. M. Lee. 2001. The Modular Ocean Data Assimilation System (MODAS), Journal of Atmospheric and Oceanic Technology, 19:240-252.

Gallaway, J. N., R. W. Howarth, A. F. Michaels, S. W. Nixon, J. M. Prospero, and F. J. Dentener. 1996. Nitrogen and phosphorus budgets of the North Atlantic Ocean and its watershed, Biogeochemistry, 35:3-25.

Garver, S. A. and D. A. Siegel. 1997. Inherent optical property inversion of ocean color spectra and its biogeochemical interpretation. 1. Time series from the Sargasso Sea, Journal of Geophysical Research, 102:18,607-18,625.

Geider, R. J. 1987. Light and temperature dependence of the carbon to chlorophyll ratio in microalgae and cyanobacteria: implications for physiology and growth of phytoplankton. New Phytology, 106:1-34.

Gill, A. E. and P. P. Niiler. 1973. The theory of seasonal variability in the ocean, Deep-Sea Research, 20:141-177.

Gordon, H. R. and D. K. Clark. 1980. Clear water radiances for atmospheric correction of coastal zone color scanner imagery, Applied Optics, 20:4175-4180.

Gordon, H. R., O. B. Brown, R. H. Evans, J. W. Brown, R. C. Smith, K. S. Baker, and D. K. Clark. 1988. A semi-analytical radiance model of ocean color, Journal of Geophysical Research, 93:10,909-10,924.

Gordon, D. C., P. R. Boudreaum K. H. Mann, J. –E. Ong, W. L. Silvert, S. V. Smith, G. Wattayakorn, F. Wulff, and T. Yanagi. 1996. LOICZ Biogeochemical Modeling Guidelines, LOICZ/R&S/95-5, VI +96 pp. LOICZ, Texel, Netherlands.

Gregg, W. W. and J. J. Walsh. 1992. Simulation of the 1979 spring bloom in the Mid-Atlantic Bight: A coupled physical/biological/optical model, Journal of Geophysical Research, 97:5723-5743.

Gross, L., S. Thiria, R. Frouin, and B. G. Mitchell. 2000. Artificial neural networks for modeling the transfer function between marine reflectance and phytoplankton pigments concentration, Journal of Geophysical Research, 105:3483-3495.

Haidvogel, D. B. and A. Beckmann. 1999. Numerical Ocean Circulation Modeling, Imperial College Press, London, 320 pp.

Harmon, R. and P. Challenor. 1997. A Markov chain Monte Carlo method for estimation and assimilation into models, Ecological Modeling, 101:41-59.

Hecky, R. E. and P. Kilman. 1984. Nutrient limitation of phytoplankton in freshwater and marine environments: A review of recent evidence on the effects of enrichment, Limnology and Oceanography, 33:796-822.

Hickey, B.M. 1993. Physical oceanography. In: Marine Ecology of the Southern California Bight, Hood, D. (ed.) Pergamon Press.

Hickey, B.M. 1998. Coastal Oceanography of Western North America from the tip of Baja California to Vancouver Is., In: Volume 11, Chapter 12, The Sea, K.H. Brink and A.R. Robinson (eds.), pp. 345-393, Wiley and Sons, Inc.

Hofmann, E. E. and J. W. Ambler. 1988. Plankton dynamics on the outer southeastern U. S. continental shelf. Part II: A time-dependent biological model, Journal of Marine Research, 46:883-917.

Hofmann, E. E. and C. M. Lascara. 1998. Overview of interdisciplinary modeling for marine ecosystems, in The Sea, Vol. 10, K. H. Brink and A. R. Robinson (eds.), John Wiley and Sons Ltd., pp 507-540.

Hoge, F. E., C. W. Wright, P. E. Lyon, R. N. Swift, and J. K. Yungel. 1999. Satellite retrieval of inherent optical properties by inversion of an oceanic radiance model: a preliminary algorithm, Applied Optics, 38: 495-540.

Hoge, F. E., C. W. Wright, P. E. Lyon, R. N. Swift, and J. K. Yungel. 2001. Inherent optical properties imagery of the western North Atlantic Ocean: Horizontal spatial variability of the upper mixed layer, Journal of Geophysical Research, 106:31,129-31,140.

Howard, K. L., and J. A. Yoder. 1997. Contribution of the subtropical ocean to global primary production, In Space Remote Sensing of the Subtropical Oceans, C. T-. Liu (Ed.), Pergamon Press, New York, 157-168.

Hurtt, G. C. and R. T. Armstrong. 1996. A pelagic ecosystem model calibrated with BATS data, Deep-Sea Research II, 43:653-683.

Hurtt, G. C. and R. T. Armstrong. 1999. A pelagic ecosystem model calibrated with BATS and OWSI data, Deep-Sea Research II, 46:27-61.

Iverson, R. L., W. E. Esaias, and K. R. Turpie. 2000. Ocean annual phytoplankton carbons and new production, and annual export production estimated with empirical equations and CZCS data, Global Change Biology, 6:57-72.

Kiefer, D. A. and B. G. Mitchell. 1983. A simple steady-state description of phytoplankton growth based on absorption cross-section and quantum efficiency, Limnology and Oceanography, 28:770-776.

King, M. D., J. Closs, S. Spangler, and R. Greenstone (eds.). 2003. EOS Data Products Handbook, Vol. 1, NASA Goddard Space Flight Center, pp. 225.

Kwiatowska, E. J. and G. S. Fargion. 2003. Merger of ocean color data from multiple satellite missions within the SIMBIOS project, In: Ocean Remote Sensing and Applications, Frouin R. J., Y. Yuan, and H. Kawamura (eds.), Proceedings of the Society of Photo-Optical Engineering (SPIE), 4892:168-182.

Lawson, L. M., Y. H. Spitz, E. E. Hofmann, and R. B. Long. 1995. A data assimilation technique applied to a predator-prey model, Bulletin of Mathematical Biology, 57:593-617.

Lawson, L. M., E. E. Hofmann, and Y. H. Spitz. 1996. Time series sampling and data assimilation in a simple marine ecosystem model, Deep-Sea Research II, 43:625-651.

LeBaron, P., P. Servais, H. Agogué, C. Courties, and F. Joux. 2001. Does the high nucleic acid content of individual bacterial cells allow us to discriminate between active cells and inactive cells in aquatic systems? Applied Environmental Microbiology 67:1775-1782.

Lermusiaux, P. F. J. and A. R. Robinson. 1999. Data assimilation via error subspace statistical estimation. Part I: Theory and schemes, Monthly Weather Review, 127:1385-1407.

Lewis, E. and D. W. R. Wallace. 1998. Program Developed for CO2 System Calculations. ORNL/CDIAC-105. Carbon Dioxide Information Analysis Center, Oak Ridge National Laboratory, U.S. Department of Energy, Oak Ridge, Tennessee.

Liu, K.-K., S. –Y. Chao, P. –T. Shaw, G. C. Gong, C. C. Chen, and T. Y. Tang. 2002. Monsoon forced chlorophyll distribution and primary productivity in the South China Sea: observations and a numerical study. Deep-Sea Research Part I, 49:1387-1412.

Lorenzen, C. J. 1970. Surface chlorophyll as an index of the depth, chlorophyll content and primary productivity of the euphotic layer, Limnology and Oceanography, 15:479-480.

Lozano, C. J., A. R. Robinson, H. G. Arango, A. Gangopadhyay, Q. Sloan, P. J. Haley, L. Anderson, and W. Leslie. 1996. An interdisciplinary ocean prediction system: Assimilation strategies and structured data sets, In: Modern Approaches to Data Assimilation in Ocean Modeling, P. Malanotte-Rizzoli (ed.), pg. 413-452, Elsevier Science, B. V.

Lynch, D. R., J. T. C. Ip, C. E., Naimie, and F. E. Werner. 1996. Comprehensive circulation model with application to the Gulf of Maine, Continental Shelf Research, 16:875-906.

Mantua, N, D. Haidvogel, Y. Kushnir, and N. Bond. 2002. Making the climate connections: Bridging scales of space and time in the U. S. GLOBEC Program, Oceanography, 15(2):75-87.

Marchesiello, P., J. C. McWilliams, and A. Shchepetkin. 2001. Open boundary conditions for long-term integration of regional oceanic models, Ocean Modelling, 3:1-20.

Marchesiello, P., J. C. McWilliams, and A. Shchepetkin. 2002. Equilibrium structure and dynamics of the California Current System, Journal of Physical Oceanography, 33, Sub judice.

Maritorena, S., D. A. Siegel, and A. R. Peterson. 2002. Optimization of a semianalytical ocean color model for global-scale applications, Applied Optics, 41(15): 2705-2714.

Matear, R. J. 1995. Parameter optimization and analysis of ecosystem models using simulated annealing: A case study at Station P, Journal of Marine Research, 53:571-607.

Matear, R. J. and G. Holloway. 1995. Modeling the inorganic phosphorus cycle of the North Pacific using an adjoint data assimilation model to assess the role of dissolved organic phosphorus, Global Biogeochemical Cycles, 9:101-119.

McClain, C. R., K. Arrigo, K.-S. Tai, and D. Turk. 1996. Observations and simulations of physical and biological processes at ocean weather station P, 1951-1980, Journal of Geophysical Research, 101:3697-3713.

McGillicuddy, D. J., J. J. McCarthy, and A. R. Robinson. 1995. Coupled physical and biological modeling of the spring bloom in the North Atlantic (I): model formulation and one-dimensional bloom processes, Deep-Sea Research Part I, 42:1313-1357.

McGillicuddy, D. J., D. R. Lynch, A. M. Moore, W. C. Gentelman, C. S. Davis, and C. J. Meise. 1998. An adjoint data assimilation approach to diagnosis of physical and biological controls of Pseudocalanus spp. in the Gulf of Maine—Georges Bank region, Fisheries Oceanography, 7:205-218.

McGowan, J. A., S. J. Bograd, R. J. Lynn, and A. J. Miller. 2003. The biological response to the 1977 regime shift in the California Current, Deep-Sea Research II, Sub judice.

Miller, A. J. and B. D. Cornuelle. 1999. Forecasts from fits of frontal fluctuations, Dynamics of Atmospheres and Oceans, 29:305-333.

Miller, A. J., E. Di Lorenzo, D. J. Neilson, B. D. Cornuelle and J. R. Moisan. 2000. Modeling CalCOFI observations during El Nino: Fitting physics and biology. California Cooperative Oceanic Fisheries Investigations Reports, 41:87-97.

Moisan, J. R. 1993. Modeling nutrient and plankton processes in the California Coastal Transition Zone. Ph.D. thesis, Old Dominion University, Norfolk, VA., 214 pp.

Moisan, J. R. and E. E. Hofmann. 1996. Modeling nutrient and plankton processes in the California Coastal Transition Zone. 1. A time- and depth-dependent model, Journal of Geophysical Research, 101:22,647-22,676.

Moisan, J. R., E. E. Hofmann, and D. B. Haidvogel. 1996. Modeling nutrient and plankton processes in the California Coastal Transition Zone. 2. A three-dimensional physical-bio-optical model, Journal of Geophysical Research, 101:22,677-22,691.

Moisan, J. R. and P. P. Niiler. 1998. The seasonal heat budget of the North Pacific: Net heat flux and heat storage rates (1950-1990), Journal of Physical Oceanography, 28:401-421.

Moisan, J. R., T. K. Moisan, and M. R. Abbott. 2002. Modeling the effect of temperature on the maximum growth rates of phytoplankton populations, Ecological Modeling, 153:197-215.

Moisan, J. R., A. Shchepetkin, E. Di Lorenzo, P. Marchesiello, K. Stolzenbach, A. J. Miller, and J. C. McWilliams. 2004. Modeling the biogeochemical processes in the coastal ocean along the U.S. Continental Margin: Calibrations using one dimensional simulations, Journal of Geophysical Research, In Prep.

Moisan, T. A. and B. G. Mitchell. 1999. Photophysiological acclimation of *Phaeocystis antarctica* Karsten under light limitation, Limnology and Oceanography, 44:247-258.

Moloney, C. L., J. G. Field, and M. I. Lucas. 1991. The size-based dynamics of plankton food webs. II. Simulation of three contrasting southern Benguela food webs, Journal of Plankton Research, 13:1039-1092.

Moore, A. M., H. G. Arango, E. Di Lorenzo, B. D. Cornuelle, A. J. Miller, and D. J. Nelson. 2004. A comprehensive ocean prediction and analysis system based on the tangent linear and adjoint components of a regional ocean model, Ocean Modeling, 7:227-258.

Mullon, C., P. Cury, and P. Penven. 2002. Evolutionary individual-based model for the recruitment of anchovy (*Engraulis capensis*) in the southern Bengula, Canadian Journal Fisheries and Aquatic Science, 59:910-922.

Nixon, S. W., J. W. Ammerman, L. P. Atkinson, V. M. Berounsky, G. Billen, W. C. Boicourt, W. R. Boynton, T. M. Church, D. M. Ditoro, R. Elmgren, J. H. Garber, A. E. Giblin, R. A. Jahnke, N. J. P. Owens, M. E. Q. Pilson, and S. P. Seitzinger. 1996. The fate of nitrogen and phosphorus at the land-sea margin of the North Atlantic Ocean. Biogeochemistry, 35:141-180.

Ocean.US. 2002. An Integrated and Sustained Ocean Observing System (IOOS) for the United States: Design and Implementation, Ocean.US, Arlington, VA, 21 pp.

Oguz, T., P. Malanotte-Rizzoli, and D. Aubrey. 1995. Wind and thermohaline circulation of the Black Sea driven by yearly mean climatological forcing, Journal of Geophysical Research, 100:6845-6863, 1995.

Oguz, T., H. W. Ducklow, and P. Malanotte-Rizzoli. 2000. Modeling distinct vertical biogechemical structure of the Black Sea: Dynamical coupling of the oxic, suboxic, and anoxic layers, Global Biogeochemical Cycles, 14:1331-1352.

Oguz, T., P. Malanotte-Rizzoli, H. W. Ducklow, and J. W. Murray. 2002. Interdisciplinary studies integrating the Black Sea biogeochemistry and circulation dynamics, Oceanography, 15:4-11.

Olivieri, R. A. and F. P. Chavez. 2000. A model of plankton dynamics for the coastal upwelling system of Monterey Bay, California, Deep-Sea Research II, 47:1077-1106.

O'Reilly, J. E., S. Maritorena, B. G. Mitchell, D. A. Siegel, K. L. Carder, S. A. Garver, M. Kahru, and C. McClain. 1998. Ocean color chlorophyll algorithms for SeaWiFS, Journal of Geophysical Research, 103:24,937-24,953.

O'Reilly, J. E., and 21 co-authors. 2000. Ocean color chlorophyll a algorithms for SeaWiFS, OC2, and OC4: Version 4. Chapter 2 in SeaWiFS Post-launch Calibration and Validation Analyses, Part 3, SeaWiFS Post-launch Technical Memorandum Series, Vol. 11, NASA.

Orr, J. C. 1999. Ocean Carbon-Cycle Model Intercomparison Project (OCMIP): Pase 1 (1995-1997), GAIM Report 7, IGBP/GAIM Office, EOS, Univ. of New Hampshire, Durham, NH.

Penven, P., C. Roy, G. B. Brundrit, A. Colin de Verdiere, P. Freon, A. S. Johnson, J. R. E. Lutjeharms, and F. A. Shillington. 2001. A regional hydrodynamic model of upwelling in the Southern Benguela, South African Journal of Science, 97:1-4.

Pickard, G. L., and W. J. Emery. 1990. Descriptive Physical Oceanography, An Introduction, Butterwork Heinmann, Oxford, 320 pp.

Platt, T. 1986. Primary production of the ocean water column as a function of surface light intensity: algorithms for remote sensing, Deep-Sea Research, 33:149-163.

Pontius, R. G. 2002. Statistical methods to partition effects of quantity and location during comparison of categorical maps at multiple resolution, Photogrammetric Engineering and Remote Sensing, 68:1041-1049.

Prospero, J. M., K. Barrett, T. Church, F. Dentener, R. A. Duce, J. N. Galloway, H. Levy II, J. Moody, and P. Quinn. 1996. Atmospheric deposition of nutrients to the North Atlantic Basin, Biogeochemistry, 35:27-73.

Prunet, P., J.-F. Minster, D. Ruiz-Pino, and I. Dadou. 1996a. Assimilation of surface data into a one-dimensional physical-biogeochemical model of the surface ocean. 1. Method and preliminary results, Global Biogeochemical Cycles, 10:111-138.

Prunet, P., J.-F. Minster, V. Echevin, and I. Dabou. 1996b. Assimilation of surface data into a one-dimensional physical-biogeochemical model of the surface ocean. 2. Adjusting a simple trophic model to chlorophyll, temperature, nitrate and pCO_2 data, Global Biogeochemical Cycles, 10:139-158.

Robinson, A. R. 1996. Physical processes, field estimation and an approach to interdisciplinary ocean modeling, Earth-Science Reviews, 40:3-54.

Robinson, A. R., H. Arango, A. Warn-Varnas, W. G. Leslie, A. J. Miller, P. J. Haley, and C. J. Lozano. 1996. Real-time regional forecasting, In: Modern Approaches to Data Assimilation in Ocean Modeling, P. Malanotte-Rizzoli (ed.), pg. 377-410, Elsevier Science, B. V.

Robinson, A. R., and P. F. J. Lermusiaux (eds.). 2000. Workshop on the assimilation of biological data in coupled physical/ecosystem models, GLOBEC Special Contribution 3, 152 pp.

Roemmich, D. 1992. Ocean warming and sea level rise along the southwest U.S. coast, Nature, 257:373-375.

Roemmich, D. and J. McGowan. 1995. Climate warming and the decline of zooplankton in the California Current, Science, 267:1324-1326.

Ryther, J. H. and C. S. Yentsch. 1957. The estimation of phytoplankton production in the ocean from chlorophyll and light data, Limnology and Oceanography, 2:281-286.

Seitzinger, S. P. and A. E. Giblin. 1996. Estimating denitrification in North Atlantic continental shelf sediments, Biogeochemistry, 35:235-260.

Semovski, S. V., B. Wozniak, and V. N. Pelevin. 1995. Multispectral ocean color data assimilation in a model of plankton dynamics, Studia I Materialy Oceanologia, Marine Physics, 68:125-147.

Semovski, S. V., and B. Wozniak. 1995. Model of the annual phytoplankton cycle in the marine ecosystem—assimilation of monthly satellite chlorophyll data for the North Atlantic and Baltic, Oceanologia, 37:3-31.

Shchepetkin, A. F., and J. C. McWilliams. 2003. The Regional Ocean Modeling System: A split-explicit, free-surface, topography following coordinates ocean model, unpublished manuscript.

Siegel, D. A., Maritorena, S., Nelson, N. B., Hansell, D. A., and Lorenzi-Kayser, M. 2003. Global distribution and dynamics of colored dissolved and detrital organic materials. Journal of Geophysical Research, 107(12):10.1029/2001JC000965.

Sieracki, M. E., E. M. Haugen, and T. L. Cucci. 1995. Overestimation of heterotrophic bacteria in the Sargasso Sea: direct evidence by flow and imaging cytometry, Deep-Sea Research I, 42:1399-1409.

Signorini, S. R., C. R. McClain, J. R. Christian, and C. S. Wong. 2001. Seasonal and interannual variability of phytoplankton, nutrients, TCO_2, pCO_2, and O_2 in the eastern subarctic Pacific (ocean weather station Papa), Journal of Geophysical Research, 106:31,197-31,215.

Smith, R. C. and K. S. Baker. 1978. The bio-optical state of ocean waters and remote sensing, Limnology and Oceanography, 23:247-259.

Smith, R. C. 1981. Remote sensing and depth distribution of ocean chlorophyll, Marine Ecology Progress Series, 5:359-361.

Smith, R. C., R. W. Eppley and K. S. Baker. 1982. Correlation of primary production as measured aboard ship in Southern California coastal waters and as estimated from satellite chlorophyll images, Marine Biology, 66:281-288.

Soetaert, K., J. J. Middelburg, P. M. J. Herman, and K. Buis. 2000. On the coupling of benthic and pelagic biogeochemical models, Earth-Science Reviews, 51:173-210.

Soetaert, K., P. M. J. Herman, J. J. Middelburg, C. Heip, C. L. Smith, P. Tett, and K. Wild-Allen. 2001. Numerical modelling of the shelf break ecosystem: reproducing benthic and pelagic measurements, Deep-Sea Research II, 48:3141-3177.

Spitz, Y. H., J. R. Moisan, M. R. Abbott, and J. G. Richman. 1998. Data assimilation and a pelagic ecosystem model: Parameterization using time series observations, Journal of Marine Systems, 16:51-68.

Spitz, Y. H., J. R. Moisan, M. R. Abbott, and J. G. Richman. 2001. Using data assimilation to configure a biogeochemical model of the Bermuda-Atlantic Time Series (BATS), Deep-Sea Research II, 48:1733-1768.

Stolzenbach, K. D., J. R. Moisan, H. Frenzel, N. Gruber, P. Marchesiello, J. C. McWilliams and J. Oram. 2004. Simulation of the plankton ecosystem in the California Current System, In Prep.

Strub, P. T., C. James, A. C. Thomas, and M. R. Abbott. 1990. Seasonal and nonseasonal variability of satellite-derived surface pigment concentration in the California Current, Journal of Geophysical Research, 95:11,501-11,530.

Strub, P. T. and C. James. 2000. Altimeter-derived variability of surface velocities in the California Current System: 2. Seasonal circulation and eddy statistics, Deep-Sea Research II, 47:831-870.

Swenson, M. S. and P. P. Niiler. 1996. Statistical analysis of the surface circulation of the California Current. Journal of Geophyical Research, 22:631-645.

Takahashi, T., R. E. Feely, R. F. Weiss, R. H. Wanninkhof, D. W. Chipman, S. C. Sutherland, and T. T. Takahashi. 1997. Global air-sea flux of CO_2: An estimate based on measurements of sea-air pCO_2 difference, Proceedings of the National Academy of Science, 94:8292-8299.

Talling, J. F. 1957a. Photosynthetic characteristics of some freshwater plankton in relation to underwater radiation, New Phytologist, 56:29-50.

Talling, J. F. 1957b. The phytoplankton population as a compound photosynthetic system. New Phytologist, 56:133-149.

Tanaka, A., M. Kishino, T. Oishi, R. Doerffer, and H. Schiller. 2000. Application of neural network to case II water, In: Remote Sensing of the Ocean and Sea Ice, Bostater, C. R., and R. Santoleri (eds.), Proceedings of the Society of Photo-Optical Engineering (SPIE), 4172:144-152.

Thomas, A. C. and P. T. Strub. 1989. Interannual variability in phytoplankton pigment distribution during the spring transition along the west coast of North America, Journal of Geophysical Research, 94:18,095-18,117.

Thomas, A. C. and P. T. Strub. 1990. Seasonal and interannual variability of pigment concentrations across a California Current frontal zone, Journal of Geophysical Research, 95:13,023-13,042.

Vallino, J. J. 2000. Improving marine ecosystem models: use of data assimilation and mesocosm experiments, Journal of Marine Research, 58:117-164.

Vichi, M. 2002. Predictability studies of coastal marine ecosystem behavior, Ph. D. Thesis, University of Oldenburg, Oldenburg, Germany.

Vichi, M., P. Oddo, M. Zavatarelli, A. Coluccelli, G. Coppini, M. Celio, S. Fonda Umani, and N. Pinardi. 2003. Calibration and validation of a one-dimensional complex marine biogeochemical flux model in different areas of the northern Adriatic shelf, Annales Geophysicae, 21:1-24.

Weingartner, T. J., K. Coyle, B. Finney, R. Hopcroft, T. Whitledge, R. Brodeur, M. Dagg, E. Farley, D. Haidvogel, L. Haldorson, A. Hermann, S. Hinckley, J. Napp, P. Stabeno, T. Kline, C. Lee, E. Lessard, T. Royer, S. Strom. 2002. The Northeast Pacific GLOBEC Program: Coastal Gulf of Alaska, Oceanography, 15(2):48-63.

Wilkin, J. L., H. G. Arango, D. B. Haidvogel, C. S. Lichtenwalner, S. M. Glenn, and K. S. Hedstrom. 2004. A regional ocean modeling system for the long-term ecosystem observatory, Journal of Geophysical Research, Sub judice.

Wunsch, C. 1996. The Ocean Circulation Inverse Problem. Cambridge University Press, 442 pp.

Zhang, T. L., F. Fell, Z. S. Liu, R. Preusker, J. Fischer, and M. X. He. 2003. Evaluating the performance of artificial neural network techniques for pigment retrieval from ocean color in Case I waters, Journal Geophysical Research, 108(C9):3286, 10.1029/2002JC001638.

Chapter 11

MONITORING BOTTOM SEDIMENT RESUSPENSION AND SUSPENDED SEDIMENTS IN SHALLOW COASTAL WATERS

[1]RICHARD L. MILLER, [2]BRENT A. MCKEE, AND [3]EURICO J. D'SA

[1]*National Aeronautics and Space Administration, Earth Science Applications Directorate, Stennis Space Center, MS, 39529 USA*
[2]*Department of Earth and Environmental Sciences, Tulane University, 208 Dinwiddie Hall, New Orleans, LA, 70118 USA*
[3]*Lockheed Martin Stennis Operations, Remote Sensing Directorate, Stennis Space Center, MS, 39529 USA*

1. Introduction

Shallow coastal waters, especially bays and estuaries, are often characterized by high turbidity generated by the resuspension of bottom sediments. Here, we define sediment as comprising both organic and inorganic materials. Bottom sediment resuspension may be caused by natural (e.g., wind or tides) or human-induced (e.g., dredging) events. The corresponding high concentration of total suspended matter (TSM) directly influences water quality, benthic (Miller and Cruise, 1995; Morton, 1996; Pilskaln et al., 1998) and phytoplankton productivity (Cloern, 1987; Wienieski, 1993; Millie et al., 2003; May et al., 2003; Schallenberg and Burns, 2004), and the redistribution and transport of pollutants and materials (Olsen et al., 1982; Simpson et al., 1998). As a result, there is considerable interest from a broad range of investigators in determining the occurrence and intensity of bottom resuspension events as well as monitoring the transport and fate of suspended sediments and associated materials.

The use of remote sensing technology to map suspended sediment concentration is well documented. Numerous investigators have established an empirical relationship between reflected solar radiance measured by airborne and satellite-based instruments with suspended sediments in a wide range of inland and coastal waters. For example, Miller and Cruise (1995) described the use of the Calibrated Airborne Multispectral Scanner (CAMS) to determine regional scale effects of suspended sediments on coral growth in Mayagüez Bay, PR USA; Carder et al. (1993) provide evidence that the Airborne Visible-Infrared Imaging Spectrometer (AVIRIS) could be used to measure water quality parameters including suspended sediments in Tampa Bay, FL USA; and high resolution (1.3 m) Compact Airborne Spectrophotographic Imager (CASI) images were used to map suspended sediment concentrations correlated to construction activities of the Øresund Link, a tunnel and bridge connecting Denmark and Sweden. Similar examples exist for the use of satellite-based instruments. Numerous investigators have successfully mapped suspended sediments in a wide range of coastal waters using the Advanced Very High Resolution Radiometer (AVHHR) (Stumpf, 1987; Stumpf and Pennock, 1998; Froidefond et al., 1993, Yan and Jing, 2000), the LandSat Thematic Mapper (TM) (Mertes, 1993; Tassan, 1993; Bilge et al., 2004) and

R.L. Miller et al. (eds.), Remote Sensing of Coastal Aquatic Environments, 259–276.

the Sea-viewing Wide Field-of-view Sensor (SeaWiFS) (Tassan, 1994; Binding et al., 2003).

It is well established that remote sensing can map suspended sediments, but the characteristics of most remote sensing instruments, or their associated costs, significantly limit their application to operational monitoring of sediment dynamics in many coastal waters. There are numerous reasons why remote sensing may not be employed operationally in studies of coastal systems. The most common limitation is the sensor's spatial or ground resolution. The nadir (just below the satellite) spatial resolution of data from the AVHRR, SeaWiFS, and MODIS (Moderate Resolution Imaging Spectrometer) is nominally 1 km, an area often too coarse in scale to adequately examine coastal processes or gradients, particularly in estuaries and bays. The National Oceanographic and Atmospheric Administration (NOAA) Landsat series of instruments (i.e., Thematic Mapper (TM), Enhanced Thematic Mapper (ETM+)) have a spatial resolution of 30 m. However, the orbital characteristics of the Landsat satellites yield a revisit time of about 16 days (i.e., the sensor images a given spot on Earth every 16 days). Consequently, Landsat sensors cannot adequately capture the temporal dynamics of coastal waters. Although airborne systems have the advantage that the user can define the deployment characteristics (e.g., time flown, area covered, spatial resolution), airborne systems are typically expensive to operate and process the data. The availability of remote sensing data and processing software has been a problem (Miller 1993). Until recently, the acquisition of data was often limited to commercial systems or processing algorithms (e.g., atmospheric correction) and software systems were absent or severely limited.

In contrast to the issues presented that describe the limitations of remote sensing, all remote sensing instruments provide a unique perspective of coastal waters unavailable through traditional field sampling techniques from ships or platforms – a large-scale synoptic view of the highly dynamic coastal environment. A major goal of efforts designed to develop the capability of operational remote sensing systems in the coastal environment is therefore to utilize a combination of sensors and techniques to overcome the limitations of sensor systems while capitalizing on the unique features of remote sensing technologies.

Few studies have examined the coupling between bottom sediment resuspension and the surface expression of resuspension through remotely sensed imagery. A recent exception is the work by Ouillon et al. (2004) where they converted a single Landsat ETM+ (Enhanced Thematic Mapper) image into a map of suspended sediment using *in situ* measurements of above-water remote sensing reflectance. The combined sediment map and *in situ* data were used to validate and refine a numerical model of fine grain suspended sediment transport in a coral reef lagoon. The major focus of their work was to demonstrate the utility of using an integrated dataset (field and remote sensing) to calibrate and validate a sophisticated numerical model of sediment transport. In this chapter we will demonstrate how a simple model of bottom sediment resuspension and recent developments in remote sensing technology can provide a rapid, widespread, and integrated approach to studying sediment dynamics in coastal waters.

In the following sections we present an approach that integrates modeling and remote sensing in a manner that is efficient and cost effective so that most students, investigators, and decision makers can monitor the processes or factors related to resuspension, suspended sediments, and transport of materials. In addition, the selection, acquisition, and processing of remotely sensed data to monitor suspended sediments and sediment transport are discussed in detail.

2. A Simple Model of Bottom Sediment Resuspension

Booth et al. (2000) present a model to determine wind-induced resuspension in shallow coastal waters. The basis of their model is the full derivation of wind-induced waves developed by the US Army Corps of Engineers Coastal Engineering Research Center (CERC, 1977, 1984). Briefly, their model was derived from a simple formulation describing surface gravity waves. In general, bottom sediment resuspension is predicted to occur when the wavelength of a wave (L) is twice the water depth (d). The wavelength of a wave is related to wave period (T) to couple the model to the wind parameters velocity (U) and fetch (F) such that (CERC, 1984):

$$gT/U_A = 0.2857(gF/U^2_A)^{1/3}, \tag{1}$$

where g is gravitational acceleration (9.8 m s^{-2}) and U_A is a wind stress factor determined by

$$U_A = 0.71(U\,R_T)^{1.23}. \tag{2}$$

The boundary layer stability factor, R_T, in (2) is taken as 1.1 (CERC, 1984). The parameter critical wind speed U_c, the threshold wind speed at which resuspension is expected to occur, is given by (Booth et al., 2000):

$$U_C = [1.2\{4127(T_C^3/F)\}^{0.813}] \tag{3}$$

where T_C, the critical wave period, is calculated as

$$T_C = (4\pi d/g)^{1/2}. \tag{4}$$

Therefore, the model requires input point measurements of wind speed, wind direction (i.e., to calculate fetch), and water depth to provide corresponding point estimates of the occurrence of bottom sediment resuspension or resuspension potential (RP). RP was defined as the difference of average daily wind speeds and derived critical wind speed (3). Bottom sediment resuspension occurs when RP is positive.

An enhanced implementation of the Booth et al. (2000) model was developed to provide greater flexibility in defining wind speed and direction (Trzaska et al., 2002); the model is now implemented in an easy-to-use computer program with a graphical interface. The Booth et al. (2000) model used 24-hr average winds grouped into a wind quadrant (e.g., N, NE, E). The new model will accommodate any combination of wind speed (m s^{-1}) and direction (0-359°). Resuspension potential is redefined as a Boolean parameter – resuspension or no resuspension. An additional parameter, resuspension intensity (RI), is introduced to represent the magnitude of the difference between measured wind speed and derived critical wind speed. That is, RP indicates areas where resuspension is predicted to occur and RI provides an index of the intensity of bottom sediment resuspension.

3. Remote Sensing of Suspended Sediment

The use of remote sensing to map suspended sediment concentration is well established for a variety of water types. Typically, the approach relates remotely-sensed measured reflectance in the red portion (ca 600-700 nm) of the visible spectrum to water column sediment concentration (see references cited above). This approach is reasonably robust in coastal and inland waters because scattering from suspended materials frequently dominates the reflectance spectrum when compared to pure water and phytoplankton absorption (Kirk, 1994; Mobley, 1994). Table 1 provides information of the red band of several remote sensing instruments.

As new sensors are developed there has been a consistent progression to use reflectance imagery to map suspended sediments. As new remote sensing instruments become operational, particularly instruments that have spectral coverage in the red or near-Infrared (ca 700 – 900 nm) part of the spectrum, numerous investigators have examined the potential use of these new instruments for mapping water quality parameters in inland and coastal waters. This effort has matured so that the latest systems now generate operational products of water quality. For example, standard products provided by the recently launched NASA EOS (Earth Observing System) MODIS series, Terra and Aqua, contain calibrated water quality parameters including suspended sediment concentration (EOS Data Products Handbook Vol. 1, 2003; Vol. 2, 2000). In addition, the standard MODIS products of normalized water leaving radiance have provided the basis of local-to-regional bio-optical algorithms (e.g., D'Sa and Miller, 2003) to quantify suspended sediments.

Since the general approach employed to derive maps of suspended sediments, or covarying variables, from remotely sensed data is basically the same regardless of sensor or environment, we will now describe in detail the acquisition, processing and analysis of MODIS data. In particular, we emphasize the application of MODIS 250 m in coastal waters using publicly available data and software implemented on low-cost computers.

The Terra spacecraft was launched as the first mission of NASA's EOS on December 18, 1999. On board the satellite are the ASTER (Advanced Spaceborne Thermal Emission and Reflection Radiometer), CERES (Clouds and the Earth's Radiant Energy System), MISR (Multi-angle Imaging Spectro-Radiometer), MODIS (Moderate-resolution Imaging Spectroradiometer), and MOPITT (Measurements of Pollution in the Troposphere) sensors. Information on Terra and the Terra instruments can be found on numerous NASA web sites. Another EOS spacecraft, Aqua, was launched in May 2002 and carries a second MODIS instrument. Terra and Aqua are in sun-synchronous orbits (cross the Equator at the same local time each day); the Terra (EOS AM) spacecraft crosses the equator at 10:30 AM local time (descending node – traveling north to south), and the Aqua (EOS PM) spacecraft crosses at 1:30 PM local time (ascending node –

Table 1. Characteristics of the red reflectance band for select instruments.

Sensor	Red Sensor Band(s)	Bandwidth (nm)
AVHRR(KLM)	1	580 - 680
AVIRIS	24-37	592 - 699
Landsat Thematic Mapper	3	630 - 690
Enhanced Thematic Mapper (ETM+)	3	630 - 690
IKONOS	3	640 - 720
MODIS (250 m)	1	620 - 670
MODIS (1 km)	13-14	662 - 683
QUICKBIRD	3	630 - 690
SeaWiFS	6	660 - 680

traveling south to north). Hence, MODIS data are available from two overpasses per day such that the solar angle is the same when a MODIS instrument crosses a given latitude.

The instruments began providing science data in February 2000 and June 2002 for the Terra and Aqua, respectively.

The MODIS instruments provide data, and subsequent operational products, for the land, ocean, cryosphere, and atmosphere. These instruments represent a continuation of a long lineage of instruments including the AVHRR, CZCS (Coastal Zone Color Scanner), SeaWiFS, and TOVS (Total Ozone Vertical Sounder). The MODIS records data in 36 spectral bands at a spatial resolution of 250 m (bands 1-2), 500 m (bands 3-7) and 1 km (bands 8-36). Data are collected at 1354 pixels, or ground points, for each scan to provide large-scale regional coverage. As stated earlier, the MODIS project team produces operational products that help describe the concentrations of suspended materials (e.g., products MOD 20 (chlorophyll *a*) and MOD 23 (suspended solids)). However, the current standard ocean products are produced using the 1 km data only. As shown in Table 1, there is a MODIS red band acquired at a ground resolution of 250 m thus providing a higher spatial resolution that is adequate to examine larger bays, estuaries, and coastal margin waters. Miller and McKee (2004) present an algorithm that accurately predicts concentrations of total suspended material (TSM) using MODIS Terra band 1 imagery. Hence, the moderately high resolution of MODIS 250 m, the near daily revisit time, and the two overpasses per day, render the MODIS instruments as a potentially powerful tool for studying the dynamics of suspended sediment transport and related processes in coastal waters.

4. Integrating a Resuspension Model and Remote Sensing

We now examine bottom sediment resuspension and suspended sediment transport in Lake Pontchartrain, LA USA using the numerical model and remote sensing approach described above. This case study will provide an example of how to implement a simple integrated approach to studying sediment dynamics in shallow coastal waters.

4.1 A CASE STUDY, LAKE PONTCHARTRAIN, LA USA

Lake Pontchartrain is a shallow urbanized estuary adjacent to New Orleans, LA (Fig. 1). The Lake Pontchartrain system consists of three embayments; Lake Maurepas, Lake Pontchartrain, and Lake Borgne. Lake Pontchartrain is a large estuary (ca. 1645 km2) with an average depth of about 3 to 4m; Lake Maurepas is considerably smaller (ca. 241 km2) with an average depth of 2 m (Flowers and Isphording, 1990); and, Lake Borgne is ca. 550 km2 and has an average depth of 2.7 m (Haralampides, 2000). The spatial extent of Lake Pontchartrain is large enough so that the MODIS 250 m data is adequate to investigate horizontal gradients (Miller and McKee, 2004). Similarly, coastal waters of comparable spatial extent, or larger, such as the Great Lakes, Chesapeake Bay, Pamilco Sound, NC, and coastal margins can potentially be studied well using 250 m data. Lake Maurepas and Lake Borgne are connected to Lake Pontchartrain by narrow passes. Lake Maurepas, located to the west of Lake Pontchartrain, is mainly a freshwater system. Lake Borgne located to the east of Lake Pontchartrain is directly connected to the northern Gulf of Mexico. Although the tidal passes between Lake Borgne and Lake Pontchartrain facilitate the inflow of saltwater and materials from the Gulf of Mexico, the tidal prism and incursion of materials into Lake Pontchartrain through this connection is relatively low (Haralampides, 2000). Several small rivers flow into Lake Pontchartrain and Lake Maurepas, yet the delivery of sediment to the lake system ranks among the lowest of estuaries in the region (Isphording et al., 1989; Bianchi and Argyrou, 1997). As a result of the general characteristics of Lake Pontchartrain (such as large surface area, shallow water depth, limited freshwater inflow, and small tidal

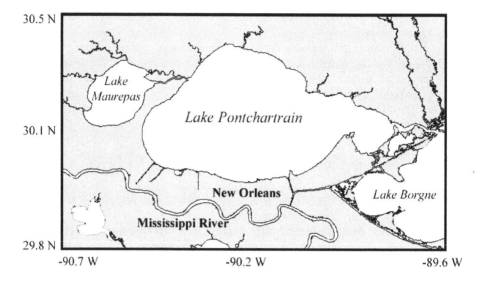

Figure 1. Location map of the Lake Pontchartrain system, a shallow urban estuary in south eastern Louisiana, USA.

Influence) the system is predominantly governed by wind-driven resuspension and circulation. Hence, Lake Pontchartrain can be considered an ideal environment in which to examine bottom sediment resuspension, the application of remote sensing to mapping suspended sediments, and the coupling of these factors.

4.2 MODELING RESUSPENSION

4.2.1 *Input variables*

Based on input requirements of the resuspension model, the bathymetry of Lake Pontchartrain and daily wind speed and direction were obtained to perform model simulations. Coastal bathymetry can be obtained from several sources including digitized charts from the NOAA (National Oceanic and Atmospheric Administration) Office of Coast Survey. For this study, a recent nautical chart was digitized on a flatbed scanner using the ELAS (Beverley and Penton, 1989) image processing system. The map was digitized at half-meter contours and resampled to 250 m pixel (spatial) resolution. The map was referenced to a geographic grid during digitization. Hence, the resulting image (Fig. 2) provides a georeferenced bathymetric map of Lake Pontchartrain at a pixel resolution of 250 m or the spatial resolution the MODIS instruments.

Additional data required for model simulations are wind speed and wind direction. These data are readily available from a wide range of sources including local airports and various meteorological stations and towers. Fortunately, a major effort is currently underway to develop a national network of state and federal coastal observing stations that include meteorological data (e.g., NOAA Coastal Ocean Observing Program (COOS) and OCEAN.US). The Gulf of Maine Ocean Observing System (GoMOOS) is a recognized leader in this effort. GoMOOS is a national pilot program to bring oceanographic and weather data to all those that need it, including the general public. Another example is the instrumented platform in Lake Pontchartrain that provided data for our analysis. The environmental monitoring station is located off the northwest shore

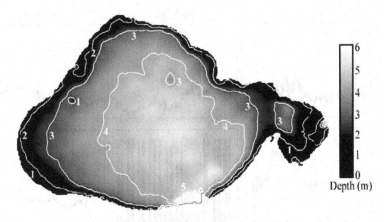

Figure 2. Digitized bathymetry of Lake Pontchartrain with 1 m contours. Pixels were resampled to 250 m to correspond to the nominal resolution of the imagery used.

(30° 18.894' N, 90° 16.831' W) and is operated and maintained by the Louisiana Universities Marine Consortium (LUMCON). LUMCON provides a web site where real time data is displayed and where users can download historical data. The platform provides data every minute for common meteorological (e.g., wind speed, wind direction, air temperature, barometric pressure) and hydrological parameters (e.g., salinity, turbidity).

4.2.2 *Numerical model*

The numerical basis of the Booth et al. (2000) model with several enhancements was implemented as a MATLAB 5.0 (MathWorks, Natick, MA) module. Developing the model in an interpretative, science-based software package such as MATLAB or RSI's IDL (Interactive Data Language, Research Systems, Inc., Boulder, CO) greatly facilitates model development and prototyping. Major enhancements of our model include a batch processing mode and more accurate of estimates of fetch. In addition, wind direction can now be from any direction rather than a direction quadrant; and images are generated for all resuspension parameters such as critical wind speed (U_C), resuspension potential (RP), and resuspension intensity (RI). An earlier version of this model was reported by Trzaska et al. (2003).

4.2.3 *Frontal passages*

To provide a comprehensive example of our integrated approach, we examined resuspension events and corresponding suspended sediments associated with the passage of several cold fronts during May 2002. Data were acquired from the LUMCON platform for 14-21 May 2002 (i.e., 2002134-2002141). The data were first examined for spurious or missing data. Select data for pre- and post frontal passage are shown in Fig. 3. These data show the typical effects of the passage of a frontal system – a switch in wind direction from the south to the north with a corresponding increase in wind speed during the front's passage. Often there is a temporary dramatic increase in wind speed as the front passes with significant gusts (Fig. 3, left panels, 2002137) and then a gradual decrease to low wind speeds during a stationary high-pressure system. As we shall see, frontal systems frequently impart sufficient energy to generate bottom sediment resuspension throughout most of the lake system.

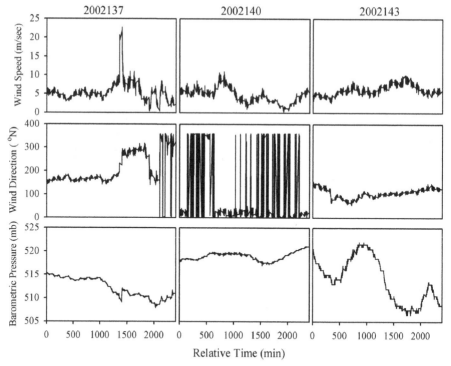

Figure 3. Select variables obtained from the LUMCON platform in Lake Pontchartrain measured during the onset of a cold front (left panels, 2002137), during a stationary high pressure system following the passage of the front (middle panels, 2002140), and during breakdown and wind reversal of the high pressure system (right panels, 2002143).

4.2.4 *Model results*

An example of the effect of increasing wind speed on the predicted occurrence of bottom sediment resuspension is shown in Fig. 4. As shown for a wind from the northeast, significant resuspension occurs with wind speeds over 4 m/sec. Trazska et al. (2002) reported that independent of wind direction, resuspension occurs in Lake Pontchartrain, except close to shore in the proximal areas of wind direction, when wind speeds exceed 4 m/sec. As wind speed in this area often exceeds 4 m/sec (Miller unpublished data), the water column of Lake Pontchartrain is usually well mixed with high turbidity resulting from bottom sediment resuspension. An example of the affect of wind speed on modeled resuspension intensity is shown in Fig. 5 for a wind speed of 1 and 6 m/sec from the northeast. Although resuspension intensity provides a relative index of the amount of wind energy that impacts the bottom, the depth to which sediment is mixed or affected depends in part on sediment grain size and consolidation of bottom sediments (Larsen et al., 1981; Richardson et al., 2002). The bottom sediment type of Lake Pontchartrain is generally homogenous dominated by small grain size particles consisting of clay, silty-clay, and silt, with only very small pockets of large grain size particles such as sand and silt-sand (Flowers and Isphording, 1990). Therefore, it is reasonable to assume that modeled images of resuspension intensity directly reflect the degree, or intensity, to which bottom sediments are impacted or resuspended in this system.

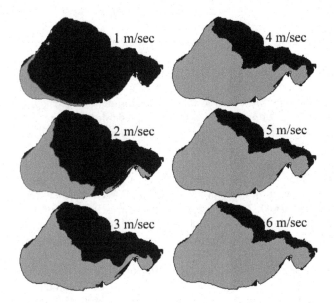

Figure 4. Resuspension potential (*RP*) obtained from model runs with an increasing wind speed of 1-6 m/sec from the northeast (45°). Black areas indicate no resuspension while gray areas indicate regions where resuspension is expected to occur (i.e., resuspension index is greater than 0).

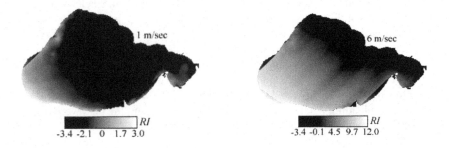

Figure 5. Resuspension intensity (*RI*) obtained from model runs with a wind speed of 1 m/sec and 6 m/sec from the northeast (45°). Note that the scales are different for each image.

4.3 MAPPING SUSPENDED SEDIMENT CONCENTRATION

4.3.1 *Acquisition and processing of MODIS data*

As stated above, to fully utilize remotely sensed data for monitoring water quality of coastal waters, particularly in applications that support environmental managers and decision makers, there is a strong need for data that has moderately high spatial resolution, a frequent revisit time, and that is cost effective. Perhaps equally important, software tools must be available for the display, processing, and analysis of data from specific instruments on low-cost computer systems. The MODIS data and available supporting software meet these requirements.

MODIS data are available without cost for downloading by accessing several data portals. For example, users can obtain data directly from the NASA EOS Data Gateway (EDG). The EOS gateway is an easy-to-use system to search for data from NASA and its affiliated centers. Access is available to most users. The EOS gateway provides a query system to determine data availability as a function of sensor, acquisition date and geographic region. MODIS data is readily available from the NASA Goddard Earth Science (GES) Distributed Active Archive Center (DAAC). Data requested can be downloaded via an ftp (file transfer protocol) or a URL-based (web-based) tool. A data granule (5 minutes time of data collection) is the smallest amount of data that can be ordered from the DAAC. A typical data set (full swath, single granule) containing data over Lake Pontchartrain is about 280 MB. The DAAC data sets required for analysis in our example are MOD02QKM (calibrated radiances level L1b full swath at 250 m) and MOD03 (geolocation fields level L1A at 1 km). MOD02QKM files contain the MODIS 250 m band 1 and 2 image data while earth geolocation points for georeferencing the image data are maintained in a MOD03 file. Data are stored in the HDF-EOS format. HDF (Hierarchal Data Format) is an efficient structure for storing multiple sets of scientific, image and ancillary data, in a single file. Individual data sets are embedded in the file as an SDS (Scientific Data Set). The SDS name for MODIS 250 m image data in a MOD02QKM file is EV_250_RefSB.

Because the EDG does not provide a preview or thumbnail image of MODIS data, we employed another site to evaluate the availability of cloud free data. We therefore accessed the DAAC's SeaWiFS HRPT (High Resolution Picture Transmission) browse image capability. A browse, or thumbnail, image is available for each SeaWiFS overpass. Although the SeaWiFS instrument passes over Lake Pontchartrain at a different time of day than MODIS, and with a different look angle, one can easily evaluate the probability of obtaining a cloud free MODIS image. This approach greatly facilitated our ordering of useable MODIS data. Fortunately, NASA recently established an 'Ocean Color Web' that provides full browse and ordering capability for MODIS Aqua data. In addition, users can establish a subscription in which specific data over a geographic area will be posted on an ftp server for download. Users are notified via email when the data is ready for downloading.

Upon downloading MODIS 250 m data to a local system, the data were displayed to assess image quality and then processed using the HDFLook 4.1 software. HDFLook was developed under collaboration between the Laboratoire d'Optique Atmospherique and Goddard Earth Science (GES) DAAC for the XWindows computer environment and has been tested on all major UNIX platforms, Linux, and MAC OS. Hence, HDFLook can be operated on a wide range of low cost computer systems. HDFLook is available free of charge from several NASA MODIS web sites.

We developed HDFLook batch scripts to extract a region of interest (29.6 to 30.5° N, -90.7 to -89.5° W) covering the Lake Pontchartrain system from the main MOD02QKM file, convert the data from calibrated radiances to surface reflectance, georeference the image, and output the image as a generic two band HDF file (Fig. 6A,B). The script also generated a graphics image file (i.e., jpg) for rapid assessment of the data using a standard image display program (Fig. 6C). In summary, HDFLook was used to review the original data ordered from the DAAC and to extract and process a specific region for subsequent analysis.

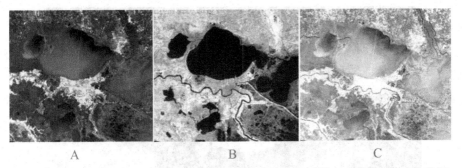

A   B   C

Figure 6. Extracted area of MODIS 250 m band 1 (620 - 670 nm) (A), band 2 (841 – 876 nm) (B) and a quick look graphics image (C) of the Lake Pontchartrain system generated using HDFLook batch processing.

HDFLook is a useful program to display and extract sub-scenes from MODIS data files, but additional software is required to fully process and analyze the HDFLook generated data files as required for this study. Most popular commercially available image processing programs now support HDF files. We used ENVI 3.4 (Research Systems, Inc., Boulder, CO) to analyze both the MODIS image files and files generated during model simulations (e.g., *RI* images).

To facilitate the use of ENVI as our primary processing software, all image files produced during a model run contained an ENVI header file. Hence, files were accessed directly. The first processing step was to access the accuracy of georeferencing by HDFLook. This was accomplished by overlaying vector-based graphics of high resolution coastal maps and adjusting manually any horizontal or vertical offsets. Land and clouds were masked using MODIS band 2 (841 – 876 nm) data. In this region of the electromagnetic spectrum, land and clouds are highly reflective whereas water, with strong absorption in this region, is very dark, resulting in a sharp contrast between these features. The varying atmospheric component of the MODIS reflectance images was removed using the clearwater, or dark pixel, subtraction technique (see for example, Gordon and Morel, 1983). Briefly, clear water is essentially a black body (i.e., no reflectance) in the spectral region of MODIS band 2 and therefore, any radiance measured by the instrument is due to atmospheric backscatter. This value is then subtracted from all pixels in the corresponding image assuming a homogenous atmospheric particle size distribution over the extent of the scene. The resulting image provides the surface reflectance values available for direct analysis of suspended sediment concentration and comparison with modeled wind-driven resuspension. A sample image derived using these processing steps is shown in Fig. 7. As can be seen, the moderately high resolution of the MODIS data is sufficient to display small scale features in reflectance or suspended sediments. These processed images alone are useful in evaluating sediment dynamics. For example, as in Fig. 7, one can access the synoptic distribution of suspended sediments, and by generating an image of the difference in reflectance between two images (e.g., Fig. 8), horizontal changes in sediment concentration can be examined. However, these changes may be due to either transport, that is, advection of sediments into or out of an area, or by localized resuspension of bottom sediments. Hence, there is a need to couple remotely sensed images of suspended sediment concentration with the output of a resuspension model to determine the sources and transport of sediments.

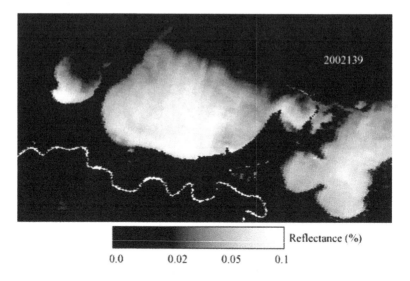

Figure 7. An example of an atmospherically corrected MODIS 250 m band 1 reflectance image acquired on 2002139. Land is masked to black.

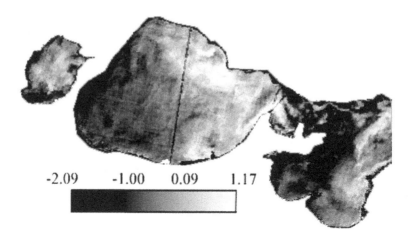

Figure 8. A difference image calculated by subtracting MODIS 250 m band 1 reflectance obtained for days 2002139 and 2002140 (2002139 minus 2002140). Image shows fine scale changes in reflectance (i.e. TSM) over a 24 hr period. Bright pixels indicate areas where TSM on 2002140 is greater than TSM concentration on 2002139; darker pixels show areas where the TSM concentration was higher on 2002139.

4.4 COUPLING MODELED RESUSPENSION AND SEDIMENT IMAGES

We now demonstrate an integrated approach to analyzing sediment dynamics in Lake Pontchartrain using results from the resuspension model and processed MODIS imagery. MODIS 250 m data were acquired for days corresponding to the passage of several frontal systems during May 2002 (2002134-2002143). The corresponding meteorological data were obtained from the LUMCON Lake Pontchartrain platform (see for example, Fig. 3). For model simulations, the average wind speed and direction were calculated during the 15 minutes prior to a MODIS overpass as it was determined that Lake Pontchartrain responded to changes in wind stress over this time period (Trzaska et al., 2002).

As discussed above, the wind-driven resuspension model provides a prediction where resuspension occurs and an estimate of resuspension intensity in Lake Pontchartrain while processed MODIS 250 m band 1 images provide maps of surface reflectance which is directly related to the concentration of total suspended matter. But how well are the results from the model related to MODIS images?

First, there is a high correlation between resuspension intensity and atmospherically corrected MODIS reflectance (Fig. 9). The data are representative of the general relationship found for all days. The data is extracted using a polygon to define a large region in western Lake Pontchartrain that contains a gradient in MODIS reflectance. As shown in Fig. 9 there is a curvilinear relationship between resuspension intensity (*RI*) and MODIS reflectance. Since no resuspension is predicted to occur at *RI* less than zero, MODIS reflectance over a negative range of *RI* indicates a *background* reflectance value (although elevated reflectance values may indicate advection of particles into a pixel). At positive values of *RI*, MODIS reflectance increases linearly with increasing resuspension intensity, (i.e. energy) until a maximum value or plateau in reflectance is reached. This upper limit in reflectance may be due in part to the depth at which resuspension can occur (e.g., maximum loading of bottom sediments). Miller and McKee (2004) reported that the dynamic range of MODIS band 1 data was sufficient to detect much higher values of reflectance and corresponding high concentrations of total suspended material in turbid coastal waters.

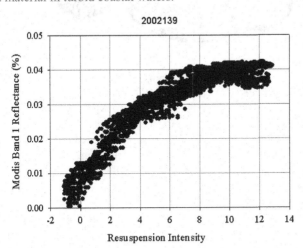

Figure 9. Sample relationship between resuspension intensity (*RI*) and MODIS band 1 reflectance data in Lake Pontchartrain on 2002139 (n=4898).

As a consequence of the relationship exhibited in Fig. 9, there is an observed similarity in the general reflectance patterns in MODIS images and patterns of predicted resuspension (*RP*) and intensity (*RI*) (Fig. 10). However, this simple comparison also indicates that differences do exist and that these differences may provide insight into the processes governing sediment dynamics.

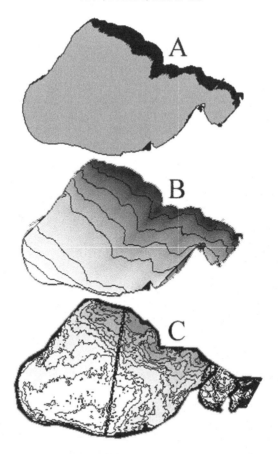

Figure 10. Comparison of modeled resuspension potential (A) and resuspension intensity (B) with surface reflectance data for MODIS 250 m band 1 (C) for 2002140. The average wind speed and direction 15 min. prior to MODIS overpass was 7.4 m/s and 45°, respectively. Resuspension potential (A) indicates that resuspension occurs (grey area) within a large region of the lake. Contours of modeled resuspension intensity (B) display near parallel contours with increasing resuspension intensity, with increasing fetch, from the northeast shore. The contours of MODIS reflectance show a similar parallel pattern near the northeast shore with an eastward movement of higher reflectance (i.e., higher suspended sediments) at the opposing southwest shore. This eastward movement may indicate the reflection or transport of water and materials as the southwesterly flow of water interacts with the southern shore.

Processed MODIS images for days 2002139-2002143 are shown in Fig. 11 and the corresponding *RP* images in Fig. 12. Barometric pressure measured at the LUMCON platform indicated that a strong frontal system passed through the area on 2002139 and a weaker front on 2002142. The *RP* images indicate that resuspension occurred over 90% of Lake Pontchartrain except on 2002141 when a high pressure system established in the region and winds from the northeast decreased significantly. However, resuspension was still predicted to have occurred over 72% of Lake Pontchartrain on 2002141 with winds at the threshold value of 4 m/sec. There is generally good agreement between the

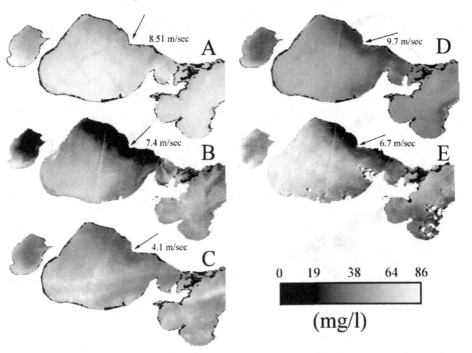

Figure 11. TSM images for days 2002139-2002143 (A-E). Arrows indicate wind direction at wind speed given at right of arrow. Bands of high cirrus clouds are shown in the bottom of C; small clouds are also shown in image E. MODIS band 1 reflectance data were converted to TSM concentration using the model of Miller and McKee (2004).

MODIS TSM images and the patterns of *RP*. The patterns and orientation of the gradients in TSM are aligned with the axis of wind direction and magnitudes are commensurate with wind speed. Here again, however, the image for 2002141 shows higher TSM concentrations than expected. A review of the wind speed data for 2002142 indicates that the average wind speed for 6 hours prior to the time averaged for the model run was over 6 m/sec. Hence, although the patterns in TSM are consistent with the *RP* image, the values may be higher as complete settling of suspended sediments has not yet occurred. It is therefore important to note that a comprehensive understanding of the events of bottom sediment resuspension and corresponding surface suspended sediment indeed requires a complete analysis of all data. The tools presented in this example however represent an excellent opportunity to thoroughly study these events and potential consequences.

5. Summary

Mapping the occurrence of bottom sediment resuspension and the distribution of suspended sediments in shallow coastal waters is critical in many scientific and environmental studies. Environmental managers and local decision makers must often know about water quality and transport of materials in coastal waters. Frequently their effort involves expensive and labor-extensive field programs to obtain the data that they

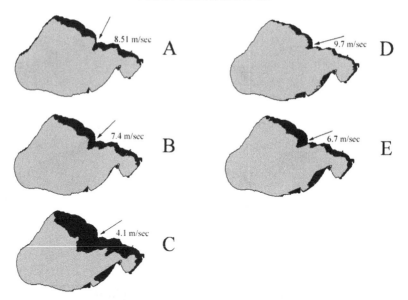

Figure 12. Resuspension potential images calculated for days 2002139-2002143 (A-E). Arrows indicate wind direction at wind speed given at right of arrow.

need at the required temporal and spatial resolutions. Although numerous investigators have shown that accurate maps of suspended sediments could be derived from remotely sensed data, the spatial resolution or frequency of remotely sensed data was insufficient to examine sediment dynamics in most in coastal waters. In addition, the software and instrument specific algorithms required to process remotely sensed data were often restricted to a limited set of users. Lastly, remotely sensed images were primarily limited to mapping suspended sediment concentration and were not coupled to the occurrence or intensity of bottom sediment resuspension.

Although the general relationship between resuspension and remotely sensed images of suspended sediments is complex, we demonstrated that an integrated approach that incorporates both a simple wind-driven resuspension model and moderately high resolution images can provide unique insight into sediment dynamics in coastal waters. We offered a preliminary example of how our modeled results and remotely sensed images could be used together to determine the source and transport of materials. Future work employing additional techniques to examine bottom sediment dynamics such as radiochemical tracers (Corbett et al., 2004) would greatly enhance an integrated approach. That is, the geochemical data would help to elucidate the small-scale processes associated with resuspension, and the results of the resuspension model and MODIS images would help extend the understanding to the larger spatial scale.

The moderately high resolution of MODIS 250 m data is useful for mapping suspended sediment concentration in a wide range of inland and coastal waters. The near daily revisit period and two MODIS instruments (morning, afternoon) can support studies of short-term events. Equally important to the potential widespread use of MODIS images in coastal studies is the significant advances in the development and open distribution of analysis software including instrument specific algorithms. Similarly, the development of effective distribution gateways of MODIS data has greatly facilitated its widespread use.

Future sensors, and sensor constellations, are expected to continue to improve the availability of data and data systems for the analysis of suspended sediments in coastal waters. As more users have access to remote sensing data, the tools to process and analyze the data, and the understanding of how to apply remotely sensed data to coastal applications, our understanding of complex coastal systems will improve and provide significant benefits to society.

6. References

Beverley, A.M., and P.G. Penton. 1989. ELAS User's Reference, Volume II, Science and Technology Laboratory, NASA, John C. Stennis Space Center, MS. [unpaginated].

Bianchi, T.S., and M.E. Argyrou. 1997. Temporal and spatial dynamics of particulate organic carbon in the Lake Pontchartrain estuary, Southeast Louisiana, U.S.A. Estuarine, Coastal and Shelf Science, 45:557-569.

Binding, C.E., D.G. Bowers, and E.G. Mitchelson-Jacob. 2003. An algorithm for the retrieval of suspended sediment concentrations in the Irish Sea from SeaWiFS ocean colour satellite imagery, International Journal of Remote Sensing, 24(19):3791-3806.

Booth, J.G., R.L. Miller, B.A. McKee and R.A. Leathers. 2000. Wind induced sediment resuspension in a microtidal estuary. Continental Shelf Research, 20(7):785-806.

Carder, K.L., R.G. Steward, R.F. Chen, S. Hawes, A. Lee and C.O. Davis. 1993. AVIRIS calibration and application in coastal oceanic environments: tracers of soluble and particulate constituents in the Tampa Bay coastal plume. Photogrammetric Engineering and Remote Sensing, 59(3):339-344.

Cloern, J.E. 1987. Turbidity as a control on phytoplankton biomass and productivity in estuaries: Continental Shelf Research, 7(11):1367-1381.

Corbett, R., B. McKee and D. Duncan. 2004. An evaluation of mobile mud dynamics in the Mississippi River Deltaic region. Marine Geology, In Press.

D'Sa, and R.L. Miller. 2003. Bio-optical properties in waters influenced by the Mississippi River during low flow conditions. Remote Sensing of Environment, 84(4):538-549.

Flowers, G.C., and W.C. Isphording. 1990. Environmental sedimentology of the Lake Pontchartrain estuary. Transactions Gulf Coast Association of Geological Societies. 11:237-250.

Froidefond, J. M., Castaing, P., Jouanneau, J. M., Prud'Homme, R., and A. Dinet. 1993. Method for the quantification of suspended sediments from AVHRR NOAA-11 satellite data, International Journal of Remote Sensing, 14(5):885-894.

Gordon, H.R. and A.Y. Morel. 1983. Remote assessment of ocean color for interpretation of satellite visible imagery: A Review, Springer-Verlag, New York, 114 pp.

Haralampides, K. 2000. A study of the hydrodynamics and salinity regimes of the Lake Pontchartrain system. Ph.D. Dissertation, University of New Orleans, 199 pp.

Isphording, W.C., F.D. Imsand and G.C. Flowers. 1989. Physical characteristics and aging of Gulf Coast estuaries: Gulf Coast Association of Geological Societies Transactions, 39:87-401.

Kirk, J.T.O. 1994. Light & photosynthesis in aquatic ecosystems. 2nd edition. Cambridge University Press, New York, NY, 509 pp.

Larsen, L.H., L. Thomas, R.W. Sternberg, N.C. Shi and M.A.H.Marsden. 1981. Field investigations of the threshold of grain motion by ocean waves and currents (Continental shelf). Marine Geology, 44(1-4):105-132.

May, C., J. Koseff, L. Lucas, J. Cloern and Dave Schoellhammer. 2003. Effects of spatial and temporal variability of turbidity on phytoplankton blooms. Marine Ecology Progress Series, 254:111-128.

Mertes, L., A.K. Smith, O. Milton, and J.B. Adams. 1993. Estimating suspended sediment concentrations in surface waters of the Amazon River wetlands from Landsat images. Remote Sensing of Environment, 43(3):281-301.

Millie, D.F., G.L. Fahnenstiel, S.E. Lohrenz, H.J. Carrick, T. Johengen and O.M.E. Schofield. 2003. Physical-biological coupling in southeastern Lake Michigan: Influence of episodic sediment resuspension on phytoplankton. Aquatic Ecology, 37:393-408.

Miller, R.L., and B.A. McKee. 2004. Using MODIS Terra 250 m Imagery to map concentrations of total suspended matter in coastal waters. Remote Sensing of Environment, In Press.

Miller, R.L., and J.F. Cruise. 1995. Effects of suspended sediments on coral growth: Evidence from remote sensing and hydrologic modeling. Remote Sensing of Environment, 53:177-187.

Miller, R.L. 1993. High resolution image processing on low cost microcomputers. International Journal of Remote Sensing. 14:655-667.

Mobley, C.D. 1994. Light and water: radiative transfer in natural waters. Academic Press, Inc.,San Diego, CA, 592 pp.

Morton, B. 1996. The subsidiary impacts of dredging (and trawling) on a subtidal benthic community in the southern waters of Hong Kong. Marine Pollution Bulletin, 32(10):701-710.

Olsen, C.R., Cutshall, N.H., and I.Larsen. 1982. Pollutant-particle associations and dynamics in coastal marine environments: a review. Marine Chemistry, 11:501-533.

Ouillon, S., P. Douillet, and S. Andréfouët. 2004. Coupling satellite data with in situ measurements and numerical modeling to study fine suspended-sediment transport: a study for the lagoon of New Caledonia. Coral Reefs, DOI: 10.1007/s00338-003-0352-z.

Pilskaln, C.H., J.H. Churchill and L.M. Mayer. 1998. Resuspension of sediment by bottom trawling in the Gulf of Maine and potential geochemical consequences. Conservation Biology, 12(6):1223-1229.

Richardson, M.D., K.B. Briggs, S.J. Bentley, D.J. Walter and T.H. Orsi. 2002. The effects of biological and hydrodynamic processes on physical and acoustic properties of sediments off the Eel River, California. Marine Geology, 182(1-2):121-139.

Schallenberg, M. and C.W. Burns 2004. Effects of sediment resuspension on phytoplankton production: teasing apart the influences of light, nutrients and algal entrainment. Freshwater Biology, 49(2):143-159.

Simpson, S.L., S.C. Apte and G.E. Batley. 1998. Effect of short-term resuspension events on trace metal speciation in polluted anoxic sediments. Environmental Science Technology, 32:620-625.

Stumpf, R.P. 1987. Application of AVHRR satellite data to the study of sediment and chlorophyll in turbid coastal waters. Washington, DC: NOAA Technical Memo NESDIS AISC 7, 50 pp.

Stumpf, R.P, and J.R. Pennock. 1989. Calibration of a general optical equation for remote sensing of suspended sediments in a moderately turbid estuary. Journal of Geophysical Research, 94(C10):14,363-14,371.

Tassan, S. 1994. Local algorithms using SeaWiFS data for the retrieval of phytoplankton, pigments, suspended sediment, and yellow substance in coastal waters. Applied Optics, 33(12):2369-2378.

Tassan, S. 1993. An improved in-water algorithm for the determination of chlorophyll and suspended sediment concentration from Thematic Mapper data in coastal waters. International Journal of Remote Sensing, 14(6):1221-1229.

Trzaska, J.R., R.L. Miller, B. McKee and R. Powell. 2002. Monitoring Sediment Resuspension Events in Shallow Aquatic Systems Using Remote Sensing and Numerical Models, Proceedings of the Seventh Thematic Conference, Remote Sensing for Marine and Coastal Environments, Miami, FL.

U.S. Army Coastal Engineering Research Center (CERC). 1984. Shore protection manual. Vol. 1:4th Edition. U.S. Army Coastal Engineering Center, Fort Belvoir, VA, 603 pp.

Wieniewski, R. 1993. The release of phosphorus during sediment resuspension. Hydrobiologia, 84:321-322.

Yan, L., and L. Jing. 2000. A suspended sediment satellite sensing algorithm based on gradient transiting from water-leaving to satellite-detected reflectance spectrum. Chinese Science Bulletin, 45(10):925-930.

Chapter 12

REMOTE SENSING OF HARMFUL ALGAL BLOOMS

RICHARD P. STUMPF AND MICHELLE C. TOMLINSON

National Oceanic and Atmospheric Administration, National Ocean Service, Silver Spring, MD 20910 USA

1. Introduction

Problems associated with blooms of unicellular marine algae, known as Harmful Algal Blooms (HABs), are global and appear to be increasing in severity and extent (Anderson et al., 1995). These phenomena have many economic, ecological, and human health impacts, such as mass mortalities of fish and marine mammals; economic loss due to reduced tourism, fish stocks, and shellfish harvests; and a suite of public health problems associated with the consumption of contaminated fish and shellfish, in addition to direct exposure to toxins. These blooms may also alter marine habitats, through shading, overgrowth induced anoxia, and adverse effects on various life stages of fish and other marine organisms (Anderson et al., 1995; Shumway, 1990, 1995).

What is a harmful algal bloom? A HAB is most accurately defined as an increase in the concentration of a phytoplankton species that has an adverse impact on the environment (Smayda, 1997). The most severe and important impacts of HABs result from toxin production. A suite of toxic syndromes have been associated with marine phytoplankton and include: Amnesic Shellfish Poisoning (ASP), Neurotoxic Shellfish Poisoning (NSP), Paralytic Shellfish Poisoning (PSP), Diarrhetic Shellfish Poisoning (DSP) and Ciguatera Fish Poisoning (CFP). Most of these syndromes occur through consumption of shellfish made toxic by ingestion of the toxin-producing phytoplankton. However, the term HAB can also describe non-toxic blooms that impact the environment indirectly through high biomass accumulation (e.g., anoxia or the alteration of trophic structure) or mechanically with such results as fish kills due to the ingestion of phytoplankton that have spines, or adverse effects on the life stages of various marine organisms including starvation (Anderson et al., 1995; Landsberg, 2002).

HABs have a broad impact. ASP has led to deaths of hundreds of California sea lions that consumed toxic anchovies (Scholin et al., 2002). The brevetoxin that causes NSP has caused massive fish kills and has led to mass mortality of dolphins in 1987, and endangered (and herbivorous) Florida manatees in 1982, 1996, and 2003 (Landsberg, 2003). Brevetoxins, when aerosolized by waves, can also cause respiratory distress in humans. Anoxia, resulting from highly concentrated phytoplankton blooms occurring in well-stratified water can lead to "jubilees" or strandings, where crabs or lobsters have walked out of the water to escape anoxic conditions (e.g. Pitcher and Calder, 2000). Blooms of *Aureococcus anophagefferens*, or "brown tides", overwhelm planktonic biomass and starve filter-feeding bivalves, who cannot use it as a food.

Although HABs are commonly referred to as "red tides" by the public, this term does not provide a good description of these phenomena. HABs are not caused by the tides; they can be many colors and are not necessarily harmful. HABs are produced by several classes of microalgae, which are characterized by different colors ranging from

R.L. Miller et al. (eds.), Remote Sensing of Coastal Aquatic Environments, 277–296.

blue-green to golden-brown to red. These classes include diatoms, dinoflagellates, raphidophytes, cyanobacteria, prymnesiophytes, pelagophytes and silicoflagellates (Landsberg, 2002). Although several of these species are found globally, they may only cause toxic events in particular regions of the world. Landsberg (2002) produced a comprehensive review of taxonomic and ecologic information on HABs and should be consulted for primary references on any specific HAB species. The roles of various species and their impacts are important from a remote sensing perspective. In the case of NSP and ASP, caused by the dinoflagellates, *Karenia* spp and *Alexandrium* spp, respectively, shellfish beds are closed when the concentration exceeds 5 cells ml^{-1}, the equivalent of 0.05 μg L^{-1}, which is below the limit for remote detection (Tester et al., 1998). Domoic acid, which causes ASP, is not always associated with high *Pseudo-nitzschia* spp. biomass, so detection of this diatom is not sufficient for detection of a domoic acid event.

Monitoring and providing early warning for toxic HABs is critical for protecting public health, wild and farmed fish and shellfish, and endangered species (such as marine mammals). Current monitoring efforts are done by measuring the concentration of toxic cells in the water and toxin levels in shellfish tissue. As these efforts are logistically demanding and labor intensive, methods which improve the efficiency of field data collection are considered essential. Accordingly, HAB monitoring programs have an intense interest in remote sensing as a tool to detect and provide the location and extent of HABs, in real-time. This chapter discusses the current and potential uses of remote sensing for HAB monitoring efforts.

2. Remote Sensing Techniques

The potential use of remote sensing for HABs was first demonstrated by Mueller (1979). An experimental ocean color sensor attached to an aircraft fortuitously flew over southwest Florida where a bloom including *Karenia brevis* (*Gymnodinium breve* at the time) was ongoing. Significant discoloration was evident in the resultant image. As the instrument was developed to simulate the Coastal Zone Color Scanner (CZCS), which was launched in late 1978, the potential for satellite detection of blooms was raised. Haddad (1982) used thermal data from the geostationary environmental satellites (GOES), to locate a tongue of cooler water which corresponded to the location of a *K. brevis* bloom off the west coast of Florida. However, the use of remote sensing for HAB monitoring became possible following the development of operational high resolution sensors such as the Advanced Very High Resolution Radiometer (AVHRR), in the mid 1980's (e.g. Tester et al., 1991; Keafer and Anderson, 1993; Gower, 1994). The general utility and characteristics of various sensors for HABs was outlined by Tester and Stumpf (1998).

The application of remote sensing to the detection and monitoring of various HAB-causing organisms has taken several approaches (Table 1). These involve detection using purely optical methods, such as Mueller (1979) presented; detection of ecological characteristics, such as the use of sea surface temperature (SST) to detect water masses conducive to growth of a particular species, and bio-physical associations such as co-occurrence with fronts or other species.

2.1 OPTICAL METHODS

Optical techniques are based on the simple premise that HABs cause discoloration of the water, hence the term "red tide". Blooms that dominate the biomass will change the

Table 1. HABs for which remote sensing is currently being used (impacts from Landsberg (2002).

HAB Species	Region (Example)	Sensing Type	Impact
Diatoms (Bacillariophyceae)			
Pseudo-nitzschia spp.	upwelling regions (West Coast-U.S.)	SST, chlorophyll	ASP, harmful to copepods, invertebrates, marine mammals, fish, birds
Dinoflagellates (Dinophyceae)			
Karenia brevis (=*Gymnodinium breve*, *Ptychodiscus breve*)	open ocean (Gulf of Mexico-U.S.)	Chlorophyll, optics, SST (in some cases)	NSP, respiratory irritation, toxic to marine organisms
Karenia mikimotoi (=*Gyrodinium aureolum*)	coastal ocean (Hong Kong, English Channel, Ireland)	SST, chlorophyll	NSP, ichthyotoxic, harmful to zooplankton and marine invertebrates
Gymnodinium catenatum	estuaries, coastal ocean, upwelling regions (Portugal, Spain, South China, Mexico)	SST, chlorophyll	PSP
Alexandrium spp.	coastal ocean (Gulf of Maine-U.S.)	SST	PSP, toxic to marine organisms
Dinophysis acuminata	estuaries, coastal ocean (South Africa, Spain)	SST	DSP
Gonyaulax spinifera	upwelling regions (British Columbia)	AVHRR band ratio	Harmful to marine organisms
Lingulodinium polyedra (=*Gonyaulax polyedra*)	coastal ocean, upwelling regions (California)	UV absorption chlorophyll	Toxic to marine organisms
Cochlodinium polykrikoides	coastal ocean (British Columbia)	SST	Ichthyotoxic, molluscicidal
Noctiluca scintillans	estuaries (Bohai Bay-China)	AVHRR visible, SST	ichthyotoxic
Prymesiophyte			
Phaeocystis pouchetti	Estuaries (Loch Striven-Scotland)	chlorophyll	ichthyotoxic
Cyanobacteria			
Nodularia spumigena	enclosed sea (Baltic Sea)	Color, AVHRR visible	Hepatotoxic, toxic to zooplankton
Microcystis spp.	estuary, lakes (Lake Erie, Lake Pontchartrain, U.S.)	Color	Hepatotoxic, ichthyotoxic
Pelagophyte			
Aureococcus anophagefferens	estuaries, upwelling regions (South Africa)	chlorophyll from aircraft	Harmful to invertebrates, ecosystem impacts

color of the water, providing an indication of the potential risk.   As a result, many monitoring programs respond to reports of water discoloration. Research on open ocean blooms beginning with the use of CZCS imagery, has shown that distinctive color patterns can aid in the detection of various types of algal blooms, including those of coccolithophores and *Trichodesmium* spp. (Balch et al., 1989; Subramaniam et al., 2002). Quantifying "discoloration" is problematic and associating a discoloration with a HAB is even more difficult.   As a result, research on HAB detection using optical techniques has varied and includes methods which examine water discoloration, chlorophyll concentration, absorption spectra, and optical characteristics.

2.1.1 *Optical theory*

As detailed discussions of ocean optics have been covered elsewhere, only the basic principals are shown here for the purpose of discussing HAB detection techniques. The irradiance reflectance of water (R), which is the ratio of light leaving the water to that entering, is determined by absorption (a) and backscatter (b_b) as follows (Gordon et al., 1988):

$$R(\lambda) \approx b_b(\lambda)/(a(\lambda) + b_b(\lambda)) \tag{1}$$

where λ is the wavelength or spectral band and the backscatter at each wavelength is represented by:

$$b_b = b_{bw} + b_{bs} + b_{bp}. \tag{2}$$

The absorption at each wavelength is represented by:

$$a = a_w + a_g + a_d + a_p \tag{3}$$

where the subscripts w, s, p, g and d indicate water, sediment, phytoplankton, gelbstoff or colored (chromophoric) dissolved organic matter (CDOM), and detritus, respectively (Morel and Prieur, 1977). Usually absorption is much greater than backscatter so that reflectance decreases with increased absorption and increases with increased backscatter. However, characteristics of both influence the spectra of the reflectance.

Remote sensing-based algorithms use either remote sensing reflectance (R_{rs}) or the normalized water-leaving radiance (nLw).   These are the same, except for a solar constant (the mean exo-atmospheric solar irradiance), F_0, and they differ from the irradiance reflectance by a distribution factor Q such that

$$nLw / F_0 = R_{rs} \approx R/Q. \tag{4}$$

There is considerable discussion regarding the value of Q, which is not relevant to this paper. For uniform water-leaving radiance, $Q = \pi$.

Several approaches are used to solve equation 1 in order to derive b_b, a, and the components of a from R_{rs}. These include empirical, semi-analytical and inversion model solutions. These are all most effective in case 1 waters. However, in all cases the accuracy and reliability depend on the stability of the spectral (λ) dependencies of b_{bs}, b_{bp}, a_g, a_d, and a_p.

For the use of quasi-analytical solutions for identifying a specific phytoplankton species, the resolution of a minimum of twelve terms is necessary. These include the magnitude and spectral shape of b_{bs}, b_{bp}, a_g, a_d, and a_p, and the R_{rs} error. The R_{rs} error is significant since the atmospheric correction of imagery is not perfect. Sea-viewing Wide Field-of-view Sensor (SeaWiFS) has gone through three significant changes in the atmospheric correction (Gordon and Wang, 1994; Siegel et al., 2000, Stumpf et al, 2003). While each has led to improvements, errors still remain, particularly due to the effect of absorbing aerosols (Schollaert et al., 2003). Finally, the solution of equation 1 presumes that the material of interest is mixed within the water column, and that the water is optically deep. Floating surface blooms cannot be determined by equation 1, nor can blooms where significant light has reflected back to the sensor from the ocean bottom. Optical methods for identifying HABs, particularly those involving *in situ* data collection, are discussed in some detail by Cullen et al. (1997) and Schofield et al. (1999).

2.1.2 *Discoloration*

Visual observations of water discoloration are used by managers for the identification of potential harmful blooms, particularly for blooms in Florida and Texas. Areas are often sampled for HABs following reports of water discoloration, based on observations aboard water or aircraft. However, HABs do not always produce water discoloration. Many coastal areas in which HABs occur are normally discolored due to sediment, CDOM from river runoff, or non-toxic blooms, therefore hindering the identification of discoloration as a HAB.

Water discoloration depends on the complicated interaction of backscattering of light by both phytoplankton and sediment, and absorption by various substances. Still, discoloration techniques for some algal blooms have worked by developing a set of rules to determine a unique reflectance pattern which is associated with a particular phytoplankton species. This approach is successful when used for the identification of monospecific blooms in case 1 water (water whose optical properties are determined by phytoplankton), and has been used to develop algorithms for *Trichodesmium* from CZCS (Subramaniam and Carpenter, 1994), SeaWiFS, and Moderate Resolution Imaging Spectrometer (MODIS) (Subramaniam et al., 2002) imagery. Haddad (1982) used a parallelepiped classifier on CZCS data to attempt to discriminate blooms of *Karenia brevis* (*Ptychodiscus brevis* in his work). However, only a few cases were examined and insufficient field data were available to validate its accuracy.

Discoloration techniques for the identification of HABs will be most effective in case 1 water, where the optical characteristics are determined exclusively by the phytoplankton. The HABs must be monospecific where variations in dissolved or detrital pigments and suspended sediment covary with the algal biomass. Also, many combinations of materials and phytoplankton can lead to the same apparent color, so confusion between HABs and other features are common.

The situation is somewhat different for cyanobacterial blooms, where water discoloration has been most effective (Lavender and Groom, 2001). These organisms often have gas vacuoles which allow them to float on the surface of the water column. As a result, they produce intensely bright surface features, which can be seen in single band or color imagery. The most dramatic examples occur in the Baltic Sea, where the frequency and extent of Nodularia blooms over several years can be characterized by single band imagery from the AVHRR (Kahru et al., 2000). *Microcystis* sp. blooms have also led to increased reflectance in the water without forming a surface layer (Budd et al., 2001).

Surface layers of cyanobacterial blooms can be discriminated from turbid water using the near-infrared (NIR) wavelengths. A temporary diversion of floodwaters from the Mississippi River into Lake Pontchartrain (an estuary in Louisiana, USA) in April 1997 resulted, several weeks later, in a massive bloom of Anabaena and Microcystis containing high levels of hepatotoxins, (Turner et al., 1999; Flocks et al., 2002). In areas where the bloom exceeded concentrations of 10^6 cells L^{-1}, the water appeared brighter in red wavelengths (Fig. 1, 04 June 1997). When the concentrations reached above 10^8 cells L^{-1}, starting before June 12, surface layers were evident in the imagery. These layers produced a much brighter signal in the NIR than the visible (Fig. 1), allowing identification of the extent of the most severe bloom. False-color infrared (NIR-band viewed as red, red-band as both green and blue) can identify the presence of surface layers resulting from the bloom. Turner et al. (1999) found the highest concentrations in the south and central lake on June 14, consistent with the distribution in the imagery.

Some other HABs have been observed due to changes in the brightness and general color (including NIR) of the water; these include *Gonyaulax spinifera* off the coast of British Columbia (Gower, 1994), and unspecified HABs in the Bohai Sea (Lin et al., 2003). These studies used AVHRR reflective visible and near-infrared bands for discrimination of water color, following techniques from Stumpf and Tyler (1988).

2.1.3 *Total chlorophyll*

The most common method for identifying a HAB is by estimating the total chlorophyll using one of the several standard remote sensing chlorophyll algorithms (O'Reilly et al., 2004; Gordon et al., 1988; Holligan et al., 1983). When a HAB dominates the phytoplankton biomass, chlorophyll concentration has the advantage of providing an estimate of the total concentration of the bloom. Tester et al. (1998) found that blooms of *Karenia brevis* could be detected with satellite at cell concentrations as low as 50 cells ml^{-1}, which is equivalent to 0.5 μg L^{-1}. Chlorophyll concentration alone does not allow for discrimination of a toxic bloom from any other bloom. On the other hand, blooms that are noxious at high cell concentrations, and therefore dominate the chlorophyll signature, may be detected through this method. For example, total chlorophyll can be quite effective for events caused by the brown tide organism *Aureococcus anophagefferens* or for blooms that may cause anoxia (Gastrich and Wazniak, 2002). Chlorophyll estimates alone have been used for the detection of toxic blooms of the genus *Karenia* (Fig. 2) (Raine and McMahon, 1998; Stumpf et al., 2003), *Gymnodinium catenatum* (Tang et al., 2003), *Lingulodinium polyedra* (Balch et al., 1989), *Chattonella* sp. (Pettersson et al, 2000b) and noxious diatoms (Ishizaka, 2003). Caution must be made in areas where these blooms lie adjacent to blooms of non-toxic species, as determining the extent of the HAB will be difficult.

2.1.4 *Optical characterization*

The development of optical models for HABs has taken two approaches: the examination of absorption spectra alone, and the examination of the entire suite of optical properties, which include both scattering and absorption characteristics. Absorption techniques focus on both photosynthetic and auxiliary pigments.

2.1.5 *Photosynthetic pigment absorption*

The analysis of absorption spectra for bloom delineation has attracted the greatest attention, as studying the unique algal pigments associated with a given species seems to be a logical approach. However, direct measurements of visible absorption spectra allow for the characterization of only some phytoplankton properties. Schofield et al. (1999) discussed the issues involved with using optical absorption methods for phytoplankton identification, through the use of *in situ* instrumentation. Millie et al. (1997) showed that a single pigment was unique between *Karenia brevis* and a diatom, allowing possible discrimination of the two species through the use of fourth-derivative analysis of the absorption spectra. Similarly, Johnsen et al. (1997) used *in situ* measurements of specific absorption bands, combined with discriminate analysis, to aid in the separation of toxic and non-toxic phytoplankton in Norwegian waters. Thus,

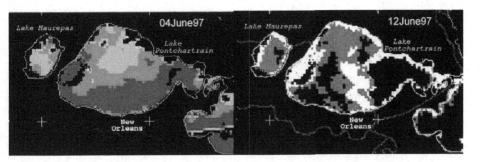

Figure 1. AVHRR imagery with red and NIR bands showing area of *Anabaena* bloom in Lake Pontchartrain, Louisiana. Areas with surface concentrations of *Anabaena* or *Microcystis* are identified where the reflectance is greater in the NIR than the red and appear white (or reddish). Areas of high water column concentrations, without surface layers, appear bright gray (or turquoise). Clouds are masked in black. The bloom intensified rapidly between 04 June and 12 June with cell concentrations increasing 1000-fold to over 2×10^8 cells L^{-1} in the central lake (Turner et al., 1999).

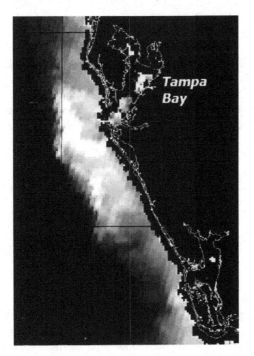

Figure 2. SeaWiFS chlorophyll image collected on 6 January 2003, during a *Karenia brevis* bloom off Tampa Bay, Florida. White (yellow to red) indicates high chlorophyll, dark (blues) indicates low. High chlorophyll off Tampa Bay identifies the HAB extent, however, patches of high chlorophyll to the south are blooms of non-harmful species. analysis of specific absorption might allow for detection of HABs. The significant questions regarding this method is whether inverse modeling of R_{rs} (Equations 1-3) can extract the absorption spectra with sufficient accuracy and whether the HAB species can be separated from the non-HAB species.

2.1.6 *Mycosporine amino acids (MAAs)*

Phytoplankton respond to ultraviolet (UV) radiation by the production of mycosporine amino acids (MAAs) (Carreto et al, 1990; Neale et al, 1998). MAAs appear to function as a sunscreen for these organisms. The concentration of MAAs and associated absorption of UV light varies depending on the amount of incident UV light, water clarity and water depth (Laurion et al., 2002). MAAs are produced rapidly and the concentration per cell changes in response to UV exposure. There is also evidence that dinoflagellates may be better at producing MAAs than diatoms (Hannach and Sigleo, 1998). The explanation for this is that dinoflagellates are motile, and can be positively phototactic which allows them to congregate in the surface water where they are exposed to a higher concentration of UV radiation. Diatoms, however, are more passive and tend to settle to the bottom, which limits their exposure. Kahru and Mitchell (1998) used shipboard R_{rs} measurements to show that UV spectra could distinguish a bloom of the potentially toxic dinoflagellate, *Lingulodinium polyedra*, from surrounding diatoms. The *Lingulodinium* bloom had lower R_{rs} in the UV relative to visible bands than did the diatoms.

While the use of UV absorption shows potential for identifying HABs, several characteristics need to be considered. Cellular concentrations of MAAs change with exposure to UV, and the relative proportions of MAAs change with their total concentration. Unlike absorption by photosynthetic pigments, MAA-absorption is not conservative within a species, so discrimination by a fixed UV-absorption spectra does not seem likely (Hannach and Sigleo, 1998; Moisan and Mitchell, 2001). In addition, other properties within the water column interfere with UV absorption. CDOM has extremely strong absorption in the UV, thus phytoplankton in water with even moderate levels of CDOM will have reduced MAAs (Laurion et al., 2002). Sufficiently high CDOM level may preclude any valid determination of MAA absorption. For use with remote sensing, the atmospheric correction becomes more problematic in the UV. Absorbing aerosols absorb most strongly at shorter wavelengths. Methods for correcting for these aerosols (e.g. Chomko and Gordon (1998) require the combination of a bio-optical model with the atmospheric aerosol model. Currently these models presume an absorption spectra relative to phytoplankton, which is not the case with the MAAs. However, low altitude remote sensing is less sensitive to atmospheric correction errors, and therefore the MAA technique may be more applicable to aircraft. If UV absorption proves to be effective in separating dinoflagellates from other algae under similar optical and atmospheric conditions, it may be a viable tool for HAB discrimination.

2.1.7 *Optical separation*

The characterization of HABs has involved optical techniques which include both absorption and scattering. Carder and Steward (1985) created a model for *Karenia brevis* (*Ptychodiscus brevis* in their work) from high resolution reflectance data collected via helicopter. This model relied on an accurate characterization of scattering components, in addition to the estimation of chlorophyll concentration. They noted that the scattering from *Karenia* appeared to be relatively low in areas of high chlorophyll concentration as compared to diatoms.

Roesler and Etheridge (McLeroy-Etheridge and Roesler, 1998; Etheridge et al., 2002; Roesler, 2003; Roesler et al., in press) have investigated inversion models to separate phytoplankton, including several HAB species, using both absorption and scattering characteristics (Fig. 3). They have found that the separation using absorption alone is insufficient. The size of the phytoplankton can produce some significant variations in the amount and spectral shape of the backscatter. It is possible, based on Mie scattering theory, to predict these backscatter characteristics. These variations in $b_b(\lambda)$ can significantly alter the spectral shape of R_{rs}, which is ignored in many bio-optical models. With a solution for spectral backscatter and known absorption characteristics Roesler and Etheridge inverted a model equivalent to equations 1-3 to estimate concentrations of dominant species in case 1 waters.

Cannizzaro et al. (in press) examined backscatter data from *Karenia brevis* and a variety of other blooms relative to the chlorophyll concentration. They found that *Karenia* had relatively low backscatter (b_{bp}) per unit chlorophyll (the latter converted to hypothetical algal backscatter, $b_{bp(chl)}$) as compared to other phytoplankton. This difference allows for the potential discrimination of *Karenia* blooms in SeaWiFS imagery (Fig. 4). They determined that the method could work in case 1 waters where chlorophyll concentration exceeds 1 μg L^{-1}.

Optical separation methods have similar limitations to those based solely on absorption spectra. Detection is most effective when the optical characteristics all co-vary with the HAB. Scattering by sediments will limit their effectiveness. While discoloration methods are not effective in the presence of CDOM, any method that can effectively separate phytoplankton absorption from CDOM absorption will allow for effective usage of optical separation techniques.

2.2 ECOLOGY

2.2.1 *Ecological characterization*

HABs occur in a variety of unique conditions. In some areas, blooms occur seasonally, in others, they respond to certain events, such as high river flow. Identification of blooms during particular times of the year and with specific ecosystem characteristics may allow for discrimination that would not be possible otherwise.

Stumpf et al. (2003) have provided a demonstration of this technique. In specific regions of the eastern Gulf of Mexico, blooms of *K. brevis* are nearly mono-specific, and dominate the biomass from late summer through early winter. New blooms occurring during that time are likely to be *K. brevis*. Stumpf (2001) devised a method for finding new blooms using a short-term anomaly. Using satellite-derived chlorophyll (from SeaWiFS in that case), a mean is determined for a two-month period beginning two weeks before the real-time image. An anomaly is taken as the difference between the image and the two-month mean. If the chlorophyll anomaly is greater than 1 μg L^{-1}, the bloom is considered "new" and is likely to be *Karenia* (Fig. 4). Tomlinson et al. (2004) evaluated the method and found that it was correct 80% of the time during the bloom season from August through the following April. Areas near river plumes, which are prone to frequent non-*Karenia* blooms, had poorer results. Fortunately, *Karenia* blooms tend to initiate in areas offshore and does not tolerate low salinity water (Steidinger et al., 1997). The temporal and spatial accuracy of this ecological characterization method was shown to be successful thus far and useful for HAB monitoring (Stumpf et al., 2003). This solution is an effective detection method and is not sensitive to the presence of sediment or other absorbing materials (provided they do not impact the chlorophyll algorithm). In addition, enhanced detection capability may

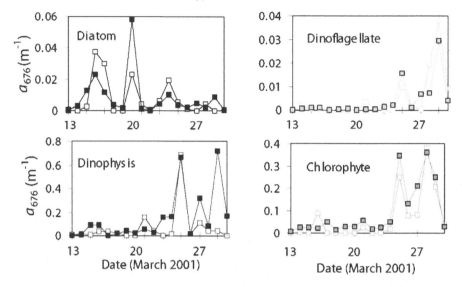

Figure 3. Modeled and measured absorption (m^{-1}) at 676 nm for taxonomic groups as derived by Roesler (in press) from inverse modeling. The modeled values are represented by filled squares while those based upon microscopic cell counts, cell size and cellular absorption efficiency are represented by open squares. Reprinted with permission of the author and publishers (Florida Fish and Wildlife Conservation Commission and the Intergovernmental Oceanographic Commission of UNESCO).

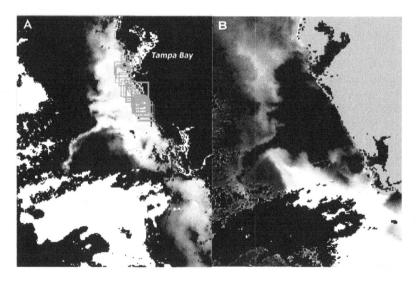

Figure 4. (A) Chlorophyll anomaly (method of Stumpf et al., 2003) and (B) the ratio of backscatter (b_{bp}) to chlorophyll specific backscatter ($b_{bp(chl)}$) method of Cannizzaro et al. (in press). White (or red) areas in the anomaly area represent chlorophyll anomalies exceeding 1 μg L^{-1}, and boxes represent the location of an ongoing *K. brevis* bloom with concentrations greater than 100 cells ml^{-1}. Dark (or dark blue) areas in the backscatter image represent low $b_{bp}/b_{bp(chl)}$ which indicate areas of probable *K. brevis*.

be possible by combining this method with other optical techniques, such as that demonstrated by Cannizzaro et al. (2002), as these are independent approaches

2.2.2 *Ecological associations*

Detecting blooms directly requires the presence of blooms that dominate the biomass. In many cases, blooms are a relatively small component of the biomass. For instance, toxic *Alexandrium* events may not "bloom" in the traditional sense (Smayda, 1997), as they rarely dominate the biomass (Keafer and Anderson, 1993). Other blooms, such as those of *Pseudo-nitzschia* occur in upwelling regions where a bloom of 1-2 μg L^{-1} chlorophyll may comprise only 10% of the total chlorophyll. In these cases, the bloom may be identified through an association with a specific ecological condition. Techniques for detecting HABs through ecological associations involve linkages with specific oceanographic or meteorological events and the timing of HAB initiation as it relates to blooms of other phytoplankton.

One of the most common methods used in this category of detection is the tracking of SST features, associated with HABs. This proves particularly useful in upwelling areas. Transient upwelling events, such as those which occur in temperate zones, may lead to patterns that initiate the development of HABs at the coast. Blooms of *Gymnodinium catenatum* are found to occur on the shoreward side of upwelling fronts, near Capes in Portugal (Figs. 4 and 5; Moita et al., 2003). On the northern Iberian coast, the occurrence of a northward current will drive these blooms to the north, and is currently monitored using SST. The *Cochlodinium* blooms in Korea develop and follow an isotherm between 21 and 25 °C, depending on the year (Suh et al., 2000). Upwelling of deep water could block the spread of the bloom along the coast in some years.

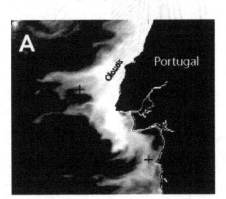 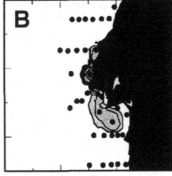

Figure 5. A comparison of (A) satellite IR imagery from September 1995 and (B) corresponding *G. catenatum* concentrations along the Portuguese coast. Lighter shades in imagery (A) represent cooler water. In (B), dots indicate sample sites, white represents cell absence; darker shades represent higher concentrations. This figure was derived from Figs. 3 and 4 of Moita et al. (2003) by permission of the author and publisher (Elsevier).

The only occurrence of *Karenia brevis* north of Florida resulted from transport via the Gulf Stream. A Florida bloom became entrained in the northwest wall of the Gulf Stream, and was transported 1000 km northward. When the bloom reached a filament that had developed off the west side of the Gulf Stream near Cape Hatteras, North Carolina, the cells were carried onto the shelf, and onshore winds brought the bloom to the coast (Tester et al., 1991). Along the southern coast of Ireland, wind shifts lead to stratification and transport that promote blooms of *Karenia mikimotoi* (formerly *Gyrodinium aureolum*). These features can be seen in the SST imagery. The HAB appears to develop near the thermocline, then due to upwelling favorable winds is transported onshore. SST imagery can show the arrival of cooler water from the pycnocline, which can be used to distinguish blooms that are HABs from those that are not (Fig. 7; Raine et al., 2001). These associations between physical features and the transport of HABs to the coast can be identified by SST and should provide a useful tool for monitoring (e.g. Gomez et al., 1999).

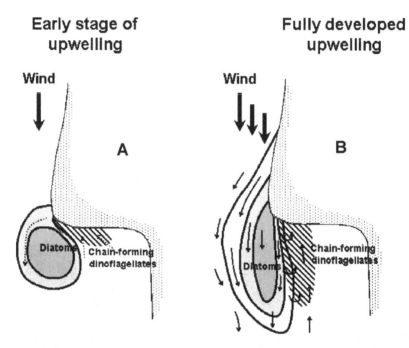

Figure 6. Conceptual model of development of a HAB associated with an equatorial trending upwelling plume along the Portuguese coast: (A) an early upwelling stage and (B) a fully developed upwelling stage. The schematic shows the relative position of diatoms vs. chain-forming dinoflagellates, relative to frontal boundaries. The arrows indicate current direction and relative speed (by length of arrow). Dotted lines represent phytoplankton distribution whereas solid lines represent the shape of the isotherms with temperatures increasing away from the upwelling core (darker shading). Derived from Moita et al. (2003) by permission of the author and publisher (Elsevier).

Conditions favoring the first occurrence of a HAB are quite important. On the west Florida coast, while *K. brevis* blooms appear to start offshore, they come onshore during upwelling favorable winds (Stumpf et al., 1998; Stumpf et al., 2003). This indicates that a new bloom that occurs during upwelling favorable conditions is more likely to be a *Karenia* bloom, somewhat similar to *K. mikimotoi* blooms along the coast of Ireland. Estuarine plumes can provide necessary nutrients to fuel certain blooms. Water mass and frontal patterns that concentrate material or lead to transport of subsurface chlorophyll blooms will reveal zones favorable for HAB development and the extent of

HABs (Pitcher et al., 1998). Therefore, SST and ocean color have been used in combination, particularly in areas where major HABs occur, to provide detection and characterization at frontal boundaries and around river plumes (Tang et al., 2003; Yin et al., 1999; Huang and Lou, 2003).

In the Gulf of Maine, Roesler (2003) has found an association between the beginning of the *Alexandrium* bloom and the seasonal shift from a spring diatom bloom to a summer dinoflagellate bloom. While *Alexandrium* rarely comprises a significant component of the biomass, the association with other dinoflagellates allows for potential remote sensing characterization. *In situ* optical techniques have allowed for the identification of this shift to dinoflagellates, and may be applied to remote sensing, leading to an infusion of optical characterization into ecological characterization.

3. Applications of Remote Sensing

Due to substantive economic losses as result of HABs worldwide, the need for remote detection of blooms, before they reach coastal areas, is critical. A single event of *Karenia brevis* in North Carolina, USA, in 1987-88 cost over US$30 million. Anderson et al., (2000) conservatively estimated the annual average loss in the US to be over US$50 million. In 1998, the value of farmed Atlantic salmon and rainbow trout in Norway was approximately US$1 billion, with an estimated loss attributed to HABs of US$2 million per year. Effective monitoring can reduce that loss (Johnsen and Sakshaug, 2000). Comparable figures exist for other regions as well.

The use of remote sensing offers a capability to routinely monitor oceanographic conditions in large geographic regions. Real-time satellite imagery of both temperature and color offers immediate tools for monitoring. Several countries have programs in place to monitor for HABs and most of these include imagery. Most monitoring efforts include a suite of operational instruments, field sampling, in addition to satellite imagery. While this is not comprehensive, the following discusses some examples.

In the western Pacific, monitoring efforts are ongoing in Korea, Japan, and China owing to extensive aquaculture in those countries. Korea has one of the most sophisticated monitoring programs in the world, with satellite imagery incorporated with a comprehensive shipboard and shoreside monitoring system (Suh et al., 2000). The emphasis in this region is on the use of AVHRR-based SST, due to the complex optical properties caused by turbidity and a diverse phytoplankton community. However, all imagery is considered (SeaWiFS and MODIS). The imagery is used routinely to aid in forecasting the movement of the HAB northward along the coast, as the bloom season progresses. The entire monitoring program has been developed to aid mitigation efforts. When a toxic bloom is found near the mouth of a bay in which important aquaculture facilities reside, mitigation efforts, such as the dispersal of clay to flocculate the HAB from the water column, are undertaken. Warnings are issued as well, so that aquaculturists can respond appropriately to an event. In Japan, ocean color imagery, including SeaWiFS (and formerly GLI) is used for monitoring the inland seas, such as the Seto Sea (Ishizaka, 2003). Since significant blooms, either of toxic or noxious HABs, can severely impact the aquaculture here as well, monitoring the extent and progression is significant. Mitigation methods are also used in Japan. China has developed the China Ecology and Oceanography of HABs (CEOHAB) program to better characterize and understand blooms (Zhu and Ye, 2003). With some of the most turbid and optically complex waters along the Chinese coast, significant demands are placed on the use of ocean color satellite imagery (e.g. SeaWiFS and AVHRR), in this area (e.g. Huang and Lou, 2003).

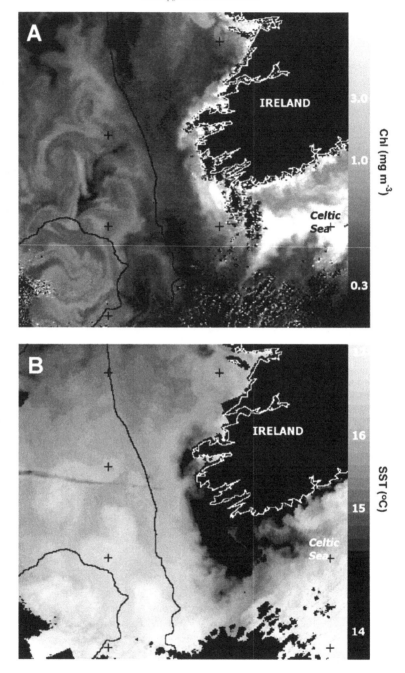

Figure 7. A) SeaWiFS ocean color and B) SST around southwestern Ireland, taken on August 4, 1998 after Fig. 5 of Raine et al., 2001. The combination of high chlorophyll and low SST along the southwestern coast shows the location of the *K. mikimotoi* bloom, which was carried onshore below the pycnocline during the upwelling event. The high chlorophyll patch offshore in the Celtic Sea is a surface bloom of *Karenia*. Reprinted by permission of the author and publisher (Kluwer).

In the United States, standard monitoring is ongoing in the Gulf of Mexico (Stumpf et al., 2003). SeaWiFS and AVHRR-SST imagery is integrated with field monitoring programs, real-time instrumentation, and basic transport modeling to identify HABs of *K. brevis* and forecast their extent and location. This effort supports managers in several states, although most blooms of *K. brevis* occur in Florida and Texas. During the bloom season, bulletins are provided to the states at least weekly. *K. brevis* is unique as a HAB in that it has significant impacts that are unrelated to the shellfish industry. *K. brevis* has killed several hundred endangered West Indian manatees in three events from 1996 (Landsberg, 2002). The toxin, when airborne at the beach, makes for unpleasant or dangerous (for those with respiratory diseases) experiences. Therefore, it has become necessary for state and local governments to provide information to the tourist industry on the status of *K. brevis* blooms.

In Europe, advanced monitoring efforts exist in Norway, Spain, and Portugal. Norway has a national monitoring program that encompasses field instrumentation, *in situ* sampling, satellite and modeling (Johnsen and Sakshaug, 2000; Petterson et al., 2000a). The information is distributed publicly, and includes the species, their locations, and the likely transport. SST is of primary interest, because it provides insights into circulation for all blooms, but color imagery (currently from SeaWiFS) is also used in the assessment of major HABs, particularly in Norway and Ireland. Spain and Portugal share blooms and situations. Monitoring involves tracking upwelling and looking specifically for conditions favorable for the appearance of the blooms. Changes in upwelling (Sordo et al., 2001) observed from AVHRR are now used for identification of conditions favorable for the various HABs (*Dinophysis acuminata*, *Gymnodinium catenatum*) to enter the estuaries. Also, as the blooms often travel north from Portugal with the season, monitoring of conditions and transport are coupled with interaction between Spanish and Portuguese fisheries authorities. SST and color imagery are used to identify the conditions favorable for HAB development, and these are monitored closely.

Significant efforts are ongoing in other countries as well. South Africa is developing a modeling effort that draws extensively on imagery (Pitcher et al., in press). Both color and temperature appear to be effective there. Ireland is applying color and temperature imagery with meteorological data to predict the onset of *Karenia mikimotoi* blooms along its southwest coast (Raine et al., 2001). These blooms do not occur annually, making the forecast all the more important. It is likely that more countries and regions will begin the routine application of remote sensing to HAB monitoring.

4. Future Prospects

Assessing future prospects is a problematic activity. Generally, the remote sensing community holds that improved optical algorithms and hyperspectral sensors will lead to the best results. However, optics will only apply where the HAB organism dominates biomass, or associates with something that does. In some cases, this works well, such as *Karenia brevis* blooms in the Gulf of Mexico. The characterization of trends in *Nodularia* blooms in the Baltic (Kahru et al., 2000) involves AVHRR visible bands; a simple technique that provides a potent analytical value. In some cases, the presence of the organisms is sufficient. For example, "brown tides" of *Aureococcus* in New Jersey and New York dominate the biomass when they occur. The state managers have found that the chlorophyll concentration alone is sufficient to identify the HAB, thus existing aircraft-based ocean color capabilities may be sufficient (Eslinger et al., 2003). For many blooms, ecological associations are critical. Therefore, future research efforts should focus on understanding the ecology of HABs and the succession of organisms as this will provide the most dramatic payback. In recent years, the major advances in applying remote sensing to monitoring have come by applying ecological characteristics and associations, for example Suh et al., 2000; Sordo et al., 2001; Raine et al., 2001; Moita et al., 2003; and Stumpf et al., 2003.

Integration of modeling with remote sensing is the next step in HAB monitoring (e.g. Petterssen et al., 2000a). This integration should naturally follow an improved understanding of HAB ecological associations. For effective use, the models will encompass a range of types. These may be heuristic, rule-based, one-dimensional, or three-dimensional. Already simple heuristic models are providing improved forecasts of HABs and better integration of imagery into the HAB monitoring realm. On the Florida coast, features identified as potential *Karenia brevis* are considered more seriously after a period of upwelling-favorable winds in early fall (Stumpf et al., 1998). Other models can play a role. Models of vertical migration (Liu et al., 2001) can provide better assessment of dinoflagellate blooms. Forecasts of the timing and response to nutrient pulses or stratification may guide hypoxia blooms. These can be synthesized with imagery to improve detection and forecasting of HABs. HABs must be monitored and their impacts forecasted, so simple detection is insufficient.

4.1 FUTURE SENSORS

HABs place stringent requirements on sensors. As the monitoring need is real, HAB detection needs sensors that are available operationally; a monitoring program cannot be built around an experimental sensor. In general, the current sensors serve the community for HAB monitoring, but they are limited in spatial resolution and accessibility. The 1-kilometer imagery found on SeaWiFS, AVHRR, and MODIS cannot resolve blooms or circulation in many bays, fjords, and estuaries. The 500-m MODIS bands may provide total pigment only, insufficient for many applications. As a commercial satellite, SeaWiFS requires a purchase, although in the US, the National Oceanic and Atmospheric Administration (NOAA) has purchased SeaWiFS data for civilian government monitoring applications since 1999. Higher resolution sensors, such as the Medium Resolution Imaging Spectrometer (MERIS), are available but have licensing concerns as well as uncertainties as to their future status. The next generation of US operational environmental satellites will have improved resolution. For example, the NPOESS (NOAA Polar Orbiting Environment Satellite System) due for launch in 2008, is expected to have color and temperature resolutions nearer to 0.7 km.

Other sensors offer some intriguing opportunities. Synthetic aperature radar (SAR), such as found on RADARSAT or the ERS-2 (European Remote Sensing Satellite), can identify surface slicks (Svejkovsky and Shandley, 2001), and so may provide additional information on certain blooms. Inexpensive aircraft-based sensors would offer significant aids to HAB monitoring, particularly in bays, lakes, and along the shore. Aircraft would also be a logical place for testing sensors that can measure UV. Improved operational satellites and standard inexpensive aircraft sensors, coupled with a well-grounded understanding of the ecology of HABs, will lead to improvements in HAB monitoring.

5. References

Anderson, D.M. (ed.). 1995. ECOHAB, The ecology and oceanography of harmful algal blooms: A national research agenda. Woods Hole Oceanographic Institution, Woods Hole, MA. 66 pp.

Anderson, D.M., P. Hoagland, Y. Kaoru, and A.W. White. 2000. Estimated annual economic impacts from harmful algal blooms (HABs) in the United States. Woods Hole Oceanog. Inst. Tech. Rept., WHOI-2000-11, 97 pp.

Balch, W.M., R.W. Eppley, M.R. Abbott, and F.M.H. Reid. 1989. Bias in satellite-derived pigment measurements due to coccolithophores and dinoflagellates. Journal of Plankton Research, 11(3):575-581.

Budd, J.W., A.M. Beeton, R.P. Stumpf, David A. Culver, and W.C. Kerfoot. 2001. Satellite observations of Microcystis blooms in western Lake Erie. Verhandlungen Internationale Vereinigung fur Limnologie, 27:3787-3793.

Cannizzaro, J.P., K.L. Carder, F.R. Chen, J.J. Walsh, Z.P. Lee, C. Heil, and T. Villareal. 2002. A novel optical classification technique for detection of red tides in the Gulf of Mexico: application to the 2001-2002 bloom event. In: Proceedings of the 10th International Conference on Harmful Algae, St.Pete Beach, Florida, 21-25 October 2002. pg. 43.

Carder, K.L. and R.G. Steward. 1985. A remote-sensing reflectance model of the red-tide dinoflagellate off west Florida. Limnology and Oceanography, 30:286-298.

Carreto, J.I., M.O. Carignan, G. Daleo, and S.G. De Marco. 1990. Occurrence of mycosporine-like amino acids in red-tide dinoflagellate *Alexandrium excavatum*: UV-photoprotective compounds. Journal of Plankton Research, 12(5):909-921.

Chomko, R.M. and H.R. Gordon. 1998. Atmospheric correction of ocean color imagery: use of the Junge power law aerosol size distribution with variable refractive index to handle aerosol absorption. Applied Optics, 37:5560-5572.

Cullen, J.J., A.M. Ciotti, R.F. Davis, and M.R. Lewis. 1997. Optical detection and assessment of algal blooms. Limnology and Oceanography, 42:1223-1239.

Eslinger, D.L., M. WanderWilt, R. Connell, R. Swift, F.Hoge. 2003. Airborne water quality measurements in shallow coastal waters. Proceedings of the 3rd Biennial Coastal GeoTools Conference, Charleston SC, January 6-9, 2003. NOAA Coastal Services Center (http://www.csc.noaa.gov/GeoTools/).

Etheridge, S. M., C. S. Roesler, H. M. Franklin, and E. Boss. 2002. Do bio-optical parameters and relationships apply to extreme algal blooms? Ocean Optics XVI Conference Proceedings (SPIE).

Flocks, J., Stumpf, R., and Kindinger, J., 2002. Satellite imagery of the 1997 Bonnet Carré spillway opening; In: Penland, S., A. Beall, and J. Kindinger (eds.) Environment atlas of the Lake Pontchartrain Basin: U.S. Geological Survey Open-File Report 02-206, 2 p.; http://pubs.usgs.gov/of/2002/of02-206/.

Gastrich, M.G., and C.E. Wazniak. 2002. A brown tide bloom index based on the potential harmful effects of the brown tide alga, *Aurococcus anophagefferens*. Aquatic Ecosystem Health & Management, 5(4):435-441.

Gomez, R.A., R. Alvarez, and O.S. Garcia. 1999. Red tide evolution in the Mazatlán Bay area from remotely sensed sea surface temperatures. Geoffisica Internacional, 38(2):63-71.

Gordon, H.R., O.T. Brown, R.H. Evans, J.W. Brown, R.C. Smith, K.S. Baker, D.K. Clark. 1988. A semianalytic radiance model of ocean color. Journal of Geophysical Research, 93(D9):10,909-10,924.

Gordon, H.R. and M. Wang. 1994. Retrieval of water-leaving radiance and aerosol optical thickness over the oceans with SeaWiFS: a preliminary algorithm. Applied Optics, 33:443-452.

Gower, J.F.R. 1994. Red tide monitoring using AVHRR HRPT imagery from a local receiver. Remote Sensing of Environment, 48:309-318.

Haddad, K.D. 1982. Hydrographic factors associated with west Florida toxic red tide blooms: An assessment for satellite prediction and monitoring. M.Sc. Thesis. University of South Florida, St.Petersburg, FL.

Hannach, G., and A.C. Sigleo. 1998. Photoinduction of UV-absorbing compounds in six species of marine phytoplankton. Marine Ecology Progress Series, 174:307-222.

Hollligan, P.M., M. Viollier, C. Dupouy, and J. Aiken. 1983. Satellite studies on the distribution of chlorophyll and dinoflagellate blooms in the western English Channel. Continental Shelf Research (vol. 2, NOS 2/3):81-96.

Huang, W.G., and X.L. Lou. 2003. AVHRR detection of red tides with neural networks. International Journal of Remote Sensing, 24(10):1991-1996.

Ishizaka, J. 2003. Detection of red tide events in the Ariake Sound, Japan. Ocean Remote Sensing and Applications, 4892:264-268.

Johnsen, G., and E. Sakshaug. 2000. Monitoring of harmful algal blooms along the Norwegian coast using bio-optical methods. South African Journal of Marine Science, 22:309-321.

Kahru, M., and B.G. Mitchell. 1998. Spectral reflectance and absorption of a massive red tide off southern California. Journal of Geophysical Research, 103(C10):21,601-21,609.

Kahru, M., J.M. Leppänen, O. Rud, and O.P. Savchuk. 2000. Cyanobacteria blooms in the Gulf of Finland triggered by saltwater inflow into the Baltic Sea. Marine Ecology Progress Series, 207:13-18.

Keafer, B.A., and D.M. Anderson. 1993. Use of remotely-sensed sea surface temperatures in studies of Alexandrium tamarense bloom dynamics. In: Toxic Phytoplanton Blooms in the Sea, Elsevier, Amsterdam (Netherlands). pg. 763-768, Dev. Mar. Biol., vol 3.

Lansberg, J.H. 2002. The effects of harmful algal blooms on aquatic organisms. Reviews in Fisheries Science, 10(2):113-390.

Laurion, I., A. Lami, and R. Sommaruga. 2002. Distribution of mycosporine-like amino acids and photoprotective carotenoids among freshwater phytoplankton assemblages. Aquatic Microbial Ecology, 26:283-294.

Lavender, S.J. and S.B. Groom. 2001. The detection and mapping of algal blooms from space. International Journal of Remote Sensing, 22(2-3):197-201.

Lin, Q., Y. Zhang, Y. Nie, and Y. Guan. 2003. Detection of harmful algal blooms over the Gulf of Bohai Sea in China at visible and near infrared (NIR) wavelengths of remote sensing. Journal of Electromagnetic Waves and Applications, 17(6):861-871.

Liu, G., G.S. Janowitz, D. Kamykowski. 2001. Influence of environment nutrient conditions on *Gymnodinium breve* (Dinophyceae) population dynamics: a numerical study. Marine Ecology Progress Series, 213:13-37.

McLeroy-Etheridge, S.L., and C.S. Roesler. 1998. Are the inherent optical properties of phytoplankton responsible for the distinct ocean colors observed during harmful algal blooms? SPIE Ocean Optics XIV, 1:109-116.

Millie, D.F., O.M. Schofield, G.J. Kirkpatrick, G. Johnsen, P.A. Tester, and B.T. Vinyard. 1997. Detection of harmful algal blooms using photopigments and absorption signatures: A case study of the Florida red tide dinoflagellate, *Gymnodinium breve*. Limnology and Oceanography, 42(5, part 2):1240- 1251.

Moisan, T.A., and B.G. Mitchell, 2001. UV absorption by mycosporine-like amino acids in *Phaeocystis antarctica* Karsten induced by photosynthetically available radiation. Marine Biology, 38:217-227.

Moita, M.T., P.B. Oliveria, J.C. Mendes, and A.S. Palma. 2003. Distribution of chlorophyll a and *Gymnodinium catenatum* associated with coastal upwelling plumes off central Portugal. International Journal of Ecology, 24:S125-S132.

Morel, A., and L. Prieur, 1977. analysis of variations in ocean color. Limnology and Oceanography, 22 (4):709-722.

Mueller, J.L. 1979. Prospects for measuring phytoplankton bloom extent and patchiness using remotely sensed ocean color images: an example. In: Toxic Dinoflagellate Blooms, Elsevier, North Holland, Inc., New York. Pp. 303-308.

Neale, P.J., A.T. Banaszak, and C.R. Jarriel. 1998. Untraviolet sunscreens in *Gymnodinium sanguineum* (dinophyceae): Mycosporine-like amino acids protect against inhabition of photosynthesis. Journal of Phycology **34**:928-938.

O'Reilly, J.E., S. Maritorena, D. Siegel, M.C. O'Brien, D. Toole, B.G. Mitchell, M. Kahru., F.P. Chavez., P. Strutton, G. Cota, S.B. Hooker., C.R. McClain, K.L. Carder, F. Muller-Karger, L. Harding, A. Magnuson, D. Phinney, G.F. Moore, J. Aiken, K.R. Arrigo, R. Letelier, and M. Culver. 2000. Ocean color chlorophyll a algorithms for SeaWiFS, OC2, and OC4: Version 4. In: SeaWiFS Postlaunch Technical Report Series, Hooker, S.B and E.R. Firestone [Eds]. Volume 11, SeaWiFS Postlaunch Calibration and Validation Analyses, Part 3. NASA, Goddard Space Flight Center, Greenbelt, Maryland. 9-23.

Pettersson, L..H., D.D. Durand, E. Svendsen, T. Noji, H. Soiland, S. Groom. 2000a. DeciDe for near real-time use of ocean colour data in management of toxic algae blooms. NERSC Technical Report no. 180-A, Nansen Environmental and Remote Sensing Center, Bergen, Norway, http://www.nersc.no/Decide-HAB/

Pettersson, L.H., D. Durand, T. Noji, H. Soiland, E. Svendsen, S. Groom, S. Lavender, P. Regner, O.M. Johannessen. 2000b. Satellite observations and forecasting can mitigate effects of toxic algae blooms. ICES CM 2000/O:07(Poster).

Pitcher, G.C., A.J. Boyd, D.A. Horstman, B.A. Mitchell-Innes. 1998. Subsurface dinoflagellates populations, frontal blooms and the formation of red tide in the southern Benguela upwelling system. Marine Ecology Progress Series, 172:253-264.

Pitcher, G.C., and D. Calder. 2000. Harmful algal blooms of the Southern Benguela Current: A review and appraisal of monitoring from 1989 to 1997. South African Journal of Marine Science, 22:255-271.

Pitcher, G., P. Montiero, and A. Kemp. (in press). The potential use of a hydrodynamic model in the prediction of harmful algal blooms in the southern Benguela.

Raine, R., and T. McMahon. 1998. Physical dynamics on the continental shelf off southwestern Ireland and their influence on coastal phytoplankton blooms. Continental Shelf Research, 18:883-914.

Raine, R., O. Boyle, T. O'Higgins, M. White, J. Patching, B. Cahill, and T. McMahon. 2001. A satellite and field portrait of a *Karenia mikimotoi* bloom off the south coast of Ireland, August 1998. Hydrobiologia, 465:187-193.

Roesler, C.S., 2003. Achievements and limitation in ocean color detection of red tides: case studies in the Benguela upwelling system and the Gulf of Maine. Proceedings of the Workshop on Red Tide Monitoring in Asian Coastal Waters., (http://fol.fs.a.u-tokyo.ac.jp/rtw/), 5 pp.

Roesler, C.S., Etheridge, S.M., Pitcher, G.C. (in press). Application of an ocean color algal taxa detection model to red tides in the Southern Benguela. Proceedings of the Tenth International Conference for Harmful Algal Blooms

Schofield, O., J. Grzymski, W.P. Bissett, G.J. Kirkpatrick, D.G. Millie, M. Moline, and C. Roesler. 1999. Optical monitoring and forecasting systems for harmful algal blooms: possibility or pipe dream? Journal of Phycology, 35:1477-1496.

Schollaert, S.E., Yoder, J.a., J.E. O'Reilly, D.L. Westphal. 2003. Influence of dust and sulfate aerosols on ocean color spectra and chlorophyll a concentrations derived from SeaWiFS off the U.S. east coast. Journal of Geophysical Research, 108(C6):22/1-22/14.

Scholin, C.A., F. Gulland, G.J. Douchette, S. Benson, M. Busman, F.P.Chavez, J. Cordaro, R. DeLong, A. De Volgelaere, J. Harvey, M. Haulena, K. Lefebvre, T. Lipscomb, S. Loscutoff, L.J. Lowenstine, R. Marin III,P.E. Miller, W.A. McLellan, P.D.R. Moeller, C.L. Powell, T. Rowles, P.Silvagni, M. Silver, T. Spraker, V. Trainer and F.M. Van Dolah. 2000. Mortality of sea lions along the central California coast linked to a toxic diatom bloom. Nature, 403:80-84.

Shumway, S.E. 1990. A review of the effects of algal blooms on shellfish and aquaculture. Journal of the World Aquaculture Society, 21:65-104.

Shumway, S.E. 1995. Phycotoxin-related shellfish poisoning: bivalve molluscs are not the only vectors. Review Fisheries in Science, 3:1-31.

Siegel, D.A., M. Wang, S. Maritorena, and W. Robinson. 2000. Atmospheric correction of satellite ocean color imagery: the black pixel assumption. Applied Optics, 39:3582-3591.

Smayda, T.J. 1997. Harmful algal blooms: their ecophysioogy and general relevance to phytoplankton blooms in the sea. Limnology and Oceanography, 45:1137-1153.

Sordo, I., E.D. Barton, J.M. Cotos, and Y. Pazos. 2001. An inshore poleward current in the NW of the Iberian Peninsula detected from satellite images, and its relation with *G. catenatum* and *D. acuminata* blooms in the Galican Rias. Estuarine, Coastal and Shelf Science, 53:787-799.

Steidinger, K.A., G.A. Vargo, P.A. Tester, and C.R. Tomas. 1997. Bloom dynamics and physiology of *Gymnodinium breve*. In: Anderson, D. M., A.E. Cembrella, and G.M. Hallegraeff. [Eds.]. The Physiological Ecology of Harmful Algal Blooms. Elsevier, Amsterdam.

Stumpf, R.P., and M.A. Tyler. 1988. Satellite detection of bloom and pigment distribution in estuaries. Remote Sensing of Environment, 24:385-404.

Stumpf, R.P., V. Ransibrahmanakul, K.A. Steidinger, and P.A. Tester. 1998. Observations of sea surface temperature and winds in association with Florida, USA red tides (*Gymnodinium breve* blooms). In: Harmful Algae. Reguera, B., J. Blanco, M.L. Fernandez, T. Wyatt. (Eds.). Xunta de Galicia and Intergovernmental Oceanographic Commission of UNESCO, Paris, France. pg. 145-148.

Stumpf, R.P. 2001. Applications of satellite ocean color sensors for monitoring and predicting harmful algal blooms. Journal of Human and Ecological Risk Assessment, 7:1363-1368.

Stumpf, R.P., M.E. Culver, P.A. Tester, M. Tomlinson, G.J. Kirkpatrick, B.A. Pederson, E. Truby, V. Ransibrahmanukul, and M. Soracco. 2003. Monitoring *Karenia brevis* blooms in the Gulf of Mexico using satellite ocean color imagery and other data. Harmful Algae, 2:147-160.

Subramaniam, A., and E.J. Carpenter. 1994. An empirically derived protocol for the detection of blooms of the marine cyanobacterium *Trichodesmium* using CZCS imagery. International Journal of Remote Sensing, 15(8):1559-1569.

Subramaniam, A., C.W. Brown, R.R. Hood, E.J. Carpenter, D.G. Capone. 2002. Detecting *Trichodesmium* blooms in SeaWiFS imagery. Deep-Sea Research II, 49:107-121.

Suh, Y.S., J.H. Kim, and H.G. Kim. 2000. Relationship between Sea Surface Temperature derived from NOAA satellites and *Cochlodinium polykrikoides* red tide occurrence in Korean coastal waters. Journal of the Korean Environmental Science Society, 9(3):215-221.

Svejkovsky, J. and J. Shandley. 2001. Detection of offshore plankton blooms with AVHRR and SAR imagery. International Journal of Remote Sensing, 22:471-485.

Tang, D.L., D.R. Kester, I.H. Ni, Y.Z. Qi, and H. Kawamura. 2003. In situ and satellite observations of a harmful algal bloom and water condition at the Pearl River estuary in late autumn 1998. Harmful Algae, 2:89-99.

Tester, P.A., R.P. Stumpf, F.M. Vukovich, P.K. Fowler, and J.T. Turner. 1991. An expatriate red tide bloom: transport, distribution, and persistence. Limnology and Oceanography, 6(5):1053-1061.

Tester, P.A., R.P. Stumpf, and K. Steidinger. 1998. Ocean color imagery: What is the minimum detection level for *Gymnodinium breve* blooms? In: Harmful Algae. Reguera. B, J. Blanco., M.L. Fernandez and T. Wyatt (Eds) Xunta de Galacia and Intergovernmental Oceanographic Commission of UNESCO.

Tester, P.A., and R.P. Stumpf. 1998. Phytoplankton blooms and remote sensing: what is the potential for early warning? Journal of Shellfish Research, 17(5):1469-1471.

Tomlinson, M.C., R.P. Stumpf, V. Ransibrahmanakul, E.W. Truby, G.J. Kirkpatrick, B.A. Pederson, G.A. Vargo, and C.A. Heil, C.A. 2004. Evaluation of the use of SeaWiFS imagery for detecting *Karenia brevis* harmful algal blooms in the eastern Gulf of Mexico. Remote Sensing of Environment, 91(3-4):293-303.

Turner, R.E., Q. Dortch, N.N. Rabalais. 1999. Effects of the 1997 Bonnet Carré opening on Nutrients and Phytoplankton in Lake Pontchartrain. Report for the Lake Pontchartrain Basin Foundation, Lakeway III Suite 2070, 3838 N. Causeway Blvd, Metairie, Louisiana 70009, USA.

Yin, K., P.J. Harrison, J. Chen, W. Huang, and P.Y. Qian. 1999. Red tides during spring 1998 in Hong Kong: is El Niño responsible? Marine Ecology Progress Series, 187:289-294.

Zhu, M. and S. Ye. 2003. Red tide monitoring in East China Sea. Proceedings of the Workshop on Red Tide Monitoring in Asian Coastal Waters., (http://fol.fs.a.u-tokyo.ac.jp/rtw/), 5 pp.

Chapter 13

MULTI-SCALE REMOTE SENSING OF CORAL REEFS

[1]SERGE ANDRÉFOUËT, [2]ERIC J. HOCHBERG, [1]CHRISTOPHE CHEVILLON, [3]FRANK E. MULLER-KARGER,[4] JOHN C. BROCK AND [3]CHUANMIN HU

[1]Institut de Recherche pour le Développement, BP A5, 98848 Nouméa, New Caledonia
[2]University of Hawaii, School of Ocean and Earth Science and Technology, Hawaii Institute of Marine Biology, P.O. Box 1346, Kaneohe, HI, 96744 USA
[3]Institute for Marine Remote Sensing, College of Marine Science, University of South Florida, 140 7th Ave. South, St Petersburg, FL, 33701 USA
[4]USGS Center for Coastal and Watershed Studies, 600 4th Street South, St. Petersburg, FL, 33701 USA

1. Introduction

Coral reefs provide an excellent case study of the application of marine remote sensing to a shallow coastal ecosystem that is spatially limited, exhibits high diversity, high productivity, and faces severe anthropogenic and climatic threats. Coral reefs, lagoons and their associated environments exhibit a high degree of natural variability in terms of water quality, benthic patchiness, and water depth. This variability presents a significant challenge for many analytical optical algorithms.

Coral ecosystems have a high intrinsic value because of their high diversity of coral, fish, and benthic species. Often described as the marine equivalent of terrestrial rainforests in terms of species richness (Hubbell, 1997), they also have significant economic value from harvesting natural resources (e.g., food, pharmacology), coastline protection, and tourism (Costanza et al., 1997). For many tropical countries or regions, coral reefs are a major or principal source of income from fisheries, aquaculture, tourism and recreation. Several island countries in the Indian and Pacific Oceans (e.g., Maldives or Tuvalu) are entirely coral reef environments.

Unfortunately, coral reefs and associated lagoons are among the most threatened coastal ecosystems worldwide (Pandolfi et al., 2003). Coral reefs react quickly to new stressors because they thrive in a narrow range of environmental conditions and are very sensitive to small changes in temperature, light, water quality and hydrodynamics. Numerous reports have documented local consequences of pollution, overfishing, urban development or coral mining (Wilkinson, 2000). Moreover, global-scale climatic changes induce new threats, even for pristine reef systems not directly under human influence (Kleypas et al, 2001). Hence, coral-based systems may serve as a unique indicator of environmental change for diagnosing the status of tropical coastal ecosystems and global change. Since remote sensing has been applied in studies of reef systems for several decades, several multidisciplinary applications and techniques have gained enough maturity to be useful for these goals. Remote sensing can potentially be used to address many reef mega-processes (sensu Hatcher, 1997) with applications in ecology, biology, biogeochemistry, geology and management of reefs.

Recently, Andréfouët and Riegl (2004) divided remote sensing of coral reefs into two categories: direct and indirect. Direct remote sensing is when the reef itself is the

R.L. Miller et al. (eds.), Remote Sensing of Coastal Aquatic Environments, 297–315.

target of remote sensing while indirect remote sensing refers to studies that focus on the oceanic and atmospheric environment around the reef. Direct studies address benthic properties and status, habitat and geomorphologic structures, bathymetry, and water circulation using satellite or airborne remote sensing data, generally using one or few coverages (Mumby et al., 1997; Hochberg and Atkinson, 2000; Andréfouët et al., 2002a; Stumpf et al., 2003; Isoun et al., 2003; Brock et al. 2004). Indirect remote sensing typically aims to describe the boundary conditions of the reefs and the spatio-temporal weather context during *in situ* surveys, or during events of interests such as coral spawning, benthic die-offs, algal blooms and other water quality events (Abram et al., 2003; Andréfouët et al., 2002b; Hu et al., 2003; Liu et al., 2003; Penland et al., 2004). Absolute measurements or anomalies in temperature, wave height and direction, sea level, chlorophyll and colored dissolved organic matter (CDOM) concentrations, aerosols, rain, solar insolation and cloud cover are parameters of interest that may be inferred from time-series analysis (Mumby et al., 2001a; Andréfouët et al., 2001a; Dunne and Brown, 2001; Liu et al., 2003; Abram et al., 2003; Otis et al., 2004).

In this chapter we present how both direct and indirect remote sensing can be integrated to address two major coral reef applications - coral bleaching and assessment of biodiversity. This approach reflects the current non-linear integration of remote sensing for environmental assessment of coral reefs, resulting from a rapid increase in available sensors, processing methods and interdisciplinary collaborations (Andréfouët and Riegl, 2004). Moreover, this approach has greatly benefited from recent collaborations of once independent investigations (e.g., benthic ecology, remote sensing, and numerical modeling).

2. Remote Sensing to Assess Coral Bleaching

2.1 CORAL BLEACHING

Within the tissues of healthy reef-building corals are populations of unicellular photosynthetic algae called zooxanthellae (Brown, 1997). Photosynthetic products from zooxanthellae contribute to coral growth and calcification, while respiration products from the coral contribute to zooxanthellae photosynthesis. Under conditions of stress, such as decreasing salinity or increasing water temperature, the coral host may expel the zooxanthellae. Without re-inoculation by a healthy population of endosymbionts, host mortality often occurs. Bleaching refers to the discoloration or whiter color of the host coral when pigmented zooxanthellae are removed. Coral bleaching was first documented in 1911 in the Florida Keys and became a well known phenomenon during the 1980-90's when large-scale massive bleaching events were reported. Massive global events occurred in 1998 and 2002 (Aronson et al., 2000; Liu et al., 2003; Berkelmans et al., 2004). This outbreak is generally attributed to positive anomalies in temperature and in ultraviolet light (Brown, 1997) in the aftermath of El Niño Southern Oscillation (ENSO) periods (Hughes et al., 2003).

With rising sea surface temperatures recorded in the world's oceans, coral bleaching is now perceived as a major threat to many reef systems (Hoegh-Guldberg, 1999; Wilkinson, 2000; Hughes et al., 2003). Some reefs impacted during the 1998 ENSO event have not recovered, potentially resulting in decreased fish abundance, phase and strategy-shifts in benthic community structures, diversity loss, and decreased overall productivity (McClanahan, 2000; Chabanet, 2002; Spalding and Jarvis, 2002). From a conservation and reef management standpoint, predicting reef vulnerability to bleaching at a scale of few tens of kilometers is important when designing monitoring programs or networks of Marine Protected Areas (MPA) since resistance to coral bleaching is an important property of areas intended to protect biodiversity (West and Salm, 2003).

2.2 REGIONAL INDIRECT ASSESSMENT OF BLEACHING

AVHRR sensors have been used for nearly a decade to monitor sea surface temperature (SST, Fig. 1) and bleaching (Liu et al., 2003). Bleaching nowcasts and forecasts are currently based on empirical SST-derived proxies, designed through trial and error, and are continuously refined. The goal of these nowcast systems is to predict the threshold above which a coral reef will be subject to bleaching. The current methods use SST, but other relevant environmental data such as sea surface height, solar insolation, cloud cover, CDOM, etc. may be used in future models. The most recent of this SST proxy, "Max3d", was defined by a statistical spatial analysis of the bleaching patterns that occurred in 1998-2002 along the Great Barrier Reef (Berkelmans et al. 2004). It was found that the maximum SST occurring over any 3-day period (hence, Max3d) during the bleaching season was a better predictor of bleaching than any other anomaly-based SST variable. These proxies have also been used in a conservation context. The goal is to determine conservation areas, resistant to bleaching. Thus, the spatial modeling combined with multivariate empirical reasoning and innovative computational techniques (e.g Baysian Belief Networks) make use of these proxies (Woolridge and Done, 2004).

Ultimately, it is the local and regional hydrodynamic regime that is the primary forcing factor of bleaching, since SST is affected by mixing and other thermodynamic processes in the upper water column. Topography, wind, low frequency currents and tidal regimes are critical oceanographic information needed to understand and forecast bleaching using numerical models of circulation. The next step will be to model water mixing and SST over scales of 1000's of km at high spatial resolution (1 km). Remote sensing data such as wind velocities and direction, wave height and direction, SST, and ocean color may help improve the parameterization and calibration of ocean circulation models by assimilation or by comparing satellite observations to model solutions.

2.3 REEF-SCALE ASSESSMENT OF BLEACHING

Bleaching can occur on reefs in a variety of spatial patterns that depends on reef geomorphology and topography, previous perturbations, the type of corals (e.g. acroporids, pocilloporids, porites), possibly coral genetics, associated zooxanthellae and their adaptation to thermal stress. To understand reef-scale heterogeneity in bleaching, reef-scale hydrodynamic models with resolutions of a few 10's of meters are required. These models will also combine physical oceanographic processes with detailed benthic community descriptions at a resolution of 10's of meters (Done et al., 2003).

At reef-scale, detailed bathymetry is required as a first step to build accurate circulation models. However, such data is not available for most reefs because of the difficulty in making sounding measurements over vast expanses of shallow waters. For instance, the topography of the Great Barrier Reef shelf and lagoon is a compilation of various data sets (mostly ship sounding), interpolated and merged in the form of a 250m-resolution grid (Lewis, 1999), which is inadequate for use in most reef-scale models.

Optical bathymetric algorithms applied to multispectral/hyperspectral satellite or airborne images such as Landsat, IKONOS or Compact Airborne Spectrographic Imager CASI) provide the most convenient way to overcome the limitations in acquiring bathymetric data for shallow clear waters (Lyzenga, 1978; Loubersac et al., 1991; Morel, 1996; Liceaga et al., 2002; Louchard et al., 2003; Stumpf et al., 2003). As a result of research conducted in the 1980's, hydrographic charts of French overseas territory (SHOM-SPT 1990) now include bathymetry derived using Satellite Pour l'Observation de la Terre (SPOT) satellite data and a multi-regression algorithm. NOAA is updating bathymetric maps of the Northwestern Hawaiian Islands using a revised ratio-algorithm applied to IKONOS data (Stumpf et al., 2003).

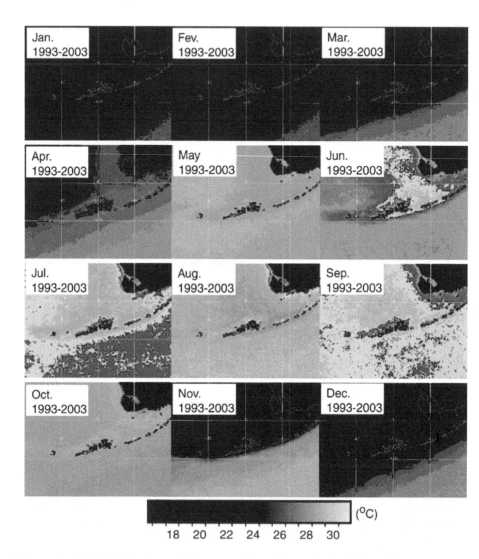

Figure 1: SST (°Celsius) monthly mean climatology obtained from AVHRR (1993-2003) for the Florida Keys. Similar time-series analysis can help detect unusually high SST values (positive anomalies) that can trigger coral bleaching events.

Active remote sensing using lasers provides an efficient alternative for detailed topography assessments of shallow coral reefs. LIDAR (LIght Detection And Ranging) altimeters such as LADS (Laser Airborne Depth Sounder) or SHOALS (Scanning Hydrographic Operational Airborne Lidar Survey) have been deployed on the Great

Barrier Reef and the Main Hawaiian Islands (Storlazzi et al., 2003). Improved systems such as the Experimental Advanced Airborne Research Lidar (EAARL), a temporal-waveform-resolving green laser altimeter constructed at NASA Wallops, was used in the northern Florida Keys during the summers of 2001 and 2002 (Brock et al., 2004). These data sets helped characterize the detailed geomorphology of a reef, such as the spur-and-grooves zone of Molokai fringing reefs (Storlazzi et al., 2003), or the rugosity of patch reef substrates in Biscayne National Park (Brock et al., 2004).

Coral bleaching has also been directly observed using aerial photographs (Andréfouët et al., 2002a), IKONOS (Elvidge et al., 2004) and Landsat data (Yamano and Tamura, 2004). The scale of interest is the community (coral assemblages) or coral colony for which sub-meter resolution is optimal (Andréfouët et al., 2002a) but beyond the current capabilities of satellite sensors. Direct sensing of bleaching is important as validation in remote sites of SST-proxy bleaching predictions (Elvidge et al., 2004), and also as a basis for planning management and monitoring activities following the bleaching event. Direct sensing is critical to surveying possible mortality or recovery that occurs within a few weeks after the peak of bleaching.

Direct remote sensing studies of bleaching have used different approaches. Change detection techniques have been applied, for example, by comparing the bleached zones detected using normalized IKONOS images acquired before and during the Great Barrier Reef 2000 event (Elvidge et al., 2004). Bleaching has been also detected using principles of radiative transfer theory applied to individual Landsat images of Ryukyus archipelago in Japan (Yamano and Tamura, 2004). Such analytical algorithms require knowledge of the reflectance of bleached corals at different depths, as well as spectral differences between bleached corals and other benthic objects (Fig. 2) (Clark et al., 2000; Holden and Ledrew, 2001; Hochberg et al., 2003).

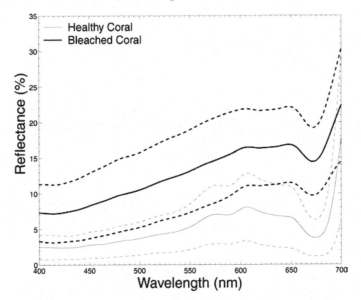

Figure 2. Statistical differences (+/- one standard deviation) between reflectance of bleached and healthy corals.

Ultimately, spectral reflectance of coral is determined by the spectral absorption and fluorescence properties of pigments residing at various locations in a coral colony, including the zooxanthellae and ectodermal and endodermal host tissues (Dove et al., 1995; Salih et al., 2000; Mazel et al., 2003a; Hochberg et al., 2004). Variability in these pigment sources contributes to the complexity in shape and magnitude of coral spectral reflectance (including the loss of pigments during coral bleaching). One goal of on-going research at the coral colony-scale is to explain reflectance according to variations in pigmentation, and thus indirectly describe the health of the colony.

The final aspect of investigating coral bleaching is to estimate benthic changes induced by the bleaching event, for example, quantifying the level of coral mortality (Mumby et al., 2001b). To date only one example of post-bleaching assessment (in Rangiroa atoll, French Polynesia) has been reported where extensive ground-truthing and airborne hyperspectral measurements were used to estimate the percent of dead corals with remarkable accuracy. Mumby et al. (2001b) reported classification results for dead and live corals within a 5% error range. There are several other recent satellite-based change detection case studies that describe benthic cover modifications, though not necessarily after bleaching-induced mortality (Dustan et al., 2001; Andréfouët et al., 2001b; Palandro et al., 2003a; Palandro et al., 2003b).

2.4   SYNTHESIS: MULTI-SCALE   APPLICATIONS   FOR   BLEACHING ASSESSMENT

Figure 3 highlights the connections between the different scales (region, reef, community, colony), remote sensing domains (indirect sensing and environmental proxies, direct sensing, spectral signatures) and non-remote sensing domains (statistical analysis, computational techniques, ecology, management) that have been integrated and merged for the bleaching application over roughly the last 8 years. As a summary, remote sensing techniques are used to forecast bleaching (e.g. using SST and statistical proxies), understand bleaching causes at regional scales (proxies), to validate SST predictions and map bleaching extent (direct high resolution change detection analysis or mapping), to calibrate/validate numerical models aimed at predicting bleaching sensitivity, and to describe the spectral signatures of coral colonies for different health state (and spectral signatures of benthic objects in general) in order to design optical radiative transfer algorithms for high resolution satellite or airborne images.

Future development will likely focus on the design of regional, high-resolution (1 km or less), optimized environmental proxies that will take advantage of SST and other remotely sensed factors and climatologies. This regional approach is very similar to the local optimization of bio-optical ocean color algorithms sought for coastal Case II waters. Numerical modelling with assimilation of remote sensing data will also help in investigating local hydrodynamic and thermodynamic processes that control bleaching. Finally, better understanding of bleaching patterns will come from detailed descriptions of coral community structures. For this goal, high resolution direct remote sensing is critical to stratify detailed quantitative surveys of benthic communities.

3. Remote Sensing to Assess Coral Reef Biodiversity

3.1 WHAT IS BIODIVERSITY ASSESSMENT?

In the introduction for a series of review papers dedicated to biodiversity published by the journal *Nature* in 2000, Tilman (2000) begins with "*The most striking feature of Earth is the existence of life, and the most striking feature of life is its diversity*". We could add "*and coral reef ecosystems harbor the highest diversity of marine life forms on Earth*". Thus, here we emphasize the application of coastal remote sensing for coral reef biodiversity assessment, a developing application with much potential.

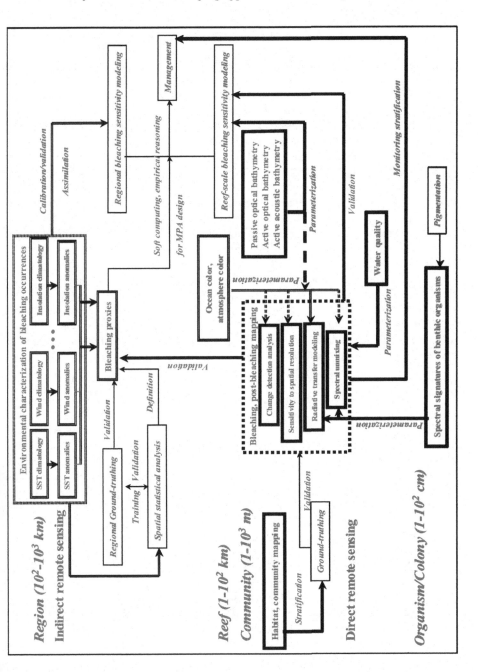

Figure 3. Network of applications for coral bleaching. Text in italics and single lines refer to non-remote sensing actions; double and bold-lines refers to indirect and direct remote sensing respectively.

First used in the early 1980's, the term "biodiversity" refers broadly to the abundance, variety, and genetic constitution of natural living communities. Purvis and Hector (2000) defined biodiversity as the sum of all biotic variation from the level of genes to an ecosystem. Practically, the field of biodiversity encompasses a vast array of scientific topics. Biodiversity science addresses the spatial and temporal patterns in biological diversity and richness, the mechanisms that control these patterns, the influence of these patterns on ecosystems functions and, conservation strategies for the preservation of these patterns. Remote sensing can help address most of these issues. Actually, the coral bleaching application previously discussed is a component of the biodiversity topic, since bleaching may result locally in a loss of biodiversity.

Several reviews of remote sensing applied to biodiversity issues have been compiled, mostly for terrestrial ecosystems (e.g. Stoms and Estes, 1993; Nagendra, 2001; Turner et al., 2003). The application of coastal remote sensing for coral reef biodiversity assessment has much potential. Turner et al. (2003) classified remote sensing applications as "direct" and "indirect". Direct applications utilize imagery with spatial and spectral resolutions adequate to describe the distributions of species or species assemblages present in a target area, thereby creating inventories of those biological units. In a coral reef context, the closest related fields are the spectral discrimination and mapping of reef habitats, communities and species. This single application has been the primary focus of coral reef remote sensing studies from the 1980's through the present. Conversely, indirect approaches, *sensu* Turner et al. (2003), seek to obtain information about diversity patterns, and patterns of ecosystems functions. Two categories emerge: 1) the definition of relevant environmental proxies that will indirectly reflect species richness patterns and will help explain processes that shape these patterns, and 2) the up-scaling of ecosystem functions and processes using habitat/community maps combined with comprehensive field data.

3.2 REEF-SCALE GEOMORPHOLOGY, HABITAT AND COMMUNITY MAPPING

The first applications of remote sensing data for reef assessment consisted of identification and mapping of reef geomorphology or habitat (Fig. 4), using aerial photographs, satellites for Earth observation, and airborne digital systems (see for instance, for recent applications, Isoun et al., 2003; Andréfouët et al., 2003, Garza-Perez et al. 2004). From these studies, there is now a good appreciation of the limits of each sensor based on their spatial and spectral resolutions. The range of map accuracies that can be expected for various complexities of habitat classification schemes has been described for representative sites where extensive ground-truthing and multi-sensor coverages exist (Mumby and Edwards, 2002; Hochberg and Atkinson, 2003; Hochberg et al., 2003a; Capolsini et al., 2003; Andréfouët et al., 2003). For instance, a compilation of results suggests that the overall accuracy (%) is related to the number of habitat classes such that $overall\ accuracy = -3.90*number\ of\ habitat\ classes + 86.38$ ($r^2=0.63$) for Landsat ETM+, and $overall\ accuracy = -2.78*$number of habitat classes $+ 91.69$ ($r^2=0.82$) for IKONOS (Andréfouët et al., 2003).

Most investigations using images with limited spectral (2-5 bands) and spatial resolution (10-80 meters) have used the statistically-oriented "sensor down" approach (*sensu* Hochberg et al., 2003a) where local knowledge of reef communities or structure and image-specific statistics drive the (generally) supervised classification of the data (Green et al., 2000). Processing image data prior to classification may be useful, if not imperative. Atmospheric, sea-surface roughness, and water column corrections have been applied to imagery of reef environments (Zhang et al., 1999; Green et al., 2000; Hu et al., 2001; Palandro et al., 2003, Hochberg et al., 2003b), though not necessarily in a consistent fashion (Andréfouët et al., 2003).

With the increasing interest in hyperspectral sensors better knowledge of spectral signatures of biotic and abiotic end-members has been achieved (Hedley and Mumby, 2002; Minghelli-Roman et al., 2002; Hochberg et al., 2003a; Kutser et al., 2003; Louchard et al., 2003). There has also been progress in understanding spectral mixing (Hedley et al., 2004), radiative transfer processes (Maritorena et al., 1994) and optimization techniques (Lee et al., 2001). These advances help justify an analytical "reef up" approach to map reef communities (Hochberg and Atkinson, 2000; Hochberg et al., 2003a). This "reef up" approach is desirable because it is physics-based, and its application is independent of site- and image-specific statistics. However, calibration accuracy, lack of adequate models, complexity of the radiative transfer processes, heterogeneity of the coral reef world, and absence of spaceborne sensors designed specifically for reef studies make this approach difficult to apply routinely.

Future work will likely exhibit a combination of the "reef up" and "sensor down" approaches depending on the application and available data. Analytical physics-based methods will certainly be effective for mapping and creating inventories of key communities with biogeographically invariant spectral properties like those described by Hochberg et al. (2003a). This likely requires very high spatial resolution imagery, at few meters resolution at most, so that the level of spectral mixing is manageable, with communities composed by few end-members (Hedley et al., 2004). Conversely, at the end of the spatial spectrum or at a geomorphological scale, "sensor down" methods will still be of interest, since the discriminating information depends on both color and topology (position and shape of the classes). While the last 10 years have seen a significant increase in papers presenting the spectral reflectance of reef objects (reviewed in Hochberg et al., 2003a), to date, there has been little focus on formalizing the spatial contextual rules that help improve spectral classification (Mumby et al., 1998; Andréfouët et al., 2000; Andréfouët al., 2003). Spatial rules are critical, but at present are empirical. Contextual rules are currently best formalized using soft-computing techniques such as fuzzy logic (Andréfouët et al., 2000; Matsakis et al., 2002) but there are still very few specific coral reef examples (Suzuki et al., 2001).

Optical data have been successfully used to assess reef communities worldwide, but always in shallow (0-30 meters at best) and relatively clear waters. There is a great deal of evidence that many deeper reef frameworks exist, along with extensive carpets of coral communities. In the Caribbean Sea, for example, the richest communities are often along deep walls and escarpments out of reach of optical data, while shallow reef flats dominate Indo-Pacific reefs. In a global warming context, medium depth coral carpets and reefs could be the only future refuge and reservoir of diversity (Riegl and Piller, 2003). Investigations of these systems will be the next frontier of direct remote sensing work for reefs.

In turbid, deep waters, shipborne acoustic remote sensing techniques are now used to complement airborne or spaceborne optical surveys. Side-scan sonars, single beam echo-sounders and multi-beam swath systems are currently under evaluation in several regions of the world. Common single-beam Acoustic Ground Discrimination Systems (AGDS) are RoxAnn®, Quester Tangent Corp. (QTC)-View™ (Collins et al., 1996; Collins and Lacroix, 1997; Tsemahman and Collins, 1997) and, more recently, ECHOplus™ (Bates and Whitehead, 2001). The respective benefits of these AGDS are compared in the context of marine habitat classification by Kenny et al. (2003). There is still a paucity of coral reef and lagoon work, but recent surveys in the Philippines (White et al., 2003), Florida (Walter et al., 2002; Moyer et al., 2002) and New Caledonia are promising. Thus far, the focus has been predominantly on geologic and sedimentologic

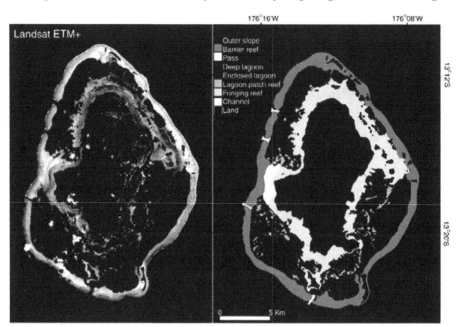

Figure 4. Geomorphological map of Wallis, a volcanic island of the South Pacific Ocean, derived from a Landsat 7 ETM+ image (left). Processing included a supervised classification and spatial contextual editing (right), thus a mix between spectral and spatial information.

characterization, but the biological diversity of deep habitats is also clearly of interest. Classification results, coupled with extensive ground-truthing, have provided results compatible with optical multispectral methods obtained in shallower waters (White et al., 2003).

The difficulty in acoustic mapping consists in translating the roughness and hardness signals acquired using a variety of frequencies into meaningful biological information. In New Caledonia, the potential of RoxAnn® for mapping complex coral reef lagoon bottoms has been tested. Acoustic responses were recorded in a wide area of the Nouméa lagoon (ca. 2 750 km²), from coastal embayment to barrier reef, in order to classify a large range of bottom types with terrestrial and carbonate sediments (Fig. 5). 267 ground-truthing sites were sampled to collect sedimentological data and habitat information to validate the acoustic classification. An example of the final AGDS product between 20 and 40 meters is presented Fig. 6. Future work will explore the coupling between the shallow optical classification with the deep acoustic classification.

Of the several multispectral imaging systems that have been deployed on reefs, only one active system provides results at centimeter scale. Mazel et al. (2003b) deployed a narrow-beam in-water line-scanning multispectral fluorescence imaging system at night on Florida and Bahamas reefs. A statistical classification exploited differences in 3 fluorescence bands and allowed a good determination of the main benthic functional groups. The Mazel et al. pilot study is successful but the required logistical support is still too cumbersome to make this technique widely available.

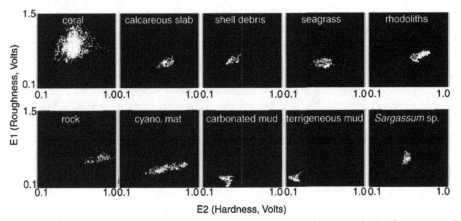

Figure 5. Acoustic AGDS (Roxann) scatter-plot signatures for a variety of bottoms of New Caledonia lagoon.

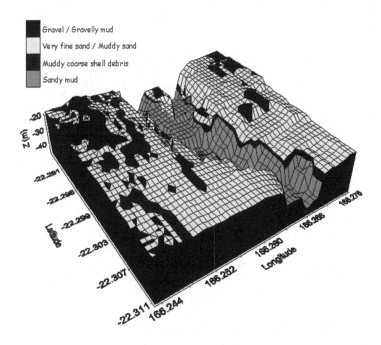

Figure 6. 3D mapping of 4 broad sedimentological classes (simplified bottom types presented in Fig. 5) near the lagoonal submarine valley of Dumbea river, New Caledonia.

Of the several multispectral imaging systems that have been deployed on reefs, only one active system provides results at centimeter scale. Mazel et al. (2003b) deployed a narrow-beam in-water line-scanning multispectral fluorescence imaging system at night on Florida and Bahamas reefs. A statistical classification exploited differences in 3 fluorescence bands and allowed a good determination of the main benthic functional groups. The Mazel et al. pilot study is successful but the required logistical support is still too cumbersome to make this technique widely available.

Geomorphology, habitat and community mapping provides explicitly the distribution and richness of species if adequate ground-truthing (species inventory) has been conducted and is then explicitly integrated into the classification scheme. In most studies, benthic classes are defined by statistical methods and result from a compromise between thematic complexity and classification accuracy (Andréfouët and Claereboudt, 2000). The inclusion of rare species or specific assemblages significantly complicates the classification scheme and most likely makes the mapping exercise statistically intractable. As a result, classification schemes are generally made of a few spectrally distinct broad classes (generally 4 to 20 at best) that explicitly include only conspicuous and dominant species or assemblages. It is difficult to reconcile this simplification with the goal of biodiversity assessment. Perhaps the solution is to consider habitat assemblages as indirect proxies for biodiversity indicators.

It has been suggested that spatial analysis of geomorphology and habitat maps could be used as predictors of benthic or fish diversity under the assumption that diversity of critical habitats will be mirrored by biodiversity patterns (Ward et al., 1999; Purvis and Hector, 2000; Mumby, 2001; Beger et al., 2003; Andréfouët and Guzman, 2004). Mumby (2001) has proposed promising theoretical methods to analyze high-resolution habitat maps, yet to be applied in the real world. Andréfouët and Guzman (2004) have opportunistically measured in the San Blas archipelago, Panama, if diversity in reef geomorphology detected with Landsat imagery could predict diversity in corals, octocorals and sponges, but the results were not totally convincing. Unfortunately, despite its strong potential, using remote sensing as an indirect way to characterize biodiversity patterns is in its infancy and more work is needed.

3.3 THE FUTURE: INDIRECT CHARACTERIZATION OF BIODIVERSITY

The natural processes that shaped modern reef biodiversity have received considerable attention. Indeed, coral reef research is driven by practical conservation goals in order to design the best possible management strategies to maintain biodiversity and ecosystem functions, services and (economic) value (Gaston, 2000; Turner et al., 2003). Gaston (2000) reviews the main ecological areas of inquiry that drive current research exploring biodiversity spatial patterns. Related to reef research, there are four domains to consider: species-energy relationships, latitudinal/longitudinal gradients in species richness, relationships between local and regional richness and taxonomic covariance in species richness. Remote sensing capabilities could be useful for most of these topics beyond providing species lists.

Species richness-energy relationships have long been emphasized since reefs thrive in nutrient poor oligotrophic oceanic environment, which seemed a paradox until relatively recently when nutrient uptake processes began to be elucidated (Atkinson and Bilger, 1992; Atkinson et al., 2001). At community scale, the metabolic standards established by Kinsey (1985) confirm that generally highly diverse coral communities have higher gross production, even if net production within the coral ecosystem is close to zero. Reef-scale productivity patterns have been up-scaled from *in situ* community metabolism measurements using remotely sensed benthic habitat maps (Atkinson and Grigg, 1984; Ahmad and Neil, 1993; Andréfouët and Payri, 2001). However, remote sensing products have not yet been used over large reef areas for inter-regional comparisons of productivity and richness.

Latitudinal/longitudinal gradients in species richness have been the focus of many regional studies (e.g. Connolly et al., 2003). The concept of distance to the center of diversity (DCD, located in the Coral Triangle, between Papua New Guinea, Indonesia and Philippine) has been key for explaining marine richness patterns in the Indo-Pacific areas (Mora et al., 2003 for fish communities). Closely linked to the influence of geographic positional variables (latitude, longitude, DCD), the question of the relationships between local and regional richness for remarkable (or well-known) groups of species, and therefore the influence of local *vs* regional/global factors, has engendered complex statistical multivariate analysis (e.g. Karlson and Cornell, 1998; Bellwood and Hughes, 2001). However, the numerous approaches, heteroclite data sets and intuitive heuristics used in coral reef biogeography analysis of biodiversity patterns are confounding and unification of the data sets would be certainly very useful.

Remote sensing will help make more comparable the different theoretical approaches by providing consistent data sets (Myers et al., 2000; Gaston 2000; Nagendra, 2001; Turner, 2003; McLaughlin et al., 2003; Guinotte et al., 2003). Today, with the wide variety of available remote sensing products (Mumby et al., 2004), there is no justification to use inaccurate, vague or arbitrary proxies of environmental factors for regional or global analyses, which has been common practice in previous studies to compensate for the lack of adequate environmental data (e.g. Bellwood and Hughes, 2001; Roberts et al., 2002). However, this implies that ecologists must master these products and collaborate with remote sensing practitioners beginning with the conceptual designs of their studies.

Geological history, eustatic variations in sea level, genetic and physical connectivity, distance to the center of diversity, temperature, turbidity, geomorphology, habitat structures, natural perturbations and human pressures are examples of factors that have influenced speciation and richness patterns in different reef regions (Galzin et al., 1994; Veron, 1995; Tomascik et al., 1997; Shulman and Bermingham 1995; Done, 1999; Fabricius and De'ath, 2001). As a result, among the standard remote sensing oceanographic products, climatologies of SST, solar insolation, water clarity, chlorophyll and suspended sediments concentrations, exposure to wind and swell, and land masses are clearly of interest. Recently, McLaughlin et al. (2003) reported on a global statistical relationship between reef occurrence and potential terrigeneous sediment sources. The methods of the analysis (k-means clustering, correlation statistics and Geographical Information System analysis) match closely the biodiversity analysis that can be now conceived.

3.4 SYNTHESIS: MULTI-SCALE BIODIVERSITY ASSESSMENTS

On one hand, biodiversity assessment includes one of the most developed and utilized remote sensing techniques, namely habitat mapping using high resolution sensors. On the other hand, it also includes one of the least developed, but equally promising techniques, namely the indirect characterization of biodiversity patterns. The efforts that have been made in a relatively short amount of time specifically for bleaching assessment have yet to be applied for these biodiversity assessments. However, we suggest that the urgency of conservation issues will challenge both the remote sensing and ecology communities and interdisciplinary work will likely improve the situation in the short term.

To establish a network of existing or potential remote sensing applications for biodiversity assessment (Fig. 7), we have considered the two domains where remote sensing will likely be considered first, namely the search for patterns (and explanations

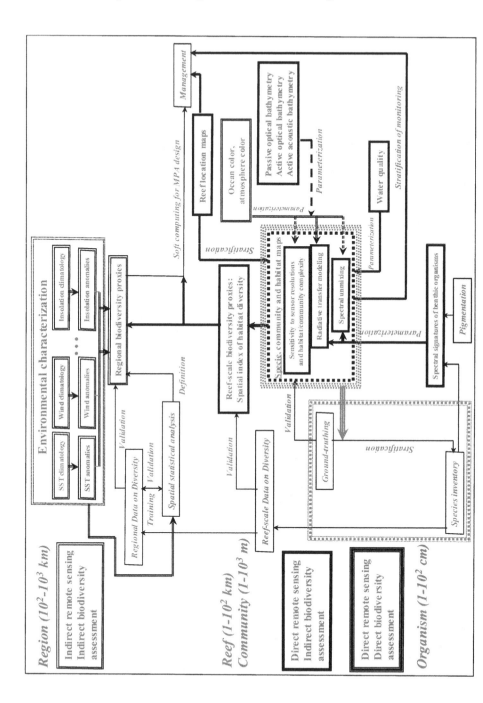

Figure 7. Network of applications for biodiversity assessment. Italicized text refers to non-remote sensing actions.

of these patterns) in gradients in species richness, and the relationships between local and regional richness. Methodologically, this translates into environmental characterization of reef regions (indirect remote sensing of reefs), mapping of reef locations, geomorphology and habitats (direct remote sensing, but indirect biodiversity assessment), and species mapping (direct remote sensing and direct biodiversity assessment). Activities non-specific to remote sensing include *in situ* surveys and spatial statistical analysis, and downstream the data flow, we added methods to design marine protected areas (soft computing) for management purposes.

4. Conclusions and Perspectives: Multi-scale, Multi-sensor, and Multi-method Approach

In this chapter, we used coral bleaching and assessing coral reef biodiversity to illustrate the integration of various remote sensing tools and techniques at several scales. This approach could be presented for other topics of coral reefs as well as other issues in coastal systems. The application of remote sensing to coral bleaching is well established while developments in the use of remote sensing to assess biodiversity are needed. These topics are representative of what can be done to describe the reef structures and their environments in order to better understand the consequences of natural or human forcing on the functioning of these ecosystems.

An analysis of coral bleaching and biodiversity suggests that no single scale (organism or reef or region or global), single method (statistical or analytical) or single sensor (airborne or satellite, multispectral or hyperspectral, active or passive) may be best for all studies. Addressing reef processes requires the capacity to streamline each component in a multi-scale, multi-sensor, multi-method approach, and take advantage of each technique. This also implies combining effectively with parallel domains such as spatial analysis or field survey designs. It is a two-way process. Remote sensing practices should be optimized according to coral reef specificities, and as a feed back, remote sensing should be used more efficiently to observe reefs and their environments.

5. Acknowledgements

This work was supported by the National Aeronautics and Space Administration (NASA).

6. References

Abram, N.J., M.K. Gagan, M.T. McCulloch, J. Chappell, and W.S. Hantoro. 2003. Coral reef death during the 1997 Indian Ocean dipole linked to Indonesian wildfires. Science, 301:952-955.

Ahmad, W., and D.T. Neil. 1994. An evaluation of Landsat Thematic Mapper (TM) digital data for discriminating coral reef zonation: Heron Reef (GBR). International Journal of Remote Sensing, 15:2583-2597.

Andréfouët, S., and M. Claereboudt. 2000. Objective class definitions using correlation of similarities between remotely sensed and environmental data. International Journal of Remote Sensing, 21(9):1925-1930

Andréfouët, S., L. Roux, Y. Chancerelle, and A. Bonneville. 2000. A fuzzy possibilistic scheme of study for objects with indeterminate boundaries: application to french polynesian reefscapes. IEEE Transactions Geoscience and Remote Sensing, 38(1):257-270.

Andréfouët, S., and C. Payri. 2001. Scaling-up carbon and carbonate metabolism in coral reefs using in situ and remote sensing data. Coral Reefs, 19:259-269.

Andréfouët, S., J. Pages, and B. Tartinville. 2001a. Water renewal time for classification of atoll lagoons in the Tuamotu Archipelago (French Polynesia). Coral Reefs, 20:399-408.

Andréfouët, S., F. Muller-Karger, E. Hochberg, C. Hu, and K. Carder. 2001b. Change detection in shallow coral reef environments using Landsat 7 ETM+ data. Remote Sensing of Environment, 79:150-162.

Andréfouët, S., R. Berkelmans, L. Odriozola, T. Done, J. Oliver, and F.E. Muller-Karger. 2002a. Choosing the appropriate spatial resolution for monitoring coral bleaching events using remote sensing. Coral Reefs, 21:147-154.

Andréfouët, S., P.J. Mumby, M. McField, C. Hu, and F. Muller-Karger. 2002b. Revisiting coral reef connectivity. Coral Reefs, 21(1):43-48.

Andréfouët, S., P. Kramer, D. Torres-Pulliza, K.E. Joyce, E.J. Hochberg, R. Garza-Perez, P.J. Mumby, .B. Riegl, H. Yamano, W.H. White, M. Zubia, J.C. Brock, S.R. Phinn, A. Naseer, B.G. Hatcher, and F.E. Muller-Karger. 2003. Multi-sites evaluation of IKONOS data for classification of tropical coral reef environments. Remote Sensing of Environment, 88:128-143.

Andréfouët, S., and B. Riegl. 2004. Remote sensing: a key-tool for interdisciplinary assessment of coral reef processes. Coral Reefs, 23:1-4.

Andréfouët, S., and H. Guzman. 2004. Coral reef distribution, status and geomorphology-biodiversity relationship in Kuna Yala (San Blas) archipelago, Caribbean Panama. Coral Reefs. In press.

Aronson, R.B., W.F. Precht, I.G. McIntyre, and T.J.T. Murdoch. 2000. Coral bleach-out in Belize. Nature, 405:36.

Atkinson, M.J., and R.W. Grigg. 1984. Model of a coral reef ecosystem. II. Gross and net benthic primary production at French Frigate Shoals, Hawaii. Coral Reefs, 3:13-22.

Atkinson, M.J., and R.W. Bilger. 1992. Effects of water velocity on phosphate uptake in coral reef-flat communities. Limnology Oceanography, 37:273-279.

Atkinson, M.J., J.L. Falter, and C.J. Hearn. 2001. Nutrient dynamics in the Biosphere 2 coral reef mesocosm: water velocity controls NH4 and PO4 uptake. Coral Reefs, 20:341-346.

Bates, C.R., and E.J. Whitehead. 2001. ECHOplus measurements in Hopavagen bay, Norway. Sea Technology, 42:34-43.

Beger, M., G.P. Jones, and P.L. Munday. 2003. Conservation of coral reef biodiversity: a comparison of reserve selection procedures for corals and fishes. Biological Conservation, 111:53-62.

Bellwood, D.R., and T.P. Hughes. 2001. Regional-scale assembly rules and biodiversity of coral reefs. Science, 292:1532-1534.

Berkelmans, R., G. De'ath, S. Kininmonth, and W. Skirving. 2004. A comparison of the 1998 and 2002 coral bleaching events on the Great Barrier Reef: spatial correlation, patterns and predictions. Coral Reefs, 23-74-83.

Brock, J., C. Wright, T. Clayton, and A. Nayegandhi. 2004. Optical rugosity of coral reefs in Biscayne National Park, Florida. Coral Reefs, 23:48-59.

Brown, B.E. 1997. Coral bleaching: causes and consequences. Coral Reefs, 16:S129-S138.

Capolsini, P., S. Andréfouët, C. Rion, and C. Payri. 2003. A comparison of Landsat ETM+, SPOT HRV, Ikonos, ASTER and airborne MASTER data for coral reef habitat mapping in South Pacific islands. Canadian J. Remote Sensing, 29(2):187-200.

Chabanet, P. 2002. Coral reef fish communities of Mayotte (western Indian Ocean) two years afer the impact of the 1998 bleaching event. Mar. Freshwater Res., 53:107-113.

Clark, C., P. Mumby, J. Chisholm, J. Jaubert, and S. Andréfouët. 2000. Spectral discrimination of coral mortality states following a severe bleaching event. International Journal of Remote Sensing, 21(11):2321-2327.

Collins, W., R. Gregory, and J. Anderson. 1996. A digital approach to seabed classification. Sea Technology, 37:83-87.

Collins, W., and P. Lacroix. 1997, Operational philosophy of acoustic waveform data processing for seabed classification. In Proceedings of COSU'97 Oceanology International '97 Pacific Rim, Singapore, CD-ROM.

Connolly, S.R., D.R. Bellwood, and T.P. Hughes. 2003. Indo-Pacific biodiversity of coral reefs: deviations from a mid-domain model. Ecology, 84(8):2178-2190.

Costanza, R., R. d'Arge, R. deGroot, S. Farber, M. Grasso, B. Hannon, K. Limburg, S. Naeem, R.V. O'Neill, J. Paruelo, R.G. raskin, P. Sutton, and M. VanDenBelt. 1997. The value of the world's ecosystem services and natural capital. Nature, 387:253-260.

Done, T.J. 1999. Coral community adaptability to environmental change at the scales of regions, reefs and reef zones. American Zoologist, 39:66-79.

Done, T.J., E. Turak, M. Wakeford, G. De'ath, S. Kininmonth, S. Wooldridge, R. Berkelmans, and M.V. Oppen. 2003. Testing bleaching resistance hypotheses for the 2002 Great Barrier Reef bleaching event. Unpublished report to The Nature Conservancy. Report Australian Institute of Marine Science.

Dove, S.G., M. Takabayashi, and O. Hoegh-Guldberg. 1995. Isolation and partial characterization of the pink and blue pigments of Pocilloporid and Acroporid corals. Biological Bulletin, 189:288-297.

Dunne, R., and B. Brown. 2001. The influence of solar radiation on bleaching of shallow water reef corals in the Adaman Sea, 1993-1998. Coral Reefs, 20(3):201-210.

Dustan, P., E. Dobson, and G. Nelson. 2001. Landsat Thematic Mapper: detection of shifts in community composition of coral reefs. Conservation Biology, 15(4):892-902.

Elvidge, C.D., J.B. Dietz, R. Berkelmans, S. Andréfouët, W.J. Skirving, A.E. Strong, and B.T. Tuttle. 2004. Satellite observation of Keppel Islands (Great Barrier Reef) 2002 coral reef bleaching using IKONOS data. Coral Reefs, 23:123-132.

Fabricius, K., and G. Death. 2001. Environmental factors associated with the spatial distribution of crustose coralline algae on the Great Barrier Reef. Coral Reefs, 19:303-309.

Galzin, R., S. Planes, V. Dufour, and B. Salvat. 1994. Variation in diversity of coral reef fish between French Polynesian atolls. Coral Reefs, 13:175-180.

Garza-Perez, J.R., A. Lehmann, and J.E. Arias-Gonzalez. 2004. Spatial prediction of coral reef habitats: integrating ecology with spatial modeling and remote sensing. Marine Ecology Progress Series, 269:141-152.

Gaston, K.J. 2000. Global patterns in biodiversity. Nature, 405:220-227.

Green, E.P., P.J. Mumby, A.J. Edwards, and C.D. Clark. 2000. Remote sensing handbook for tropical coastal management, UNESCO, Paris.

Guinotte, J.M., R.W. Buddemeier, and J.A. Kleypas. 2003. Future coral reef habitat marginality: temporal and spatial effects of climate change in the Pacific basin. Coral Reefs, 22(4):551-558.

Hatcher, B.G. 1997. Coral reef ecosystems: how much greater is the whole than the sum of the parts? Coral Reefs, 16:S77-S91.

Hedley, J.D., and P.J. Mumby. 2002. Biological and remote sensing perspectives of pigmentation in coral reef organisms. Advances in Marine Biology, 43:277-317.

Hedley, J.D., P.J. Mumby, K.E. Joyce, and S.R. Phinn. 2004. Spectral unmixing of coral reef benthos under ideal conditions. Coral Reefs, 23:60-73.

Hochberg, E., and M. Atkinson. 2000. Spectral discrimination of coral reef benthic communities. Coral Reefs, 19:164-171.

Hochberg, E., and M. Atkinson. 2003. Capabilities of remote sensors to classify coral,algae and sand as pure and mixed spectra. Remote Sensing of Environment, 85(2):174-189.

Hochberg, E.J., M.J. Atkinson, and S. Andréfouët. 2003a. Spectral reflectance of coral reef bottom-types worldwide and implications for coral reef remote sensing. Remote Sensing of Environment, 85(2):159-173.

Hochberg, E.J., S. Andréfouët, and M.R. Tyler. 2003b. Sea surface correction of high spatial resolution Ikonos images to improve bottom mapping in near-shore environments. IEEE Transactions Geociences and Remote Sensing, 41(7):1724-1729.

Hochberg, E., M. Atkinson, A. Apprill, and S. Andréfouët. 2004. Spectral reflectance of coral. Coral Reefs, DOI: 10.1007/s00338-003-0350-1

Hoegh-Guldberg, O. 1999. Climate change, coral bleaching and the future of the world's coral reefs. Marine and Freshwater Research, 50(8):839-866.

Holden, H., and E. LeDrew. 2001. Hyperspectral discrimination of healthy versus stressed corals using in situ reflectance. Journal of Coastal Research, 17(4):850-858.

Hu, C., F. Muller-Karger, S. Andréfouët, and K. Carder. 2001. Atmospheric correction and cross-calibration of LANDSAT-7/ETM+ imagery over aquatic environments: multi-platform approach using SeaWiFS/MODIS. Remote Sensing of Environment, 78:99-107.

Hu, C., K.E. Hackett, M.K. Callahan, S. Andréfouët, J.L. Wheaton, J.W. Porter, and F. Muller-Karger. 2003. The 2002 ocean color anomaly in the Florida Bight: a cause of local coral reef decline? Geophysical Research Letters, 30:art. 1151.

Hubbel, S.P. 1997. A unified theory of biogeography and relative species abundance and its aplication to tropical rain forests and coral reefs. Coral Reefs, 16:S9-S21.

Hughes, T.P., and 16 authors. 2003. Climate change, human impacts, and the resilience of coral reefs. Science, 301:929-933.

Isoun, E., C. Fletcher, N. Frazer, and J. Gradie. 2003. Multi-spectral mapping of reef bathymetry and coral cover; Kailua Bay, Hawaii. Coral Reefs, 22(1):68-82.

Karlson, R.H., and H.V. Cornell. 1998. Scale-dependant variation in local vs regional effects on coral species richness. Ecological Monographs, 68(2):259-274.

Kenny, A., I. Cato, M. Desprez, G. Fader, R. Schuttenhelm, and J. Side. 2003. An overview of seabed-mapping technologies in the context of marine habitat classification. 60, 411-418):

Kinsey, D.W. 1985, Metabolism, calcification and carbon production. I. System level studies. In Proceedings of 5th Int. Coral Reef Congr., Tahiti, 4:505-526.

Kleypas, J.A., R.W. Buddemeier, and J.-P. Gattuso. 2001. The future of coral reefs in an age of global change. International Journal of Earth Sciences, 90:426-437.

Kutser, T., A. Dekker, and W. Skirving. 2003. Modelling spectral discrimination of Great Barrier Reef benthic communities by remote sensing instruments. Limnology Oceanography, 48(1,part 2):497-510.

Lee, Z.P., K.L. Carder, R.F. Chen, and T.G. Peacock. 2001. Properties of the water column and bottom derived from Airborne Visible Infrared Imaging Spectrometer (AVIRIS) data. JGR, 106(C6):11639-11651.

Lewis, A. 1999. Depth model accuracy: a case study in the Great Barrier Reef Lagoon. In: Lowell K., and Jaton A. (eds), Spatial accuracy assessment. Ann Arbor Press, Chelsea, Michigan, pp. 71-78.

Liceaga-Correa, M.A., and J.I. Euan-Avila. 2002. Assessment of coral reef bathymetric mapping using visible LANDSAT TM data. International Journal of Remote Sensing, 23(1):3-14.

Liu, G., A. Strong, and W. Skirving. 2003. Remote sensing of Sea Surface Temperature during the 2002 (Great) Barrier Reef coral bleaching. EOS Transactions AGU, 84(15):137, 144.

Loubersac, L., P.Y. Burban, O. Lemaire, H. Varet, and F. Chenon. 1991. Integrated study of Aitutaki's lagoon (Cook Islands) using SPOT satellite data and in situ measurements: bathymetric modelling. Geocarto Int., 6(2):31-37.

Louchard, E., P. Reid, F. Stephens, C. Davis, R. Leathers, and T. Downes. 2003. Optical remote sensing of benthic habitats and bathymetry in coastal environments at Lee Stocking Island, Bahamas: A comparative spectral classification approach. Limnology Oceanography, 48(1, part 2):511-521.

Lyzenga, D.R. 1978. Passive remote sensing techniques for mapping water depth and bottom features. Applied Optics, 17(3):379-383.

Maritorena, S., A. Morel, and B. Gentili. 1994. Diffuse reflectance of oceanic shallow waters: influence of water depth and bottom albedo. Limnology and Oceanography, 39(7):1689-1703.

Matsakis, P., J.M. Spalding, and L.M. Szatandera. 2002. Applying Soft Computing in Defining Spatial Relations, Springer-Verlag.

Mazel, C.H., M.P. Lesser, M.Y. Gorbunov, T.M. Barry, J.H. Farrel, K.D. Wyman, and P.G. Falkowski. 2003a. Green-fluorescent proteins in Caribbean corals. Limnology and Oceanography, 48:402-411.

Mazel, C.H., M.P. Strand, M.P. Lesser, M.P. Crosby, B. Coles, and A.J. Nevis. 2003b. High-resolution determination of coral reef bottom cover from multispectral fluorescence laser line scan imagery. Limnology and Oceanography, 48:522-534.

McClanahan, T. 2000. Bleaching damage and recovery potential of Maldivian coral reefs. Marine Pollution Bulletin, 40(7):587-597.

McLaughlin, C.J., C.A. Smith, R.W. Buddemeier, J.D. Bartley, and B.A. Maxwell. 2003. Rivers, runoff and reefs. Global and Planetary Change, 39(1-2):191-199.

Minghelli-Roman, A., J.R.M. Chisholm, M. Marchioretti, and J.M. Jaubert. 2002. Discrimination of coral reflectance spectra in the Red Sea. Coral Reefs, 21:307-314.

Mora, C., P. Chittaro, P.F. Sale, J.P. Kritzer, and S.A. Ludsin. 2003. Patterns and processes in reef fish diversity. Nature, 421:933-936.

Morel, Y. 1996. A coral reef lagoon as seen by SPOT. In Proceedings of 8th Australasian Remote Sensing Conference, Canberra, Australie, pg. 51-62.

Moyer, R.P., B. Riegl, R.E. Dodge, B.K. Walker, and D.S. Gilliam. 2002. Acoustic remote sensing of reef benthos in Broward County, FLorida (USA). In Proceedings of 7th Int. Conf. Remote Sensing for Marine and Coastal Environments, Miami, FL, CDROM.

Mumby, P.J., E.P. Green, C.D. Clark, and A.J. Edwards. 1997. Coral reef habitat mapping: how much detail can remote sensing provide? Marine Biology, 130:193-202.

Mumby, P., C.D. Clark, E.P. Green, and A.J. Edwards. 1998. Benefits of water column correction and contextual editing for mapping coral reefs. International Journal of Remote Sensing, 19(1):203-210.

Mumby, P.J. 2001. Beta and habitat diversity in marine systems: a new approach to measurement, scaling and interpretation. Oecologia, 128(2):274-280.

Mumby, P., J. Chisholm, A. Edwards, S. Andréfouët, and J. Jaubert. 2001a. Cloudy weather may have saved Society Island reef corals during the 1998 ENSO event. Marine Ecology Progress Serie, 222:209-216.

Mumby, P.J., J.R.M. Chisholm, C.D. Clark, J.D. Hedley, and J. Jaubert. 2001b. A bird-eye view of the health of coral reefs. Nature, 413:36.

Mumby, P.J., and A.J. Edwards. 2002. Mapping marine environments with IKONOS imagery: enhanced spatial resolution can deliver greater thematic accuracy. Remote Sensing Environment, 82:248-257.

Mumby, P.J., W. Skirving, A.E. Strong, J.T. Hardy, E.F. LeDrew, E.J. Hochberg, R.P. Stumpf, and L.T. David. 2004. Remote sensing of coral reefs and their physical environment. Marine Pollution Bulletin, 48(3-4):219-228.

Myers, N., R.A. Mittermeier, C.G. Mittermeier, G.A.B. deFonseca, and J. Kents. 2000. Biodiversity hotspots for conservation priorities. Nature, 403:853-858.

Nagendra, H. 2001. Using remote sensing to assess biodiversity. International Journal of Remote Sensing, 22(12):2377-2400.

Otis, D., K. Carder, D. English, J. Ivey, and H. Warrior. 2004. CDOM transport from the Bahamas Banks. Coral Reefs, 23:152-160.

Palandro, D., S. Andréfouët, F. Muller-Karger, P. Dustan, C. Hu, and P. Hallock. 2003. Detection of changes in coral communities using Landsat 5/TM and Landsat 7/ETM+ data. Canadian J. Remote Sensing, 29(2):201-209.

Palandro, D., S. Andréfouët, P. Dustan, and F.E. Muller-Karger. 2003. Change detection in coral reef communities using the Ikonos sensor and historic aerial photographs. International Journal of Remote Sensing, 24(4):873-878.

Pandolfi, J.M., R.H. Bradbury, E. Sala, T. Hughes, K. Bjorndal, R. Cooke, D. McArdle\, L. McClenachan, M. Newman, G. Paredes, R. Warner, and J. Jackson. 2003. Global trajectories of the long-term decline of coral reef ecosystems. Science, 301:955-958.

Penland, L., J. Kloulechad, D. Idip, and R.v. Woesik. 2004. Coral spawning in the western Pacific Ocean is related to solar insolation: evidence of multiple spawning events in Palau. Coral Reefs, 23:133-140.

Purvis, A., and A. Hector. 2000. Getting the measure of biodiversity. Nature, 405:212-219.

Riegl, B., and W. Piller. 2003. Possible refugia for reefs in times of environmental stress. Int. J. Earth Sciences, 92:520-531.

Roberts, C.M., C.J. McClean, J.E.N. Veron, J.P. Hawkins, G.R. Allen, D.E. McAllister, C.G. Mittermeier, F.W. Schueler, M. Spalding, F. Wells, C. Vynne, and T.B. Werner. 2002. Marine biodiversity hotspots and conservation priorities for tropical reefs. Science, 295:1280-1284.

Salih, A., A. Larkum, G. Cox, M. Kuhl, and O. Hoegh-Guldberg. 2000. Fluorescent pigments in corals are photoprotective. Nature, 408:850-853.

SHOM-SPT. 1990, Les dossiers de spatio-préparation des campagnes hydrographiques aux Tuamotu-Gambier. In Proceedings of Pix'Iles 90: Int. Workshop on Remote Sensing and Insular Environments in the Pacific: integrated approaches, Nouméa-Tahiti, pp. 593-595, 6 pl.

Shulman, M.J., and E. Bermingham. 1995. Early life histories, ocean currents, and the population genetics of Caribbean reef fishes. Evolution, 49(5):897-910.

Spalding, M., and G. Jarvis. 2002. The impact of the 1998 coral mortality on reef fish communities in the Seychelles. Marine Pollution Bulletin, 44(4):309-321.

Stoms, D.M., and J.E. Estes. 1993. A remote sensing research agenda for mapping and monitoring biodiversity. Int. Journal of Remote Sensing, 14(10):1839-1860.

Storlazzi, C., J. Logan, and M. Field. 2003. Quantitative morphology of a fringing reef tract from high resolution laser bathymetry: Southern Molokai, Hawaii. Geological Society of America Bulletin, 115(11):1344-1355.

Stumpf, R., K. Holderied, and M. Sinclair. 2003. Determination of water depth with high-resolution satellite imagery over variable bottom types. Limnology Oceanography, 48(1(Part2)):547-556.

Suzuki, H., P. Matsakis, S. Andréfouët, and J. Desachy. 2001, Satellite image classification using expert structural knowledge : a method based on fuzzy partition computation and simulated annealing. In Proceedings of Annual Conf. of the Int. Ass. for Mathematical Geology, Cancun, Mexico, CDROM.

Tilman, D. 2000. Causes, consequences and ethics of biodiversity. Nature, 405:208-211.

Tomascik, T., A.J. Mah, A. Montji, and M.K. Moosa. 1997. The ecology of the Indonesian seas, Periplus Editions, Dalhousie.

Tsemahman, A.S., W.T. Collins, and B.T. Prager. 1997, Acoustic seabed classification and correlation analysis of sediment porperties by QTC View. In Proceedings of Oceans'97, Halifax.

Turner, W., S. Spector, N. Gardiner, M. Fladeland, E. Sterling, and M. Steininger. 2003. Remote science for biodiversity science and conservation. Trends in Ecology and Evolution, 18(6):306-314.

Veron, J.E.N. 1995. Corals in space and time: the biogeography and evolution of the Scleractinia, Comstock/Cornell,

Walter, D., D. Lambert, and D. Young. 2002. Sediment facies determination using acoustic techniques in a shallow-water carbonate environment , Dry Tortugas, Florida. Marine Geology, 182:161-177.

Ward, T.J., M.A. Vanderklift, A.O. Nicholls, and R.A. Kenchington. 1999. Selecting marine reserves using habitats and species assemblages as surrogates for biological diversity. Ecological Applications, 9(2):691-698.

West, J., and R. Salm. 2003. Resistance and resilience to coral bleaching: implications for coral reef conservation and management. Conservation Biology, 17(4):956-967.

White, W.H., A.R. Harborne, I.S. Sotheran, R. Walton, and R.L. Foster-Smith. 2003. Using an acoustic ground discrimination system to map coral reef benthic classes. Int. Journal Remote Sensing, 24(13):2641-2660.

Wilkinson, C. 2000. Status of the coral reefs of the world: 2000, Australian Institute Marine Sciences, Townsville.

Woolridge, S., and T. Done. 2004. Learning to predict large-scale coral bleaching from past events: A Bayesian approach using remotely sensed data, in-situ data, and environmental proxies. Coral Reefs, 23:96-108.

Yamano, H., and M. Tamura. 2004. Detection limits of coral reef bleaching by satellite remote sensing: simulation and data analysis. Remote Sensing of Environment, 90:86-103.

Zhang, M., K.L. Carder, F.E. Muller-Karger, Z. Lee, and D.B. Goldgof. 1999. Noise reduction and atmospheric correction for coastal applications of Landsat Thematic Mapper imagery. Remote Sensing of Environment,70:167-180

Chapter 14

REAL-TIME USE OF OCEAN COLOR REMOTE SENSING FOR COASTAL MONITORING

ROBERT A. ARNONE AND ARTHUR R. PARSONS

Naval Research Laboratory, Code 7330, Ocean Sciences Branch, Stennis Space Center, MS, 39529 USA

1. Introduction

Advances in ocean optics research over the last five years have yielded an improved understanding of how different water components affect the reflectance spectra of the ocean, or ocean color. This increased understanding has enabled the development of algorithms to separate the reflectance spectra acquired remotely by ocean color satellites into estimates of select in-water constituents. The application of new algorithms to ocean color data has dramatically increased our understanding of the in-water optical environment and, as a consequence, our understanding of optically active materials and associated biogeochemical properties. Ocean color technology and algorithm development has progressed to where ocean color remote sensing is now suitable to detect and monitor coastal ocean processes. Continued advances in our understanding of the complex nature of optically active in-water particles (the composition, size, and index of refraction, which can affect the reflectance spectra significantly) and dissolved substances will provide new and improved algorithms to better derive estimates of these constituents from remotely sensed ocean color spectra.

Unfortunately, most of the research tools available for ocean color remote sensing are used by a limited science community and have not been transferred to general applications or operational users. Based on the unique utility of ocean color remote sensing for monitoring coastal and ocean processes, a specific plan to transfer these new capabilities is essential to better support future decisions about these environments (Arnone, 1999).

Since the launch of the first weather satellite, the Television and Infrared Observation Satellite (TIROS), on April 1, 1960, the growth of operational satellite products has revolutionized weather prediction. The rapid increase in satellite weather products, and their assimilation into numerical models, has made satellite sensing an integral part of weather forecasting. The evolution of ocean remote sensing is occurring at a similarly rapid pace; operational ocean monitoring is building on the heritage of operational meteorology. The oceanographic community is at a crucial point where the availability of real-time data, coupled with recent advances in algorithm development and image processing, can revolutionize our ability to monitor the ocean.

Real-time remote sensing for operational decision-making means different things for different applications. For some operational applications, the best data needed to make a decision must be available within an hour of collection. In this context, however, the decision maker must understand the limitations and potential errors associated with the data products generated to establish how these factors affect their decision. In other

R.L. Miller et al. (eds.), Remote Sensing of Coastal Aquatic Environments, 317–337.

situations, data provided within days, weeks, or months is adequate to make effective decisions. The increase in time may be related to the time scales of the processes affecting the decision or, more importantly, may be related to a more rigorous requirement for product accuracy or validation. These expanded requirements translate into increased data processing and analysis; hence, the time until data products are available after collection is increased. For example, most scientific studies require highly accurate products and as such must wait until the data products are optimized. In contrast, most operational users, such as environmental managers, require more timely data and therefore cannot afford the time it takes to produce research-quality products.

Satellite sensors such as the Sea-Viewing Wide Field-of-View Sensor (SeaWiFS), the Moderate Resolution Imaging Spectroradiometer (MODIS, on both the Terra and Aqua satellites), the Medium Resolution Imaging Spectrometer (MERIS), Hyperion, and the Landsat Thematic Mapper (TM) are providing a near real-time capability to monitor the biogeochemical, optical, and physical processes of coastal and open oceans. Current research has advanced well beyond the Coastal Zone Color Scanner (CZCS) of the 1980's, where chlorophyll *a* was the primary data product. Researchers are now closely examining the spectral signatures of remote sensing reflectance to develop new methods to extract a variety of in-water optical properties. As a result of advances in atmospheric correction (Gordon and Wang, 1994; Arnone et al., 1998; and Ruddick et al., 2000), rigorous controls on sensor calibration, and improvements in ocean optical instrumentation, it is now possible to accurately measure the reflectance spectrum over water. However, spaceborne spectroscopy of the ocean is in its initial stages of development. To prepare for satellite spectroscopy of the future, the ocean community must identify the pathway to translate remote sensing ocean color data into products that support the decision-making process, are easy to use by the operational community, and are cost-effective to obtain on a routine basis.

There is a suite of data sources and data products available to environmental managers to aid in their decisions regarding coastal issues. These include *in situ* measurements, numerical models, ocean climatologies, and remote sensing data. Most coastal managers require highly reliable data as input to the decision-making process. Hence they may not use remote sensing products as a primary data source, opting to rely on other data types. However, as remote sensing products are validated and become more reliable, the use of remote sensing data (in combination with other data sources) will provide improved products to coastal decision makers.

Our effort is to identify and demonstrate how ocean color products can be applied to operational applications. Exploitation of these products requires the ocean community to: 1) identify and maintain continuous streams of operational products (best available products from reliable data sources); 2) train the operational community in the use of these products; 3) establish interactions between the scientist and operational manager or decision maker to identify the need for new or improved products; and, 4) develop new products for testing and use in monitoring and decision making.

2. Why Real-time Monitoring?

Coastal biogeochemical and optical processes occur on time scales of hours (or less) in response to physical processes such as mixing, due to local winds or tides, and freshwater river discharge. The ability to monitor and predict changes in these processes over large spatial scales is possible only through real-time satellite observations. These data can then be used to support such decisions as when to close a shellfish bed due to a Harmful Algal Bloom (HAB, Stumpf and Tomlinson, Chapter 12) or how to best mitigate a polluted river plume. For many operational applications, products supporting such decisions must be available within an hour from when the data are collected by the satellite. This capability is well within reach, given today's computer processing power and online distribution networks. In the near future, geostationary satellite platforms such as the Geostationary Operational Environmental Satellite-R (GOES-R) will provide near-continuous (hourly) coverage of our coastal waters, thereby creating a new paradigm in coastal monitoring, and enabling mangers to track changes along our coasts at higher spatial, temporal, and spectral resolution.

However, even mature products derived from well-understood remotely sensed data from such satellites as SeaWiFS and MODIS are not routinely used by the operational community. To fully leverage the suite of products derived from these satellites in decision-making, they must be integrated into the flow of data already available to coastal managers and other decision makers. Additionally, coastal managers must be trained in the application of these products to the specific decision to be made.

Streamlined access to satellite-based products and their availability to a more public audience are becoming more possible as a result of the internet. Additionally, satellite datasets and derived products are being integrated with numerical models through advances in data fusion. Satellite datasets are being used as data layers within Geographic Information Systems (GISs) through new integration techniques. Multiple types of data (*in situ* oceanographic and meteorological measurements, ocean model output, and satellite data) are now being coupled through web-based distributed data servers. Technology advances such as these will provide new data services that support the routine use of remotely sensed data and products by decision makers within the next five years.

The objective of this chapter is to illustrate how ocean color remote sensing products can be used for real-time monitoring and tracking of coastal biogeochemical and physical processes. Recently-developed capabilities in ocean color remote sensing supporting the operational community are discussed. Here we discuss mechanisms that have been successfully used to bridge the research-to-operational gap, that enable rapid product generation, and assist in decision making.

3. Coastal and Ocean Products from Ocean Color

The major development in ocean color remote sensing since the CZCS era (1978-1986) is that satellite optical signals have been theoretically coupled with multi-channel atmospheric aerosol models and complex oceanic optical models to derive in-water properties. Previously, atmospheric correction algorithms based on near-infrared aerosol models and simple empirical channel (band) ratios were used to determine ocean chlorophyll concentration. Recent developments leverage the radiometric properties of the satellite optical signature to apply radiative transfer theory to atmospheric aerosol models and in-water optical physics to determine a universal method to derive ocean properties. These advances were facilitated by more accurate radiometric measurements that resulted from improved stability of optical sensors on satellites. The improved consistency and accuracy of satellite data enabled the effective use of radiative transfer models to derive optical properties from measured reflectance spectra. Advances in in-water optical instrumentation (Twardowski et al., Chapter 4) enabled the results of these models to be validated and provided methods to vicariously calibrate remote sensing sensors. These sensors are now ready to quantitatively monitor in-water optical properties from space.

Additionally, significant advances in the ability of ocean color satellites to accurately detect and measure biological, geological, and optical processes have taken place in the last five years. Recent papers (for example, Arnone and Gould, 1998; Arnone et al., 2002; Carder et al., 1999; Lee et al., 2002) provide detail semi-analytical and optimization approaches to separate the in-water signal (i.e., remote sensing reflectance) into fundamental optical components. The ability to resolve optical properties has extended the utility of ocean color remote sensing beyond research applications, enabling their use in such applications as monitoring river plume discharge. The ability to detect different and changing optical properties over the plume provides a means to trace marine and terrigenous particles (in this case by deriving the backscattering coefficient using techniques described in Lee et al., 2002 and Carder et al., 1999). Pollution discharge can be monitored by its association with Colored Dissolved Organic Matter (CDOM). The spatial variation of optical signatures is used to understand the mixing and dispersion processes of suspended sediments and fresh water plumes in coastal regions. Furthermore, these properties can be used to provide products supporting diver underwater visibility assessment and underwater laser system performance (Arnone, 1999).

Until recently, ocean color sensors were used to measure water turbidity or clarity. The optical research community has drawn away from these terms, since they were never clearly defined nor were they quantitative. Instead of turbidity or clarity, one now can refer to more precise optical measurements, such as the Inherent Optical Properties (IOPs); those optical properties of the water and water constituents that are independent of the ambient light field. IOPs consist of the absorption (a), scattering (b), and attenuation (c) coefficients that more quantitatively define in-water optical conditions. Furthermore, IOPs can be used to measure the light field and monitor its effect on biological, geological, and chemical processes. Over the past five years, ocean color remote sensing has been used to quantify and monitor an expanding number of in-water biological, geological, and optical properties. Table 1 is a listing of common bio-optical properties derived from ocean color remote sensing and products, under development for their use in ocean and coastal decision making. Note, that many of these products previously validated for less dynamic ocean environments, are now being validated for coastal waters (Ladner et al., 2002).

3.1 APPLICATIONS USING THE BACKSCATTERING COEFFICIENT

Remote sensing reflectance has been linked with two IOPs, backscatter and absorption (Gordon et al., 1975), which can be used to derive a spectral backscatter coefficient image (Lee and Carder, Chapter 8). The backscattering coefficient (b_b) results from the interaction of the light field with particles in the backward direction (90-180 degrees). Total scattering (b) represents the integration of scattering over 0 to 180 degrees.

The backscattering coefficient is closely associated with suspended particle concentration; however, the specific relationship varies based on the type of particles present in the water. Marine particles are extremely diverse and can be composed of organic (phytoplankton and organic detritus) and/or inorganic (clays, silts, silicates, calcites tests, etc.) materials. These particle types are widely different optically and are the result of a complex assemblage of coastal processes that includes bottom resuspension, particle flocculation and agglomeration, phytoplankton growth and decay, and river discharge. Backscatter represents a bulk measurement of the interaction of the light field with these diverse particle types. The backscattering coefficient is influenced by the different properties of these particles such as size, index of refraction, shape, composition, etc., as well as whether they are organic or inorganic in nature. Additionally, the specific dynamics of the coastal area determines not only the types of particles present within the water column, but their associated properties as well. Therefore, it is difficult to develop universal relationships between the backscattering coefficient and particle concentration. There are, however, some assumptions that allow an estimate of particle concentration. If it is assumed that particles across a given region (the Mississippi River plume, for example) have similar type and size, then the magnitude of the backscattering coefficient at 550 nm is directly proportional to the particle concentration. Regional algorithms that derive total particle concentration from the backscattering coefficient are generally not necessary, as the absolute measure of the backscattering coefficient is sufficient to track relative particle concentrations.

Table 1. Common bio-optical and satellite-derived ocean properties from SeaWiFS, and MODIS (Terra and Aqua). The wavelength dependence of products is indicated by λ.

Validated products

Chlorophyll concentration	biological processes such as algal (harmful and non-harmful) blooms and decay
Spectral backscattering coefficient - $b_b(\lambda)$	90 to 180 degree particle scattering linked to concentration, composition, index of refraction of organic (marine) and inorganic (terrigenous) particles; resuspension of particles
Spectral absorption coefficient - $a_t(\lambda)$	total absorption, changes in water quality
Spectral absorption colored dissolved organic matter - $a_{CDOM}(\lambda)$	conservative tracer of river plumes, linked with coastal salinity, photo-oxidation processes
Spectral particle absorption coefficient - $a_p(\lambda)$	particle composition, (organic and inorganic particles)
Spectral phytoplankton absorption coefficient - $a_\phi(\lambda)$	absorption linked to differences in chlorophyll packaging within phytoplankton cells
Remote sensing reflectance - $R_{RS}(\lambda)$	spectral absolute water color and water signature
Diffuse attenuation coefficient - ($k532$, $k490$)	light penetration depth, light availability at depth
Aerosol concentration - Epsilon	type and distribution, affects visibility, atmospheric correction methods
Beam attenuation coefficient - $c(\lambda)$	total light attenuation using a collimated beam
Diver visibility	horizontal visibility, average target size, target contrast, solar overhead illumination
Laser penetration	underwater performance of lasers (imaging or bathymetry systems)
Sea surface temperature	skin temperature / bulk temperature (MODIS)

Table 1. Continued.

Exploratory products

Surface Salinity	absorption at 412 nm or CDOM absorption in coastal areas with high surface gradients
Particle size distribution (junge distribution) (<1 um - >100 um) and concentration	spectral backscattering coefficient in coastal waters, concentration at different sizes
Particulate Organic Matter (POM)	a(detritus @443nm) to estimate carbon flux
Particulate Inorganic Matter (PIM)	a(detritus @ 412 nm) to estimate particle flux
Total Particle Concentration	particle composition (organic / inorganic particles), regional dependent
Particle organic / inorganic ratio for each size class	particle fluxes in surface water, settling. and resuspension
Satellite water mass optical classification	- identification of specific water masses using optical signature and tracking movements
Satellite products integrated with numerical models of currents	interpreting how physical processes (advection) affect the bio-optical response (e.g., advection of chlorophyll blooms)
Vertical profiles of bio-optical properties	determined by assimilating modeled mixed layer depth with the satellite surface chlorophyll
Primary Production	determined through linked seasonal SST and chlorophyll fluorescence

However, if one seeks to compare dynamically different regions, it may be necessary to modify or "tune" an existing algorithm to accurately compare different regions.   For most operational applications regional tuning is not required, since it is the bulk measurement that is most useful. For example, the backscattering coefficient can be used to trace the Mississippi River sediment plume or locate the turbidity maximum in an estuary.

Twardowski et al. (2001) examined methods to determine differences in particle type, size, and index of refraction from spectral backscatter and total scatter data. Haltrin and Arnone (2003) applied these spectral backscatter relationships to SeaWiFS data, in an attempt to estimate particle size distribution and the index of refraction. The results indicated that geospatial differences exist in the distribution of large and small particles in Mississippi River plume surface waters.   Similarly, Stavn et al. (2002), Stavn and Gould (2003), and Gould et al. (2002) demonstrated the use of spectral backscattering, combined with spectral particle absorption, to determine particulate organic carbon (POC) and particulate inorganic matter (PIM). These regional algorithms demonstrate how ocean color remote sensing can provide invaluable information to monitor and track carbon of terrestrial origin and ocean-generated carbon as they are distributed and dispersed along the coast.   Over longer time scales, sequential spectral backscattering data can be collected and these measurements can be used to determine the origin of in-water particles and their fate within the coastal zone. Currently, *in situ* sampling methods do not provide the spatial coverage afforded through remote sensing to determine particle distributions over large regions effectively.

3.2 APPLICATIONS USING THE ABSORPTION COEFFICIENT

Historically, ocean color remote sensing has been used most frequently to derive the absorption coefficient in open ocean areas to estimate the chlorophyll concentration of surface waters. Open ocean chlorophyll algorithms are based on the spectral absorption properties of phytoplankton (Gordon and Clark, 1977; O'Reiley et al., 1998), as determined by applying an empirical relationship of channel ratios. Ratio algorithms minimize the influence of the backscattering signal and employ the assumption that the ocean color signature is a direct measure of chlorophyll absorption. Although this assumption holds true in the open ocean, the ocean color signature in complex coastal waters is a result of a variety of constituents, in addition to those associated with phytoplankton chlorophyll.

Semi-analytical algorithms developed in the last few years (Lee et al., 2002; Arnone et al., 2000; and Carder et al., 1999) partition the total in-water absorption ($a_t(\lambda)$) into four basic absorption coefficient components: water ($a_w(\lambda)$); phytoplankton ($a_\phi(\lambda)$); CDOM ($a_{cdom}(\lambda)$); and, detritus ($a_{detritus}(\lambda)$). These individual absorption components are extremely important in coastal areas, as they are highly variable spatially and temporally and can be used to determine different coastal properties. Therefore, to determine phytoplankton concentration in the coastal zone, it is important to separate out the other absorption components. The spatial extent of remote sensing image products of different absorption components allows for an improved capability to monitor coastal processes on a regional basis. For example, differences in phytoplankton and CDOM absorption coefficients are used to suggest areas where phytoplankton growth and decay may occur.

Images of the CDOM absorption coefficient can provide a monitoring capability for characterizing the dispersion of river discharge into coastal waters, as CDOM in the coastal zone may act as a conservative tracer. The long organic molecules of CDOM provide active ionic bond sites for chemical radicals and agents associated with terrestrial-based pollutants. CDOM absorption can be used to track pollutants and heavy metals associated with organic matter in near-coastal plumes

In particle-rich coastal waters, non-living particles can provide a significant contribution to light absorption. These particles can be composed of inorganic clays, silts, etc., or decayed organic matter (degraded organic debris from seagrasses and terrestrial sources) and are not associated with living phytoplankton. The detritus absorption coefficient contributes significantly to an ocean color signal in the presence of high concentrations of inorganic particles. Through the application of semi-analytical algorithms, detritus absorption and phytoplankton absorption coefficients can be used to characterize the distribution of living and non-living particles in coastal regions. For example, these absorption-based products can be used to distinguish between particles in the water resulting from bottom resuspension and those resulting from phytoplankton blooms.

3.3 APPLICATIONS USING THE BEAM ATTENUATION COEFFICIENT

The beam attenuation coefficient can be used to estimate the suspended particle concentration in coastal waters and to determine underwater diver visibility. The spectral beam attenuation coefficient (c) defines how a collimated beam of light is attenuated within the water column and is most closely coupled to the term turbidity. The coefficient is the sum of the total absorption coefficient and the total scattering coefficient. The attenuation coefficient is derived from the ocean color spectrum through the application of semi-analytical algorithms that determine the total absorption and backscattering coefficients (b_b). The backscattering coefficient is then converted to total scattering using the Volume Scattering Function (VSF) (Petzold, 1972). However, Haltrin et al. (2002) and Rossler and Boss (2003) suggest that major differences exist in the b_b-to-b relationship for different water types and that these differences are dependent on the type of in-water particles (mineral or organic) and their index of refraction. The attenuation coefficient is used to estimate horizontal underwater diver visibility as $4.8/c$ for a fixed target size and contrast, with solar light directly overhead (Davies-Colley, 1988; Bowers, 2003).

3.4 APPLICATIONS USING THE DIFFUSE ATTENUATION COEFFICIENT

Ocean color remote sensing can be used to determine the diffuse attenuation coefficient at different wavelengths (Austin and Petzold, 1981; 1986; and Mueller et al., 1990). The diffuse attenuation coefficient ($k(\lambda)$ or $k_d(\lambda)$) is an apparent optical property (AOP) that is dependent on local lighting conditions and represents the rate at which diffuse light will be attenuated with depth in the ocean. For open ocean waters, k is estimated at 490 nm, the wavelength at which maximum penetration occurs. However, the spectral dependence of k has been modeled such that attenuation coefficients at 490 nm is extended to other wavelengths (Austin and Petzold, 1986). The e-folding depth, represented by $1/k$ is the attenuation length (or the depth) where the surface radiation decreases by $1/e$ or approximately 37% of its initial surface value. In clear ocean waters such as the Gulf Stream, $1/k$ occurs at a depth of approximately 25 meters. In coastal turbid water $1/k$ is usually 1 to 2 meters. In fact, properties retrieved from ocean color remote sensing represent the light integrated from within this first attenuation depth.

If the water column is assumed homogeneous, the subsurface light field can be estimated at any depth based on the surface irradiation and $k490$. The subsurface light field can then be used to determine the Photosynthetic Available Radiation (PAR), the measurement of the light field from 400 to 700 nm, the portion of the spectrum that is available to support biological activity. $k490$ is used to determine whether conditions are suitable for plant growth. Additionally, the subsurface light field is used to determine how much visible light can heat subsurface waters and affect subsurface heat fluxes (Rochford et al., 2001).

The diffuse attenuation coefficient is also used to estimate the penetration depth of laser systems. Neodymium - Yttrium/Aluminum/Garnett (Nd:YAG) lasers operate at a wavelength of 532 nm and are used to obtain bathymetric soundings or for underwater laser imaging. The performance of these systems is assessed by the number of attenuation lengths the system can operate, which can be obtained from the $k532$ image product. For example, the $k532$ product is used to determine if the waters along a coastline are too turbid for a laser bathymetry system to obtain reliable depth measurements. The $k532$ product provides significant cost savings as it helps operators identify when and where their sensors will function optimally.

3.5 EXAMPLES OF REMOTE SENSING OCEAN COLOR PRODUCTS

Operational users of remote sensing ocean color products typically require that a product be available almost immediately (defined anywhere from a few hours to several days). Many times, operational decisions cannot wait until the imagery is cloud free, sensor calibration is validated, reprocessing is performed with the most current software updates, or products are completely validated. The current aim for ocean operational products is to provide the best, most recent ocean data in as timely a basis as possible using the best software processing methods (sensor calibration and algorithms) available.

3.5.1 *Models and imagery*

Clouds and haze restrict the utility of imagery in the visible portion of the spectrum. Latest pixel composite (LPC) methods can provide accurate operational products for real-time ocean applications (Sandidge et al., 2002), even when clouds affect the derivation of in-water properties. The most current and valid image pixel values for each location collected over a one-week period are used to create a daily operational cloud-free product. These methods produce a real-time product that is almost cloud free and provides the best estimate of ocean properties available for each day. Examples of image products generated using these techniques are shown in Figs. 1 and 2 for the Gulf of Mexico. Figure 1 is a SeaWiFS chlorophyll image from January 24, 2004. Figure 2 is a MODIS SST image. Note that black regions occur where there are no valid pixel values over the entire week.

In general, one product per day is adequate for the longer response time of open ocean processes. Using the seven-day LPC, in combination with 12-hour model runs, provides a new capability to monitor shelf and open ocean processes in the time scales they typically occur. However, in coastal areas shorter response times require multiple satellite 'looks' per day to remove clouds and capture coastal dynamics. By increasing satellite coverage to several times per day over a week period, such as that possible from MODIS Terra and Aqua, our ability to generate operational products to support decision making in coastal areas is enhanced.

The clear oligotrophic (lower chlorophyll) waters of the Loop Current and spin-off eddies are shown in the darker gray scales of Fig. 1. The Mississippi River plume is shown extending to the east in light (higher chlorophyll) gray shades. Surface currents (arrow vectors) obtained using the Navy Coastal Ocean Model (NCOM) (Martin, 2000; and Ko et al., 2003) illustrate the response of chlorophyll to local circulation patterns. Surface currents provide excellent tools to enhance our understanding and interpretation of image-based bio-optical information. For example, patches of chlorophyll along the West Florida shelf show little coupling to the model-derived currents in Fig. 1; however, if we animate a sequence of images, our ability to locate the origin of these chlorophyll patches, track them over time, and predict their future movement is enhanced.

The position of temperature fronts along the shelf break in Fig. 2 illustrates how ocean color remote sensing can capture the interaction of coastal and offshore waters. The Mississippi River plume has a specific temperature and chlorophyll signature that characterizes its associated physical and biological processes. The contours on this figure represent sea surface height (SSH) from NCOM model runs. SSH is used to identify the position of the Loop Current and the warm core eddy in the central Gulf. By combining image products and model output, model results can be validated and problem areas located. Animated sequences of these real-time surface fields of bio-optical and physical properties improve our ability to monitor ongoing processes. In turn, models are now being used to provide consistency between image time steps when cloud cover limits satellite coverage.

Figures 1 and 2 also illustrate how remote sensing products can be used to identify locations in the Gulf where biological response is driven by physical forcing (i.e., where circulation contributes to phytoplankton distribution). For example, the anti-cyclonic eddy shown in the northeastern Gulf of Mexico is advecting chlorophyll-rich coastal waters offshore. One can also observe elevated chlorophyll between the eddy and Loop Current, which was responsible for transporting West Florida shelf water into the central

Gulf. Surface currents provide excellent methods to determine where this phytoplankton bloom originated and how it changes in response to advection.

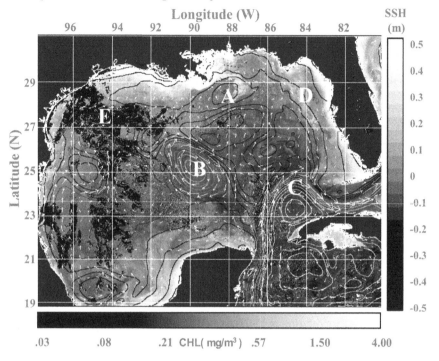

Figure 1. Real-time chlorophyll image of the Gulf of Mexico derived from SeaWiFS data obtained January 24, 2004 showing an eddy (A), a warm core ring (B), the loop current (c), and pockets of high chlorophyll (D). Clouds are masked black (E).

Additionally, to support the operational user, a real-time moving average is accumulated daily and used to calculate weekly, monthly, and yearly means of in-water properties. These statistics, that are provided in real time as they are re-calculated, help the decision maker to observe trends and to determine whether present ocean conditions are similar or dissimilar to an average over the time scale of interest. For example, by taking the difference between today's chlorophyll values and the seasonal mean chlorophyll values identifies residuals or anomalies that may be linked to Harmful Algal Blooms.

3.5.2 *Coastal Scales of Variability*

Channel 1 of MODIS (Aqua and Terra) provides greatly improved spatial resolution required to resolve coastal processes in estuaries, harbors, and bays. Channel 1 has a 250-meter Ground Sample Distance (GSD) resolution with a wide spectral range (620-670 nm) that is capable of monitoring the attenuation coefficient (Arnone et al., 2002). However, channel 1 is not designed as an ocean monitoring channel and has less dynamic range and less sensitively than the spectral ocean channels (channels 8 through 16), typically used for ocean sensing at approximately 1 km GSD. Unlike channel 1, the ocean channels 8-16 are spectrally separated with high sensitivity and dynamic range and are capable of detecting the subtle changes in open ocean waters.

Following the removal of Rayleigh scattering and aerosols from MODIS channel 1, (Arnone et al., 1998, 2002; and Casey et al., 2002), remote sensing reflectance is used to determine the beam attenuation coefficient using an algorithm similar to the one applied to NOAA's Advanced Very High Resolution Radiometer (AVHRR) (Gould and Arnone, 1996). In turbid coastal waters, where high particle scattering occurs around 620-670 nm, there is sufficient remote sensing reflectance. However, in open ocean waters, channel 1 does not have the sensitivity to resolve the remote sensing reflectance.

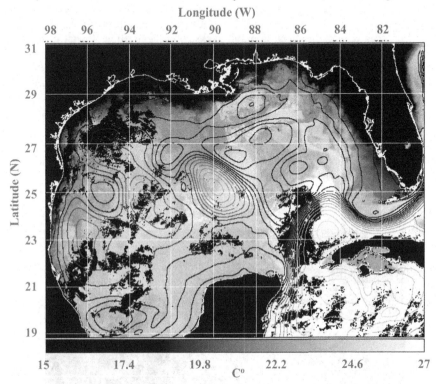

Figure 2. Coincident to the SeaWiFS image shown in Fig. 1, this MODIS (Aqua) sea surface temperature image represents a seven-day latest pixel composite with contours of SSH (same scale as Fig. 1).

Figure 3 shows an example of the beam attenuation coefficient at 670 nm derived from MODIS channel 1 and illustrates the spatial variability of particle concentration in the Mississippi Sound and the Northern Gulf of Mexico. Note that the finer spatial scales observed in this image also have short temporal variability. Also shown on this image are current vectors from a coincident tidal model (PCTides, Posey et al., 2002). The addition of these vectors demonstrates how tidal currents strongly influence observed spatial patterns in optical properties. A strong westward tidal flow is strongly coupled with the signature of the Mobile Bay plume moving westward. Similar gradients of particle distribution associated with terrestrial-borne sediments are evidenced between the coast and the barrier islands.

3.5.3 *Spectral Ocean Optical Properties of Coastal Waters*

The semi-analytical algorithm of Lee et al. (2002) is used to uncouple the backscattering and absorption coefficients from SeaWiFS spectral reflectance imagery for the Mississippi Bight and River plume (Fig. 4). Notice that the Mississippi River plume is transported eastward forming a "green" river off the Gulf Coast, that is visible in the images of the backscattering and total absorption coefficients. An offshore anti-cyclonic eddy is observed between offshore and shelf waters and interacts with this "green" river. The complex interaction of a river plume and eddy field on the coastal shelf is characterized by the differences in the bio-optical properties of the backscattering coefficient at 555 nm and the absorption coefficients from CDOM, detritus, and chlorophyll. Initially, these properties appear as co-varying responses within the imagery (the eddy is identified by similar absorption properties). However, it is the differences with these absorption and backscattering coefficients that identify the processes occurring in these waters. For example, the backscattering image illustrates the strong particle scattering of the Mississippi River plume and decreases from the river mouth as particles settle offshore. These river plume particles also show a strong detritus absorption. However, detritus absorption is not observed along the Gulf Coast to the northeast. The Gulf Coast region is characterized by high CDOM and low chlorophyll and detritus absorption. This suggests that these areas are impacted by high sources of CDOM-rich river fluxes. Similarly, high CDOM absorption defines the periphery of the offshore eddy suggesting a degrading phytoplankton bloom within the eddy's biological cycle. Note also the elevated chlorophyll within the eddy center.

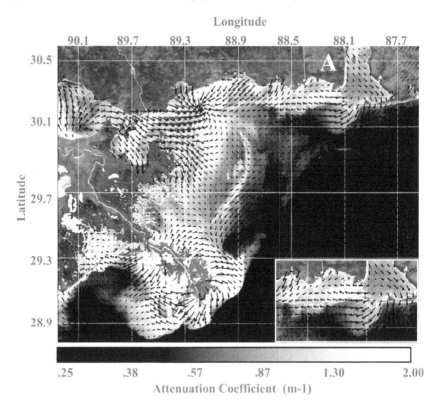

Figure 3. MODIS Aqua image of the beam attenuation coefficient of the Mississippi River plume using the 250 m channel 1. Arrows represent tidal currents. Insert is an enlargement of the Mobile Bay Plume (A).

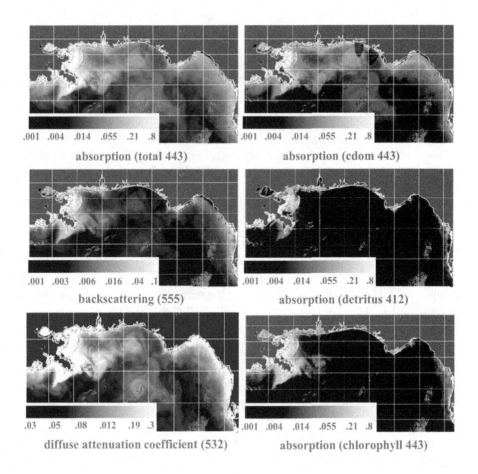

Figure 4. SeaWiFS image of the Mississippi Bight for August 17, 2000 that has been uncoupled from the spectral ocean color signature into six optical coefficients.

3.5.4 *Vertical Ocean Properties*

Subsurface optical properties may be very different than those observed in ocean color imagery, which represents an integrated value over the first attenuation length. This equates to a depth that is often shallower than subsurface optical layers associated with chlorophyll maxima, nepheloid layers, and other optical boundaries found in coastal areas. The vertical structure of bio-optical properties is currently obtained using *in situ* vehicles (gliders or Remote Piloted Vehicles, etc.) or ship-based measurements (profilers or tethered arrays). In many cases, operational applications (such as estimates supporting diver visibility or the penetration depth of a laser system) require information about the vertical profile of the bio-optical properties at or near the bottom.

Additionally, by knowing the vertical optical structure of the water column, near-surface remote sensing data can be better interpreted. For example, the disappearance of a chlorophyll bloom observed in the remote sensing product may not be associated with chlorophyll decay, but with downwelling of bloom below the 'view' of the satellite. These processes occur at convergence zones and can only be determined when the imagery is combined with surface current models to recognize the convergent and divergent regions.

In open ocean waters, the chlorophyll maximum layer frequently occurs deeper than one attenuation length and cannot be observed using satellite imagery. This chlorophyll maximum typically occurs at the base of the mixed layer. By combining the satellite surface chlorophyll data with model outputs of the mixed-layer depth (MLD) and wind stress, three dimensional bio-optical properties can be estimated. Remotely sensed chlorophyll values can be extended to depth by combining the shape of the vertical profile with surface chlorophyll. The shape of the vertical optical profile has been represented as a Gaussian distribution to characterize the chlorophyll maximum (Lewis et al., 1983) and combined with an exponential decay based on the satellite diffuse attenuation coefficient (k490):

$$b(z) = b_0 e^{(kb_0 z)} + \frac{h}{s(2\pi)^{0.5}} e^{\frac{-(z-z_m)^2}{2s^2}}$$

The vertical chlorophyll profile is a unimodal Gaussian distribution over the baseline chlorophyll signal b_0, with the parameters h, s, and z_m controlling the chlorophyll peak height over the baseline (h), the thickness of the deep chlorophyll maximum (DCM) (s), and the depth of the DCM (z_m), respectively (Bowers et al., 2002).

In open ocean systems, the peak in the Gaussian distribution is assumed to be co-located with the mixed layer depth. Elevated nutrients below the MLD enhance growth of the chlorophyll when sufficient light is available. The width of the Gaussian shape is assumed as linked to ocean mixing or wind speed. This distributes the intensity of the chlorophyll maximum within the water column in response to the intensity of surface winds. The strength and shape of the chlorophyll maximum changes as the upper ocean mixes and redistributes the chlorophyll in the upper water column. As winds increase, the chlorophyll maximum becomes destabilized and is less intense at depth as the subsurface chlorophyll is dispersed within the upper water column, which is also accompanied by a change in profile shape. With reduced winds, the chlorophyll maximum intensifies at depth.

Initially, this base profile is calculated at each pixel location over the image domain with b_0 set equal to the SeaWiFS-derived surface chlorophyll concentration. For each pixel, the Gaussian shape and thickness of the maximum layer are parameterized by local bathymetry, nowcast hydrographic data of MLD and surface winds from NCOM and the Navy Operational Global Atmospheric Prediction System (NOGAPS) numerical models. The profile shape is then linked to the SeaWiFS chlorophyll value by setting the integrated chlorophyll profile within the first attenuation length equal to the SeaWiFS chlorophyll. The vertical profile shape is computed and matched for each chlorophyll pixel in the image to determine the vertical distribution of chlorophyll concentration (Fig. 5). The figure illustrates the surface SeaWiFS chlorophyll and the modeled vertical cross section of chlorophyll. Note the subsurface chlorophyll maximum that extends offshore from the coast.

3.5.5 *Temporal Variability of Ocean Color*

Monitoring bio-optical processes in dynamic areas such as the coastal zone requires multiple satellite observations per day. Biological processes can occur on short time scales in response to the light field (PAR) and nutrient availability. Phytoplankton growth has been shown to double within a day and degradation rates in the surface ocean respond similarly. In addition to the rapid change associated with biological processes, physical processes also change the color of surface water on short time scales. Coastal jets and tidal currents change water mass composition through mixing and advection. Bottom sediment resuspension changes the types of particles and their concentration in response to tidal movements and local wind events. The dispersion of river plumes and coastal jets have been shown to change on the time scale of hours. Real-time ocean color data obtained from multiple satellites provides methods to track these changing bio-optical properties at short time scales.

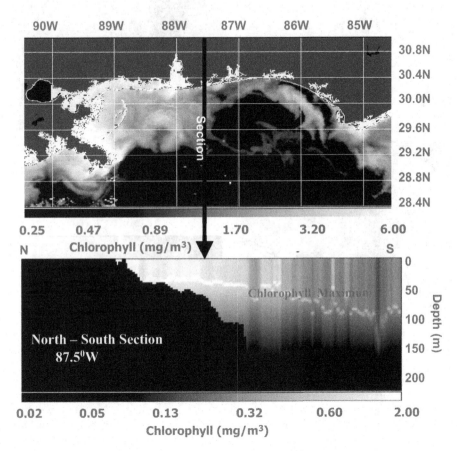

Figure 5. Surface chlorophyll derived from SeaWiFS on December 20, 2003 and the modeled vertical chlorophyll structure offshore near the Mississippi Bight.

Figure 6 shows changes in the diffuse attenuation coefficient at 532 nm in the Persian Gulf from approximately 0700 to 1000 GMT using data from three satellites (MODIS (Terra), SeaWiFS, MODIS (Aqua)). These $k532$ examples represent how the estimated penetration depth and performance of laser systems changes throughout the day. Using multiple satellites to track bio-optical changes requires precise radiometric inter-sensor calibration. The sensor is required to maintain a 0.5% radiometric precision at the top of the atmosphere to achieve a 5% water reflectance that is used in image processing algorithms. Observing daily changes in bio-optical properties from multiple satellites assumes the sensor inter-calibration, atmospheric correction, and the in-water bio-optical algorithms are correct. Differences in $k532$ values between these images are on the order of 15% and verify that multiple satellites can be used effectively for operations.

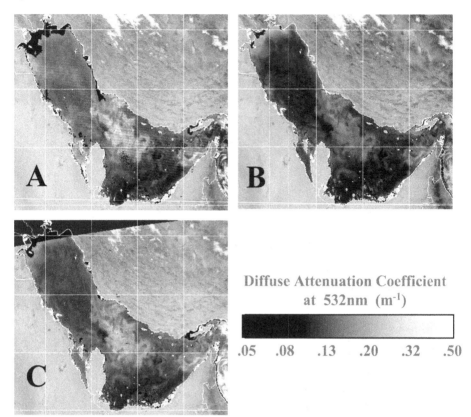

Figure 6. Daily changes in the diffuse attenuation coefficient at 532 nm in the Persian Gulf are shown using three satellites on February 7, 2003. MODIS Terra image acquired at ~0700 GMT (A). SeaWiFS image acquired at ~0900 GMT (B). MODIS Aqua image acquired at ~1000 GMT (C).

Data from three or more images per day can provide sequential products that can be used for planning tactical operations in the face of a frequently changing surface ocean. These data can provide new sources of useful information to a new source of users who must deal with rapidly changing coastal environments, even though the data might be perishable. These users require rapid, real-time delivery of products that require a commensurately rapid processing capability and immediate data availability. During Operation Iraqi Freedom, MODIS Aqua and Terra high resolution 250 m diver visibility products were used to support diving operations. Near real-time products of the Persian Gulf were sent to mission planners within three hours of collection. Examples of this product illustrating several spatial scales of diver visibility products are shown in Fig. 7.

4. Ocean Color Products for Operational Use

Operational real-time ocean color products are different from those products used for research applications. Operational products require:

1) Highly accurate data. Assessment to determine the best product available must be a continual process. This also assumes that products are closely linked with advances in research.

2) Data be available. As operational users and customers begin to rely on image products in their decision making, a break in the availability of these products will cause users to rely on less informative data sources simply because they are available when needed.

3) Continuous collaboration between operations and research. Researchers must understand how these products are used and how they can be improved. Users must understand product capabilities and limitations.

4) A format that is easily combined with other data. Methods to fuse and integrate bio-optical products with other tools such as model output or ship observations must be available to operational users.

5) The user understand how the bio-optical products are applied to their mission.

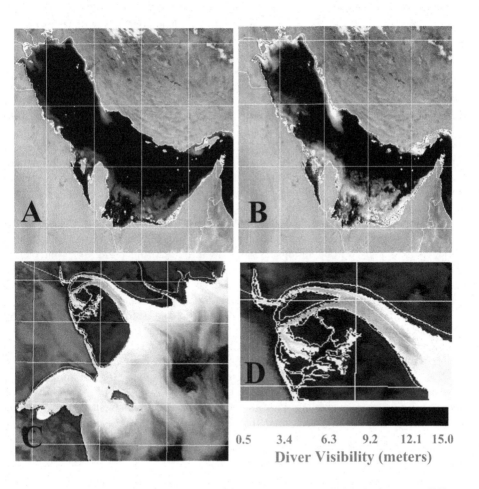

Figure 7. Products of the Persian Gulf from MODIS and SeaWiFS shown at different spatial resolutions for March 14, 2003. SeaWiFS True Color image (A). SeaWiFS diver visibility at 1 km (B). MODIS diver visibility at 250 m (C). MODIS diver visibility of the Khawr Abd Allah waterway (D).

Users must be trained in the interpretation of new products and know their limitations. This is the most difficult requirement as it bridges the gap between the focus of the science community developing the next generation of products and users that must quickly interpret and apply the products with little prior exposure to similar types of products.

5. Supporting and Maintaining Operations

Ocean color products are advancing rapidly with the advent of new satellite sensors, algorithms, and *in situ* observations. These advances are providing a new capability that directly supports real-time decision making. It is well understood that the majority of research products will not be transitioned into operational use because of difficulties in exploiting the product (i.e., lack of robustness, poorly understood, not quantitative, uncertainty in validation, inconsistency between sensors, etc.). In fact, history within the Navy suggests that less than 10% of ocean color basic research will likely be used in operations or decision making. However, these products can be used to support additional applications beyond their original intent. This hopefully will lead to direct connectivity (one-to-one) of research and operational products, which is desirable but many times difficult to establish.

Operational ocean monitoring using satellites requires that products be directly coupled with the decision making process and that the researcher understands how these products are being used. The traditional paradigm of having the decision maker identifying requirements to be addressed by researchers who do not understand how these requirements fit with the application is a slow and tedious effort, often giving raise to a series of trial-and-error products. Researchers need to know the broader aspect of operational issues the decision makers are trying to solve so that creative solutions can be quickly implemented. Operational users are not familiar with the extent to which monitoring the ocean has evolved as a result of ocean color remote sensing nor are they knowledgeable of how to use the wealth of products that are available.

Initially, pre-operational decision making tools can be made available to the operational users on an experimental basis. Note that some of these experimental products may be used in operations and others will not. At this point, direct interaction between the researcher and the users is required to determine if the product can be modified to improve its utility. Only the products that are truly useful in the decision making process should be transitioned into operational products.

The key to exploiting the full potential of satellite data is collaboration between the research and operational communities. To address even the most pressing operational problems, researchers and decision makers can work together to identify potential solutions. Communication is further required to fine tune and adjust new products to more effectively support the decision-making process. It is normally a rapid process for a researcher to modify algorithms used in decision making tools, but it may be a more laborious and time-consuming process to get changes implemented into existing operational products. Improving communications will accelerate what is always an iterative process and will produce improved operational products.

6. Conclusions

The research community has reached new levels in ocean color remote sensing and is continually addressing new challenges. The community is moving beyond the historical focus on chlorophyll products. They now have a suite of new and exciting products that support coastal and ocean monitoring, as well as help improve our understanding of in-water biological, geological, and chemical properties.

These new capabilities have resulted from an expansion of ocean optics research and improved instrumentation. Our new understanding enables answers to "what comprises the ocean color signature" and even a partial answer to "what causes the ocean color signature." We are now able to quantitatively assess the constituents of coastal and ocean surface waters (such as CDOM and other dissolved substances, as well as particle distribution and type). Our advances were facilitated directly as a result of improvements in radiometric precision and sensor stabilization, providing reliable and consistent datasets from which to work.

Ocean color can provide a suite of new optical products that can be effectively used to monitor the ocean and coastal waters in real-time. Our next step is to use our new found understanding to answer inportant questions, such as:

How is CDOM distributed in coastal areas?
How does river-borne CDOM evolve and change as it disperses along the coast?
What processes change particle distributions in the coastal ocean?
How are organic and inorganic particles distributed in coastal and shelf waters?

Research is taking us beyond qualitative looks at the sea surface. Future directions will refine our methods to uncouple the in-water constituents as we progress beyond the multi-spectral signatures available from SeaWiFS and MODIS satellites to the hyperspectral satellites of the near future. These new suites of optical properties are enabling us to characterize and track watermasses by their optical signatures. These classification methods begin to develop new approaches to recognize changes in biogeochemical cycles (Arnone et al., 2004). For the first time, methods are available to truly monitor changing optical signatures that will begin to help determine the anthropologic impact that populations have on coastal waters, and ultimately, the ocean.

Future applications of hyperspectral sensors are well beyond the real-time applications that are available to decision makers today. Research efforts continue to identify new ways to exploit present ocean color tools to meet the needs for today's real-time monitoring. These tools will pave the way for the more seamless integration of new sensors and products being designed for the future.

Even now, the operational community is moving beyond the static views of the ocean provided by current ocean color remote sensing products. Operational support has evolved from climatologies and historical databases to the ability to retrieve data multiple times per day. Similarly, the use of satellite-based products has evolved to animated time series presentations (movie loops) that allow for the visualization of ocean color changes over time. The capability now exists to exploit multiple ocean color sensors to quantitatively monitor changing biogeo-optical properties of the ocean at different time and space scales. Additionally, the ability to fuse ocean color products and numerical model output provides new capabilities to monitor and predict the dynamics and biogeochemical cycles in coastal waters. These capabilities are available in real-time to help the researcher with adaptive sampling of the ocean or to help coastal managers make enlightened decisions regarding one of our most valued resources.

7. References

Arnone, R.A., P. Martinolich, R.W. Gould, Jr., M. Sydor, R. Stumpf, and S. Ladner. 1998. Coastal optical properties using SeaWiFS. Ocean Optics XIV.

Arnone, R.A. and R.W. Gould. 1998. Coastal Monitoring Using Ocean Color. Sea Technology, 39(9):18-27.

Arnone, R.A. 1999. Integrating satellite ocean color into navy operations. Backscatter. 5pg.

Arnone, R.A. and R.W. Gould. 2001. Mapping Coastal Processes with Optical Signatures. Backscatter, 12(1):17-24.

Arnone, R.A., R.W. Gould, Jr., A.D. Weidemann, S.C. Gallegos, and V.I. Haltrin. 2000. Using SeaWiFS ocean color absorption, backscattering properties to discriminate coastal waters. Ocean Optics XV.

Arnone, R.A. Z.P. Lee, P. Martinolich, and S.D. Ladner. 2002. Characterizing the optical properties of coastal waters by coupling 1 km and 250 m channels on MODIS – Terra. Ocean Optics XVI.

Arnone, R.A., A.M. Wood, R.W. Gould, Jr. 2004. The Evolution of Optical Watermass. The Oceanographic Society Fall . 17:2.

Austin, R.W. and T.J. Petzold. 1981. The determination of the diffuse attenuation coefficient of seawater using the Coastal Zone Color Scanner. In: Oceanography from Space, J.F.R. Gower, (Ed.), Plenum Press, New York, 18 pg.

Austin, R.W. and T. J. Petzold. 1986. Spectral dependence of the diffuse attenuation coefficient of light in natural water. Optical Engineering, 25:471-479.

Bowers T.E., S.D. Ladner, R.A. Arnone, R.W. Gould, D.N. Fox, L.F. Smedstad, and C.N Barron. 2002. Modeling the Vertical Distribution of Bio-optical Properties Based on SeaWiFS Imagery. 7th International Conference on Remote Sensing for Marine and Coastal Environments.

Bowers, T.E. 2003. Directionally dependent threshold visual detection range of submerged non-self-luminous objects. Master's Thesis, University of Southern Mississippi, 71 pp.

Carder, K. C., F.R. Chen, Z.P. Lee, S.K. Hawes, D. Kamykowski. 1999. Semianalytic moderate-resolution imaging spectrometer algorithms for chlorophyll a and absorption with bio-optical domains based on nitrate-depletion temperatures, Journal of Geophysical Research, 104(C3):5403-5421.

Casey, B., R.A. Arnone. P. Martinolich, S.D. Ladner, M. Montes, D, Kohler, W.P. Bissett. 2002. Characterizing The Optical Properties Of Coastal Waters Using Fine And Coarse Resolution. Ocean Optics XVI.

Davies-Colley, R.J. 1988. Measuring water clarity with a black disk, Limnology and Oceanography, 33:616-623

Gordon, H.R., O. B. Brown, and M.M. Jacobs. 1975. Computed relationships between the inherent and apparent optical properties of a flat homogenous ocean. Applied Optics, 14:417-427.

Gordon, H.R. and D. K. Clark. 1977. Clear water radiances for atmospheric correction of coastal zone color scanner imagery. Applied Optics, 20(24):4175-4180.

Gordon, H. R. and M. Wang. 1994. Retrieval of water leaving radiance and aerosol optical thickness over the ocean with SeaWiFS: a preliminary algorithm. Applied Optics, 33(3):443-452.

Gould, R. W and R. A. Arnone. 1996. Estimating the beam attenuation coefficient in coastal water from AVHRR Imagery. Coastal Shelf Research, 17(11):1375-1387.

Gould, R.W., Jr., R.H. Stavn, M.S. Twardowski, and G.M. Lamela. 2002. Partitioning optical properties into organic and inorganic components from ocean color imagery. Ocean Optics XVI.

Haltrin, V.I., M.E. Lee., E.B. Shybanov, R.A. Arnone, A.D. Weidemann, and W.S. Pagau. 2002. Relationship between backscattering and beam scattering coefficients derived from new measurements of light scattering phase Functions. Ocean Optics XVI.

Haltrin, V.I., and R.A. Arnone. 2003. An algorithm to estimate concentrations of suspended particles in seawater from satellite optical images (2003). 2nd International Conference Current Problems in Optics of Natural Waters (ed. Levin, I and G. Gilbert) 5pg.

Ko, Dong S., R.H. Preller, and P. J. Martin. 2003. An Experimental Real time Intra Americas Sea Ocean Nowcast/Forecast System for Coastal Prediction. AMS 5th Conference on Coastal Atmospheric & Oceanic Prediction & Processes.

Ladner, S., R.A. Arnone, R.W. Gould, Jr., and P.M. Martinolich. 2002. Evaluation of SeaWiFS optical products in coastal regions. Sea Technology, 43(10):29-35.

Lee, Z., K.L. Carder, and R.A. Arnone. 2002. Deriving inherent optical properties from water color: a multiband quasi-analytical algorithm for optically deep waters. Applied Optics, 41(27):5755-5772.

Lewis, M.R., J.J. Cullen, T. Platt. 1983. Phytoplankton and thermal structure in the upper ocean – Consequences of nonuniformity in chlorophyll profile. Journal of Geophysical Research, 88:2565-2570.

Martin, P.J. 2000. A description of the Navy Coastal Ocean Model Version 1.0. NRL Report: NRL/FR/7322-009962. 39 pg.

Mueller, J., C. Trees and R.A. Arnone. 1990. Evaluation of the CZCS diffuse attenuation coefficient algorithms in coastal waters Proceeding in SPIEE Ocean Optics X, Orlando, FL.

O'Reilly, J. S. Maritorena, B. Mitchell, D. Siegel, K. Carder, S. Garver, M. Kahru, C. McClain. 1998. Ocean color chlorophyll algorithms for SeaWiFS. Journal of Geophysical Research, 103:24,937-24,953.

Petzold, T.J. 1972. Volume scattering functions for selected ocean waters. Report SIO ref. 72-78 Scripts Institute of Oceanography Visibility Laboratory. 79pg.

Posey P.G., R. H. Preller, and G. M. Dawson, S.N. Carroll. 2002. Software Test Description (STD) for the Globally Relocatable Navy Tide/Atmospheric Modeling System (PCTides). NRL-MR-8618. 45p.

Rochford, P. A., A. B. Kara, A.J. Walcraft and R.A. Arnone. 2001. The Importance of Solar Subsurface Heating in Ocean General Circulation Models. Journal of Geophysical Research, 106:30923-30938.

Rossler, C.S and E. Boss. 2003. Spectral Beam Attenuation Coefficient retrieved from ocean color inversion, Geophysical Research Letters, 30(9):1498.

Ruddick, K.G., F. Ovidio, and M. Rijkeboer. 2000. Atmospheric correction of SeaWiFS imagery for turbid coastal and inland waters. Applied Optics, 39(6):897-912.

Sandidge, J. C., P. Martinolich, R.A. Arnone, S.D. Ladner, and R.W.Gould. 2002. Optimizing Coverage by Compositing Ocean Color Imagery. 7th International Conference on Remote Sensing of Coastal and Marine Environments.

Stavn, R.H. and R.W. Gould, Jr. 2003. Biogeo-optics: Organic Matter Scattering in Case 2 Waters. Oceanography Society Annual Meeting.

Stavn, R.H., R.W. Gould, Jr., and G.M. Lamela. 2002. The biogeo-optical model: The database and testing. Ocean Optics XVI.

Twardowski, M.S., E. Boss, J. B. Macdonald, W. S. Pegau, A. H. Barnard, and R. V. Zaneveld. 2001. A model for estimating bulk refractive index from the optical backscattering ratio and the implications for understanding particle composition in case I and case II waters. Journal of Geophysical Research, 106:4129-14142.

Remote Sensing and Digital Image Processing

1. A. Stein, F. van der Meer and B. Gorte (eds.): *Spatial Statistics for Remote Sensing.* 1999   ISBN: 0-7923-5978-X
2. R.F. Hanssen: *Radar Interferometry.* Data Interpretation and Error Analysis. 2001   ISBN: 0-7923-6945-9
3. A.I. Kozlov, L.P. Ligthart and A.I. Logvin: *Mathematical and Physical Modelling of Microwave Scattering and Polarimetric Remote Sensing.* Monitoring the Earth's Environment Using Polarimetric Radar: Formulation and Potential Applications. 2001   ISBN: 1-4020-0122-3
4. F. van der Meer and S.M. de Jong (eds.): *Imaging Spectrometry.* Basic Principles and Prospective Applications. 2001   ISBN: 1-4020-0194-0
5. S.M. de Jong and F.D. van der Meer (eds.): *Remote Sensing Image Analysis.* Including the Spatial Domain. 2004   ISBN: 1-4020-2559-9
6. G. Gutman, A.C. Janetos, C.O. Justice, E.F. Moran, J.F. Mustard, R.R. Rindfuss, D. Skole, B.L. Turner II, M.A. Cochrane (eds.): *Land Change Science.* Observing, Monitoring and Understanding Trajectories of Change on the Earth's Surface. 2004   ISBN: 1-4020-2561-0
7. R.L. Miller, C.E. Del Castillo and B.A. McKee (eds.): *Remote Sensing of Coastal Aquatic Environments.* Technologies, Techniques and Applications. 2005   ISBN: 1-4020-3099-1